Springer Geology

The book series Springer Geology comprises a broad portfolio of scientific books, aiming at researchers, students, and everyone interested in geology. The series includes peer-reviewed monographs, edited volumes, textbooks, and conference proceedings. It covers the entire research area of geology including, but not limited to, economic geology, mineral resources, historical geology, quantitative geology, structural geology, geomorphology, paleontology, and sedimentology.

More information about this series at https://link.springer.com/bookseries/10172

Tatiana Chaplina
Editor

Processes in GeoMedia—Volume IV

Springer

Editor
Tatiana Chaplina
Ishlinsky Institute for Problems in Mechanics
of the Russian Academy of Sciences (IPMech RAS)
Moscow, Russia

ISSN 2197-9545 ISSN 2197-9553 (electronic)
Springer Geology
ISBN 978-3-030-76327-5 ISBN 978-3-030-76328-2 (eBook)
https://doi.org/10.1007/978-3-030-76328-2

This Springer imprint is published by the registered company Springer Nature Switzerland AG
The registered company address is: Gewerbestrasse 11, 6330 Cham, Switzerland

Contents

Determination of Cavities Under the Concrete Slab Anchoring the Upper Slopes of the Novosibirsk Hydroelectric Power Station Using Standing Waves

K. V. Fedin, Y. I. Kolesnikov, and L. Ngomayezwe

Abstract The possibility of a resonant method for detecting defects in the slab anchoring the embankment dams and levee slopes, using Novosibirsk hydroelectric power station as the study object is shown. The accumulation of amplitude spectra of acoustic noise recordings makes it possible to determine the frequencies of the first few modes of standing compressional waves generated by noise in concrete slab fasteners. A sharp increase in the frequency of the lowest mode of standing waves is an indicator of the appearance of voids or De compaction of the base soils under the slabs.

Keywords Hydroelectric dam · Upper slopes fastenings · Cavities · Diagnostics · Standing waves method · Areal observations

1 Introduction

The construction of hydroelectric power stations (HEPs) in many cases is accompanied by the construction of embankment dams and levees, which during operation can be exposed to various processes such as pressure of waves and ice, internal erosion due to filtration processes, seasonal freezing and thawing, etc., leading to changes in their design characteristics. Statistics show that embankment dams account for more than half of all accidents that occur on hydraulic structure dams (Yuzbekov 2004).

K. V. Fedin (✉) · Y. I. Kolesnikov
Trofimuk Institute of Petroleum Geology and Geophysics SB RAS, Novosibirsk, Russia
e-mail: fedinkv@ipgg.sbras.ru

Novosibirsk State University, Novosibirsk, Russia

Y. I. Kolesnikov
e-mail: kolesnikovYI@ipgg.sbras.ru

K. V. Fedin
Novosibirsk State Technical University, Novosibirsk, Russia

L. Ngomayezwe
Scientific and Industrial Research and Development Centre, Harare, Zimbabwe

T. Chaplina (ed.), *Processes in GeoMedia—Volume IV*, Springer Geology,
https://doi.org/10.1007/978-3-030-76328-2_1

Various types of fastenings are used to protect the upper slopes of embankment dams and levees exposed to the dynamic effects of waves and ice. In particular, at the Novosibirsk hydroelectric power station, the upper slopes are reinforced with concrete slabs. The appearance of defects in the upper slopes occurs due to changes in the state of the soil and the loss of concrete strength in the reinforcing concrete slabs. Identification of these defects allows one to timely take appropriate measures to eliminate them.

Inspection of dams is usually carried out by visual inspection or observation, for example, using aircraft or spacecraft. However, given that changes in the state of concrete slabs and soils beneath them are usually hidden, indirect diagnostic methods can be used to search for defects, in particular, various geophysical methods. For the Novosibirsk hydroelectric power station, an important practical task was the detection of voids under the concrete slabs fastening the upper slopes (Fig. 1). In addition, it is also desirable to control the thickness of the concrete slabs at the measuring points. The most common method used for solving this kind of problems, for example, inspection of road surfaces, is based on electromagnetic sounding on road surfaces and their bases by GPR (Derobert et al. 2001; Khamzin et al. 2017; Saarenketo and Scullion 1994). This method is characterized by high performance and fairly good accuracy, but with a large variability of the electrical parameters of the study medium, it may require periodic calibration of the equipment. In addition, concrete slabs are usually reinforced with steel meshes, which also limits the capabilities of GPR.

Acoustic methods, such as echolocation, are also used to examine hard surfaces (Baker et al. 1995; Choi and Park 2019). This method gives good results in controlling the thickness of the cover, but is ineffective in identifying voids and weakened zones underneath. Below we show an example of the application for detecting voids under concrete slabs and controlling their thickness using an acoustic method based on the extraction of standing waves from acoustic noise recorded on the surface of a concrete slab.

Fig. 1 Fastening slabs for the upper slope of the Novosibirsk hydroelectric power station

2 Method

This method has already been applied earlier, for example, during physical modeling and field experiments to determine voids in soil sediments, inspect the road surface and determine the thickness of ice (Kolesnikov and Fedin 2015, 2018; Kolesnikov et al. 2018; Fedin et al. 2019). The method is reduced to the registration of acoustic noises on the surface of the object under study and the accumulation of amplitude spectra of a large number of noise records. This makes it possible to extract the standing waves formed under the object from the noise.

In our case, the study objects are concrete layers (slabs), under which there is either a more rigid base (sandy soil with a rock outline), or, if there are defects in the base, a softer soil or cavities filled with air or water. In such layers (slabs), during the formation of standing waves between the lower and upper boundaries, depending on the conditions of reflection on them, either an integer number of half-lengths or an odd number of quarters of the length of standing waves (like standing waves in rods with loose ends or one loose end (Khaykin 1971)).

The standing waves frequencies of vertical compression-tension in the layer in these two cases are determined, respectively, by the formulas,

$$f_n = \frac{nV_p}{2h} \tag{1}$$

for a concrete slab under which there is a cavity or non-rigid soil, or

$$f_n = \frac{(2n-1)V_p}{4h} \tag{2}$$

for a concrete slab lying on top of a hard ground. Here n is the mode number of standing waves, V_p is the velocity of longitudinal waves, h is the plate thickness.

In both cases, the interval between adjacent natural frequencies $\Delta f = (f_{n+1} - f_n)$ equals $V_p/2h$. This is manifested in the appearance of a regular sequence of resonant peaks at the standing waves frequencies on the amplitude spectra averaged as a result of accumulation. Thus, at known velocity V_p and frequencies of any two adjacent (in mode order) standing waves of vertical compression-tension, the thickness of the concrete slab can be determined by the formula

$$\mathrm{h} = \frac{V_p}{2\Delta F} \tag{3}$$

Note also that for Formula (1) $\Delta f = f_1$, and for Formula (2) $\Delta f = 2f_1$ That is, over water, air, or loose soil, the interval Δf is equal to the frequency of the lowest mode, and over hard ground its value double.

3 Experimental Technique and Results

To assess the possibilities of using the acoustic method of standing waves for examining the fastenings of the embankment dam upper slopes of the Novosibirsk Hydroelectric Power Station, field experiments were carried out on four presumably "problematic" slabs. Slab size area was 10×10 m^2. Registration of noise records was carried out on the entire surface of the slabs along a square grid with a step of 1 m. A two-channel digital oscilloscope B-423 with a sampling frequency of 100 kHz was used for recording. The experimental setup is shown in Fig. 2.

Two wide-band piezoceramic piston-type sensors with a vertically directed axis of maximum sensitivity were used as receivers. These receivers were installed directly on a clean slab surface during measurements. The signals from the sensors were recorded with a digital oscilloscope and then recorded on the hard disk of a computer (laptop) for further processing. To speed up the measurements near the observation points, the concrete slabs were subjected to additional noise exposure using a hard-bristled brush driven by an electric screwdriver. This made it possible to reduce the registration time at each point from 5 to 10 min required when working with natural acoustic noises to about 30 s.

During processing, the recordings were divided into fragments with a duration of 8192 counts (approximately 82 ms), after which the amplitude spectra of these fragments were accumulated. Figure 3 shows examples of averaged spectra of noise recordings recorded at two observation points. It can be seen that even with a relatively short recording duration of noise, at least two regular resonance peaks can be confidently identified in their spectra, which agree with Formulas (1) or (2).

In particular, for a concrete slab lying on a more rigid base (a rock outline on a sandy ground), which was the case in most observation points, these peaks are located on the frequency axis in accordance with Formula (2). This is illustrated by the spectrum shown in Fig. 3a, where the peak frequencies corresponding to the first two modes of standing compressional waves are 3.33 and 9.99 kHz. At the same time, the distribution of the peaks is consistent with Formula (1) the cavity formed under the slab (Fig. 3b), their frequencies are 6.65 and 13.3 kHz.

The fact that we observe exactly standing waves of vertical compression-tension of a concrete layer and not other types of waves is due to the use of sensors in measurements that record mainly the vertical component of acoustic noise. Confirmation of

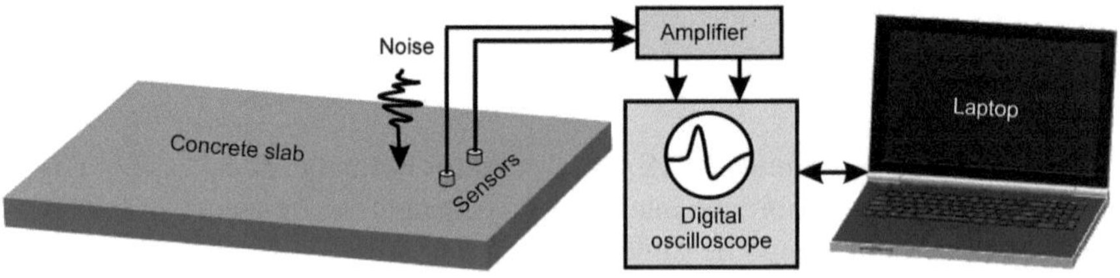

Fig. 2 Experimental setup

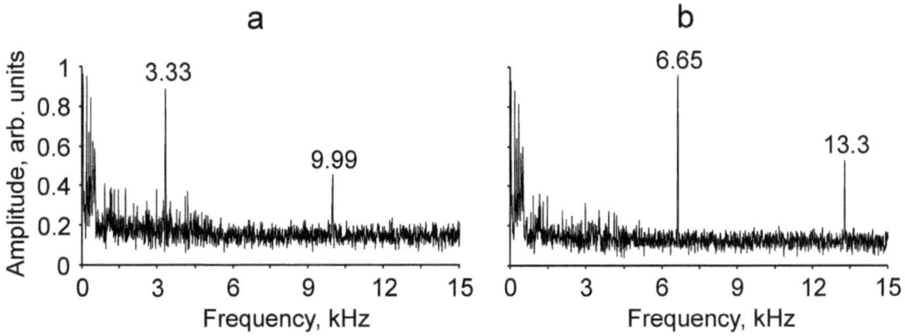

Fig. 3 Examples of averaged amplitude spectra of noise records recorded on the slabs of the upper slope of the dam **a** above the rigid base and **b** above the cavity

this nature of standing waves is provided by comparing the results of direct measurement of the concrete slab thickness at the control point and calculating the thickness using the Formula (3). The concrete thickness measured with a tape measure along the slab with a loose end at the bottom was approximately 30 cm. The lowest mode frequency determined from the noise recorded near the bottom side is $f_1 = \Delta f =$ 6.94 kHz, which, when measured by the pulse method using concrete velocity V_p = 4150 m/s in accordance with the Formula (3), gives almost the same thickness of 29.9 cm.

Figures 4 and 5 show the distribution of the lowest mode frequency over the surface of the investigated slabs according to the results of measurements in 2019 summer and repeated measurements on the same slabs in 2020 summer. It can be seen from the figures that in some zones of the slabs adjacent to their ends, an approximately two-fold increase in the frequency of the lowest mode is observed. This indicates that in these zones the contact of the slab with the base is broken (a washout has formed) or there has been a significant decrease in the rigidity of the base soils. As a result, the standing wave frequencies in these zones are determined by Formula (1), in contrast to zones where there is a more rigid base under the slabs, there these frequencies are determined by Formula (2). Weaker frequency fluctuations are probably related to the heterogeneity of slabs in thickness or changes in the concrete properties.

It should be noted that Figs. 4a, b and 5a, b shows the results for two adjacent slabs from the upper row to which other slabs also adjoin from below, and Figs. 4c, d and 5c, d, results for two slabs from different parts of the lower row under which there are no other slabs. This difference is manifested in the fact that in the second case immediately near the lower edge of the slabs, there is also a two-fold decrease in the lowest mode frequency of standing waves, which is obviously associated with the gradual washout of a part of the foundation soil in these places.

Comparing Figs. 4 and 5, it can be seen that over the year, the size of anomalies observed earlier increased as well as new although still small, high-frequency anomalies have appeared. This indicates further erosion of the soil under the investigated “problematic” slabs fastening the upper slope of the Novosibirsk hydroelectric

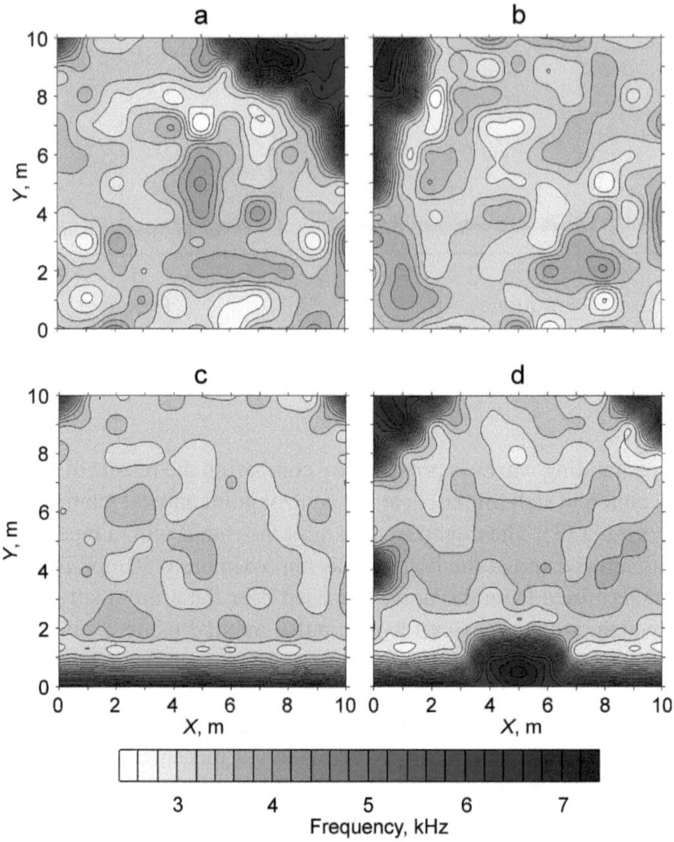

Fig. 4 Frequency distribution of the lowest mode of standing waves on the concrete slab surfaces of the upper (**a**, **b**) and lower (**c**, **d**) rows (2019 summer)

station. Thus, the results of the experiments showed that the method of standing waves based on recording acoustic noise signals can be successfully used for diagnosing and monitoring the slope fastenings quality of the embankment dams and levees.

4 Conclusion

Using Novosibirsk Hydroelectric Power Station as the study object, the possibilities of using the standing waves method for detecting voids and soil decompaction zones under concrete slabs fastening the slopes of the embankment dam and levees are investigated. Experiments have shown that the accumulation of amplitude spectra of noise records allows one to confidently determine the natural frequencies of standing

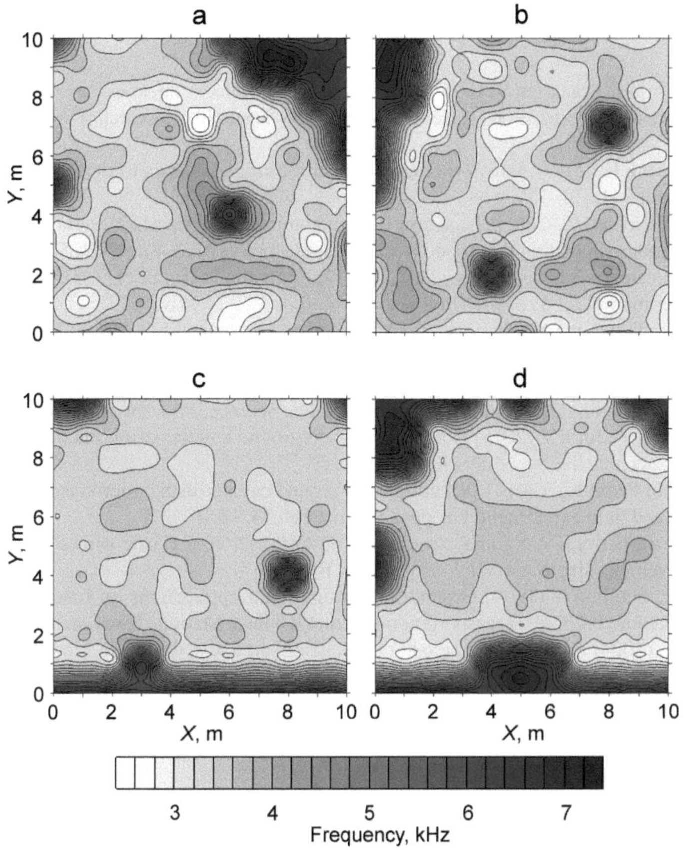

Fig. 5 Frequency distribution of the lowest mode of standing waves over concrete slab surfaces of the upper (**a**, **b**) and lower (**c**, **d**) rows (2020 summer)

compressional waves generated by noise in concrete slab fastenings. A sharp increase in the frequencies of standing waves, in particular their lowest mode, is an indicator of the appearance of voids under the slabs or decompaction of the base soils.

The proposed standing waves method makes it possible to reliably diagnose changes in the concrete base soils of the concrete slabs fastening the slopes of the embankment dams and levees, in particular, to identify the voids formed and loose soils zones. The performance of the method when searching for defects and the reliability of their detection are competitive in comparison with other methods used to solve similar problems.

Acknowledgements The availability of the obtained data is limited by the requirements of the project in the framework of which it was obtained (FNI project No 0331-2019-0009 "Dynamic analysis of seismic data for building realistic models of the geological environment based on mathematical and physical modeling").

References

Baker MR, Crain K, Nazarian S (1995) Determination of pavement thickness with a new ultrasonic device. Research Report 1966–1. Center for Highway Materials Research, University of Texas, El Paso

Choi WY, Park KK (2019) Array type miniaturized ultrasonic sensors to detect urban sinkholes. Measurement 141:371–379

Derobert X, Fauchard C, Côte P, Le Brusq E, Guillanton E, Dauvignac JY, Pichot C (2001) Step-frequency radar applied on thin road layers. J Appl Geophys 47(3–4):317–325

Fedin KV, Kolesnikov YI, Ngomayezwe L (2019) Determination of ice thickness by standing waves. Process Geomed 4(22):528–533 (In Russian)

Khamzin AK, Varnavina AV, Torgashov EV, Anderson NL, Sneed LH (2017) Utilization of air-launched ground penetrating radar (GPR) for pavement condition assessment. Constr Build Mater 141:130–139

Khaykin SE (1971) Physical foundations of mechanics. Science, Moscow (In Russian)

Kolesnikov YuI, Fedin KV (2015) Detection of underground cavities using microtremor: physical modelling. Seismic Technol 4:89–96 (In Russian)

Kolesnikov YI, Fedin KV (2018) Detecting underground cavities using microtremor data: physical modelling and field experiment. Geophys Prospect 66:342–353

Kolesnikov YuI, Fedin KV, Ngomayezwe L (2018) Application of elastic waves for the diagnosis of rigid pavement. Eng Surv XII 7–8:84–91 (In Russian)

Saarenketo T, Scullion T (1994) Ground penetrating radar applications on roads and highways. Research Report 1923–2F. Transportation Institute, College Station, Texas, p 36

Yuzbekov NS (2004) Problems of assessing the state of soil dams. Civil Sec Technol 2(6):62–65

Acoustic Measurements on Synthetic Fractured Samples Made Using FDM 3D Printing Technology

G. A. Dugarov, Yu. I. Kolesnikov, K. V. Fedin, Yu. A. Orlov, and L. Ngomayezwe

Abstract This paper presents the acoustic measurements results on synthetic samples containing a system of oriented disc-shaped cracks simulating a fractured medium. An effective model for such media is a model with a transversely isotropic symmetry system under the condition that the host medium is isotropic. FDM 3D printing technology with molten plastic thread was used to form synthetic samples. The results of the recorded acoustic measurements have shown the occurrence of anisotropy of the host medium due to horizontal layering of plastic threads. In the case of printing internal cracks using this technology, the samples contain inhomogeneities due to the greater sintering of plastic threads around the cracks. It was also shown that the use of SLA technology (printing from liquid photopolymer) potentially avoids the indicated effects, but requires developing the technique for printing samples with cracks. Acoustic measurements on the sample without cracks formed by this technique confirmed its isotropy.

Keywords 3D printing · Synthetic samples · Fractured media · Acoustic properties · Velocities · Experimental measurements

---

G. A. Dugarov · Yu. I. Kolesnikov · K. V. Fedin (✉) · Yu. A. Orlov
Trofimuk Institute of Petroleum Geology and Geophysics SB RAS, Novosibirsk 630090, Russia
e-mail: fedinkv@ipgg.sbras.ru

K. V. Fedin · Yu. A. Orlov
Novosibirsk State University, Novosibirsk 630090, Russia

K. V. Fedin
Novosibirsk State Technical University, Novosibirsk 630073, Russia

L. Ngomayezwe
Scientific and Industrial Research and Development Centre, 1574 Alpes Road/Technology Drive, Harare, Zimbabwe

T. Chaplina (ed.), *Processes in GeoMedia—Volume IV*, Springer Geology,
https://doi.org/10.1007/978-3-030-76328-2_2

1 Introduction

One of the main factors that determine the physical properties of hydrocarbon reservoirs is the structure of the void space, namely the shape and orientation of cracks and pores, their fluid filling and connectivity. Evaluation of the influence of these cracks and pores on the elastic properties of rocks is one of the most important tasks of seismic exploration, which, in turn makes it possible to evaluate only the macroscopic properties of rocks. One of the approaches to solving this problem is physical modeling, the creation of synthetic samples. When modeling a fractured environment, there are different approaches in creating such samples. A mixture of sand with epoxy resin (Rathore et al. 1994; Wei and Di 2008; Tillotson et al. 2012), or sand with cement (Santos et al. 2017), or quartz sand, clay minerals, and sodium silicate (Ding et al. 2014, 2018) can be used for manufacturing. For the formation of cracks, aluminum discs or discs made of polymeric materials are used, which are preliminarily placed in the sample in certain positions and then chemically or thermally destroyed and washed out, leaving behind disc-shaped cracks. The cracks can also be made of rubber and then placed in the samples, for example, in epoxy resin (Santos et al. 2015). The disadvantages of these approaches are limitations caused by manual production of samples, errors in positioning accuracy and crack shape. In the work (Stewart et al. 2013), the authors also introduce a new method of laser etching to create cracks in an isotropic homogeneous glass block.

3D printing technology makes it possible to form complicated structures of the void space of synthetic samples. The work (Huang et al. 2016) shows the possibility of creating such fractured synthetic samples. One test sample with disc-shaped cracks was printed and its anisotropy was studied. The fracture direction was horizontal which corresponds to the VTI medium (transversely isotropic with a vertical symmetry axis). But at the same time, in addition to disc-shaped cracks, voids were also observed in the space between the cracks in the sample (see Fig. 2c in Huang et al. (2016)), which may be due to low print accuracy. In the work (Ndao and Do 2017), the authors also printed one sample and measured velocities of compressional (P) and shear (S) waves. Authors compared the results with estimation by different effective models. But 16 areas with different crack densities were printed in one sample. Acoustic measurements was carried out in such a way that the rays crossed different combinations of these areas at different angles, which made it impossible to separate the effects from different areas. No comparison with the isotropic case without voids cracks was made. In this paper, a series of synthetic samples with disc-shaped cracks were made and acoustic experiments to study their properties was also carried out.

2 Synthetic Sample Parameters

There are three main technologies that are used in 3D printing, these are Fused Deposition Modeling (FDM, layer-by-layer molten plastic thread), Stereo Lith Aparatus (SLA, from photopolymer hardened by laser) and Selective Laser Sintering (SLS, laser sintering of powder material), see Fig. 1. The FDM technology was used for the manufacture of synthetic fractured samples in the works (Huang et al. 2016; Ndao and Do 2017). The advantage of this technology is the ability to print cavities inside bodies. SLA printing technology is suitable for simulating samples containing fluid-saturated voids in the case of sample protection from UV radiation. With SLS technology, when attempting to print closed voids, they will be filled with unsintered powder. In this paper, FDM and SLA 3D printing technologies were used to make synthetic samples.

To select the parameters of the printed fracturing, a preliminary analysis of works on the production of synthetic fractured samples was carried out. Samples with disc-shaped/penny-shaped cracks arranged in an orderly manner were considered (Ding et al. 2014, 2018; Amalokwu and Best 2016; Figueiredo et al. 2018, 2019). The cracks were flattened along one of the axes, $a = h$, h—crack thickness, and with the same two other radiuses, $b = c = 2r$, $2r$—crack diameter. The aspect ratio f for the crack is calculated as

$$f = \frac{h}{2r}.$$

The crack density e is defined as (Figueiredo et al. 2018):

$$e = N\frac{\pi h r^2}{V},$$

where N—number of cracks, V—the volume occupied by the cracks.

Three models without cracks, with a small number of cracks ($e = 0.024$), and with a large number of cracks ($e = 0.048$) were selected. The crack thickness and

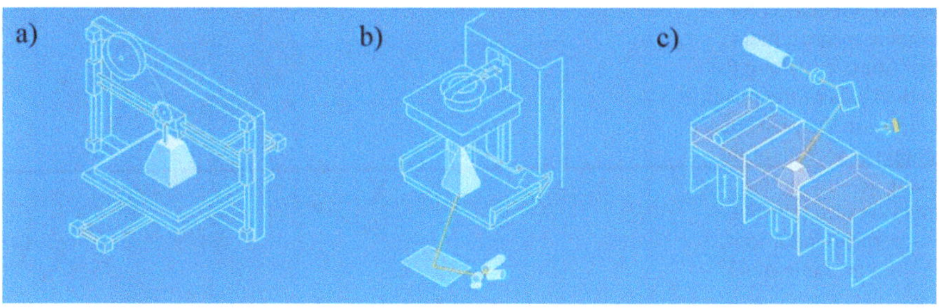

Fig. 1 Schematic illustration of printing samples by different technologies: **a** FDM, **b** SLS, **c** SLA (https://formlabs.com/blog/fdm-vs-sla-vs-sls-how-to-choose-the-right-3d-printing-technology/)

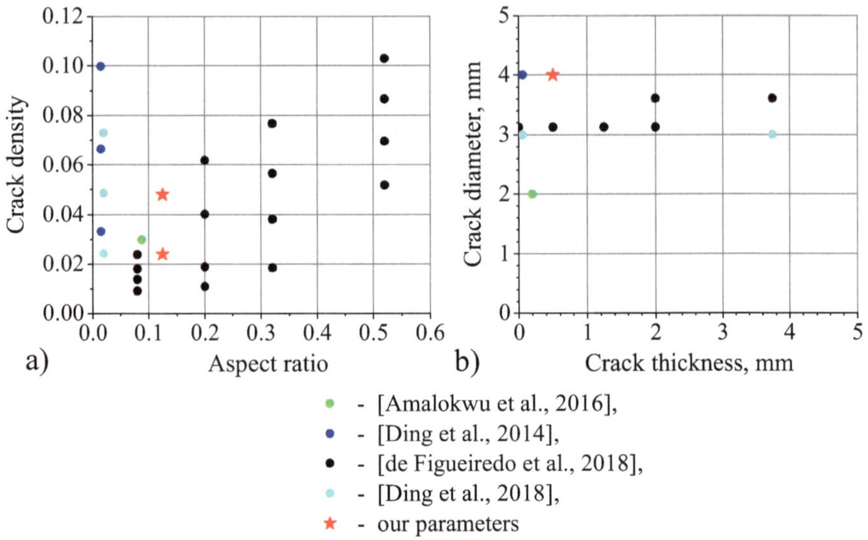

Fig. 2 Comparison of selected crack parameters: **a** aspect ratio and crack density, **b** crack thickness and diameter in synthetic samples, with data from other works

diameter for the two models with a small and large number of cracks are identical, 0.5 mm and 4 mm, respectively. A comparison of crack parameters with data from other works is shown in Fig. 2.

The samples were shaped in the form of a regular octagonal prism, which provides good acoustic contact on the lateral faces and the ability to accurately position the sensors on the samples for measurements along the direction of fracturing, perpendicular to it and at an angle of 45°, see Fig. 3. Three samples were made using FDM technology, these are: samples without voids, as well as with a small and large number of voids. A Prusa i3 Steel printer was used for printing with the following

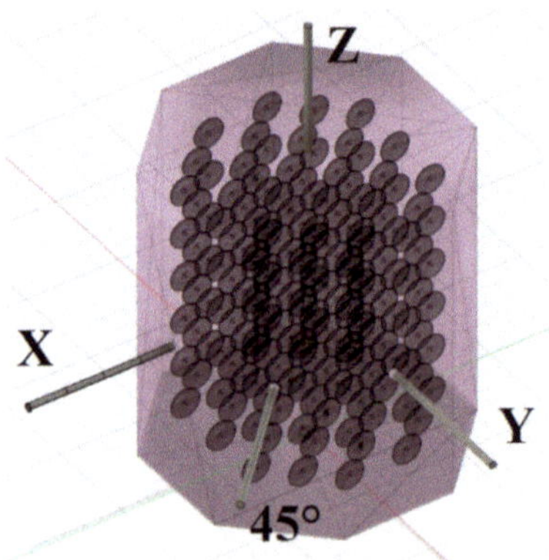

Fig. 3 Model of the sample with internal voids and marked orientation for acoustic measurement directions: Z—along the vertical axis of the prism, X, Y, 45°—in the plane of the octagonal section of the prism, X—along the direction of fracturing, Y—perpendicular to it, 45°—at an angle of 45°

Fig. 4 Sections of samples illustrating the implementation of internal voids made by FDM (left) and SLA (right) technologies

specifications: nozzle—0.2 mm, print width—0.24 mm, layer height—0.12 mm, material—PLA plastic. Printing accuracy is about 200 μm. One sample was also made without cracks using SLA technology. An Anycubic Photon S printer was used for printing with the following specifications: XY accuracy—47 μm, Z—1.25 μm, UV wavelength—405 nm, material-FunToDo NanoClear photopolymer. The printing accuracy was about 50 μm.

For quality comparison of internal voids printed by different technologies, sample horizontal slices through the centers of the voids was printed (along the plane of an octagonal prism), see Fig. 4. When using the FDM technology, mutually perpendicular interlayering of threads were observed (at an angle of 45° to the direction of internal voids), which should lead to the presence of anisotropy due to horizontal layering, in addition to anisotropy due to the presence of a system of vertical voids. A greater sintering of the threads around the voids is also observed, which can lead to heterogeneity of the samples printed by this technology.

3 Results of Acoustic Measurements

Excitation and reception of P-wave was carried out by piezoelectric piston-type transducer with an operating frequency range from 50 to 500 kHz. For S-wave, a sensor with an adjustable directional pattern were used (Bobrov et al. 1987), which were switched on in the mode of excitation and reception of S-wave with a frequency range from 30 to 250 kHz. The source was supplied with a rectangular pulse from the G5-54 generator with a duration of 0.8 microseconds and an amplitude of 60 V. The signal from the receiver was preliminarily amplified by a factor of 10–100 and recorded with B-423 digital oscilloscope.

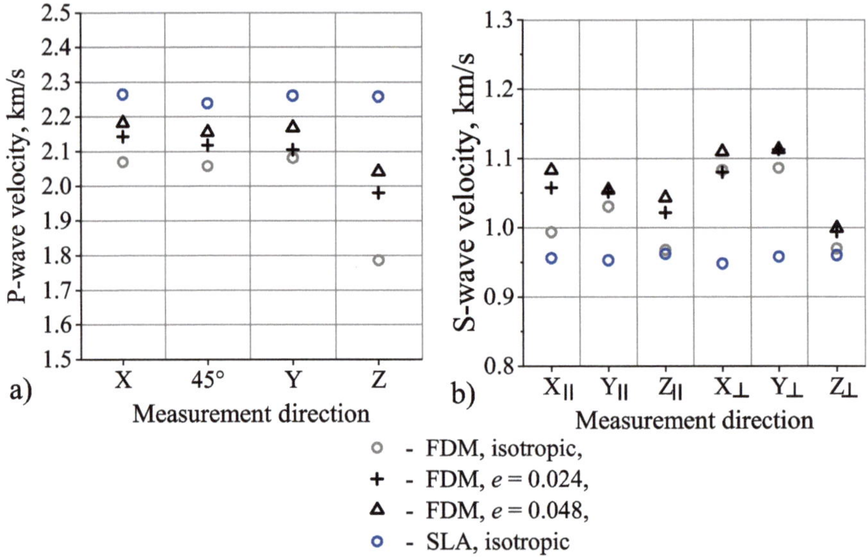

Fig. 5 Velocities of P- (**a**) and S-waves (**b**) for different measurement directions (see Fig. 3) on synthetic fractured samples

Acoustic measurements on samples printed by FDM technology showed the presence of anisotropy in the sample without cracks. The measurement data of P- and S-wave velocities are shown in Fig. 5, the measurement directions correspond to those shown in Fig. 3. The P-wave velocity along the vertical axis (Z direction) is less than in the plane of the octagonal section of the prism (X, 45°, Y directions). This velocity ratio is typical for the VTI model, which is effective for horizontally thin layered media. The anisotropy was caused by horizontal stratification of the threads. The difference in the P-wave velocities in the vertical and horizontal directions is about 14%, which corresponds to a strong anisotropy.

This effect could be taken into account by using the orthorhombic symmetry system model, which is effective for horizontally thin-layered media with subvertical fracturing. However, for samples with voids, the P-wave velocity along the vertical axis (Z direction) is higher than for a sample without these voids. This may be due to the previously mentioned fact of different sintering of threads around voids, which leads to the formation of inhomogeneity of samples. Such samples cannot be described by efficient models that imply approximation of the medium by a homogeneous model.

Using SLA technology potentially avoids the occurrence of these effects. For the sample without cracks, the velocities of both P-and S-waves are almost identical for different measurement directions, with deviations of less than 1%, see Fig. 5. However, when trying to print a sample using SLA technology with cracks, an isotropic sample is formed by using a transparent photopolymer, in which cracks are visually observed, but in fact filled with a solid photopolymer. Further development of the technique for printing synthetic fractured samples using SLA technology containing dry or filled with liquid photopolymer cracks is required.

4 Conclusion

The paper presents two 3D printing technologies FDM and SLA which were applied in printing of synthetic fractured samples. Three samples without and with small and large number of disc-shaped cracks were made using the FDM technology. Their acoustic properties were studied. It is shown that horizontal layering due to layer-by-layer printing with thread leads to strong VTI anisotropy. Also, higher P-wave velocities are observed in the samples with cracks than in the sample without them. This may be due to the appearance of inhomogeneities in the samples during the implementation of internal voids due to greater sintering of the thread around them. It is shown that the use of SLA technology can potentially avoid these effects, but requires the development of a technique for printing samples with cracks.

Acknowledgements This work was carried out with the financial support of the RFBR grant No. 19-05-00730.

References

Amalokwu K, Best AI, Chapman M (2016) Geophys Prospect 64:942
Bobrov BA, Geek LD, Orlov YA (1987) Ultrasonic broadband sensor with adjustable directivity pattern. USSR Patent No. SU 1290137 A1 (in Russian)
de Figueiredo JJS, do Nascimento MJS, Hartmann E, Chiba BF, da Silva CB, de Sousa MC, Silva C, Santos LK (2018) Geophysics 83(5):C209
de Figueiredo JJS, Chiba BFF, do Nascimento MJ, da Silva CB, Santos LK Can (2019) J Appl Geophys 161:255
Ding P, Di B, Wang D, Wei J, Li X (2014) J Appl Geophys 109:1
Ding P, Di B, Wang D, Wei J, Zeng L (2018) Wave Motion 81:1
https://formlabs.com/blog/fdm-vs-sla-vs-sls-how-to-choose-the-right-3d-printing-technology/
Huang L, Stewart RR, Dyaur N (2016) Baez-Franceschi J. //. Geophysics 81(6):D669
Ndao B, Do D-P, Hoxha D (2017) Geophys Prospect 65:181
Rathore JS, Fjaer E, Holt RM, Renlie L (1994) Geophys Prospect 43:711
Santos LK, de Figueiredo JJS, Omoboya B, Schleicher J, Stewart RR, Dyaur N (2015) J Appl Geophys 114:81
Santos LK, de Figueiredo JJS, Macedo DL, de Melo AL, da Silva CB (2017) J Pet Sci Eng 156:763
Stewart RR, Dyaur N, Omoboya B, de Figueiredo JJS, Willis M, Sil S (2013) Geophysics 78(1):D11
Tillotson P, Sothcott J, Best AI, Chapman M, Li X-Y (2012) Geophys Prospect 60:516
Wei J, Di B, Wang Q (2008) Pet Sci 5(2):119

Variational Identification of the Transport Model Parameters in the Azov Sea Based on Remote Sensing Data

Kochergin Vladimir Sergeevich and Kochergin Sergey Vladimirovich

Abstract The model of passive impurity transport in the Azov Sea is used as a constraint to minimize the quadratic functional of forecast quality, which characterizes the deviations of model estimates from the measured data of the suspended matter concentration in the upper sea layer. Variational algorithms are built to search for optimal parameters of the model, which can be either constant or variable in space and time. The obtained results of numerical modeling are analyzed taking into account satellite information about the concentration of suspended matter in the upper sea layer.

Keywords Concentration of suspended matter · Variation algorithm · Assimilation · Conjugate problem · Azov sea · Assimilation of measurement data · Substance flows

1 Introduction

To obtain adequate estimates of the dynamic characteristics of suspended matter in the sea, it is necessary to use not only modern dynamic models (Blumberg and Mellor 1987; Fomin 2002; Ivanov and Fomin 2008), models of transport and transformation of impurities, but also algorithms for searching for input parameters of these models based on the assimilation of measurement data. The quality criterion for a particular model is objective reality, i.e. measurement data. In recent years, with the development of satellite Oceanography, modern methods of image processing and converting them into figures have appeared (Kremenchutsky et al. 2014). Such information can be received online, so the development and testing of methods for determining certain input parameters of numerical modeling is an important and urgent task. This work is devoted to the construction of parameter identification algorithms of suspended matter flows at the bottom for the passive impurity transport model. The variational approach based on solving adjoint problems is actively used for solving a wide class of problems (Marchuk 1982; Marchuk and Penenko 1978; Shutyaev 2019;

K. V. Sergeevich (✉) · K. S. Vladimirovich
Marine Hydrophysical Institute RAS, Sevastopol 299029, Russian Federation

T. Chaplina (ed.), *Processes in GeoMedia—Volume IV*, Springer Geology,
https://doi.org/10.1007/978-3-030-76328-2_3

Shutyaev et al. 2018; Zalesny et al. 2016; Shutyaev and Parmuzin 2019; Kochergin and Kochergin 2015).

In numerical implementation of algorithms for searching for optimal estimates, gradient methods are used, which are implemented by minimizing the quadratic functional of the forecast quality. The solution of the adjoint problem is necessary for constructing the gradient of the functional in the direction of which the iterative descent occurs. When implementing the variation procedure, the solution of the main, adjoint problems and the problem in variations is performed. The latter is used when defining an iterative parameter for gradient descent. In the process of integrating these problems, TVD approximations are applied (Harten 1984). To implement the assimilation procedure, the used flow fields and turbulent diffusion coefficients were obtained using the model (Fomin 2002; Ivanov and Fomin 2008) in Sigma coordinates for the Azov Sea area under East wind influence.

2 Methods

The model of passive impurity transfer in σ-coordinates has the form:

$$\frac{\partial DC}{\partial t} + LC = 0, \tag{1}$$

with the conditions on the lateral boundaries:

$$\Gamma : \frac{\partial C}{\partial n} = 0, \tag{2}$$

boundary conditions on the surface and at the bottom:

$$\sigma = 0 : \frac{\partial C}{\partial \sigma} = 0, \tag{3}$$

$$\sigma = -1 : \frac{\partial C}{\partial \sigma} = Q(x, y),$$

and initial data:

$$C(x, y, \sigma, 0) = C_0(x, y, \sigma). \tag{4}$$

In (1)–(4), the following designations are accepted:
$L = \frac{\partial DU}{\partial x} + \frac{\partial DV}{\partial y} + \frac{\partial W}{\partial \sigma} - \frac{\partial}{\partial x} A_H \frac{\partial D}{\partial x} - \frac{\partial}{\partial y} A_H \frac{\partial D}{\partial y} - \frac{\partial}{\partial \sigma} \frac{K}{D} \frac{\partial}{\partial \sigma}$, $t \in [0, T]$—time. C—admixture concentration; D—dynamic depth; σ—dimensionless vertical coordinate that varies from -1 (on the bottom) to 0 (on the sea surface); $Q(x, y)$—admixture flow

at the bottom; U, V, W—components of the speed field; A_H and K—coefficients of horizontal and vertical turbulent diffusion, respectively; n – normal to the side border; Γ—the border of the region M; $M_t = M \times [0, T]$.

3 Variational Assimilation Algorithm

To solve the problem of assimilation of measurement data it is necessary to find a minimum convex quadratic functional of forecast quality

$$I_0 = \frac{1}{2}(P(RC - C_{me}), (P(RC - C_{me})))_{M_t}. \tag{5}$$

where R is the operator for projecting to observation points, and P is the operator for filling in the field of forecast residuals with zeros in the absence of measurement data. Write (5) for linear relations (1)–(4) as:

$$I = I_0 + \left(\frac{\partial C}{\partial t} + LC, C^*\right)_{M_t} + \left(\frac{\partial C}{\partial n}, C^*\right)_{\Gamma_t} + \left(C - C_0, C^*\right)_M + \left(\frac{\partial C}{\partial \sigma}, C^*\right)_{\sigma_t^{-1}} - \left(Q(x, y), C^*\right)_{\sigma_t^{-1}}, \tag{6}$$

We integrate the corresponding (6) expression for the variation of the functional in parts, taking into account the analog of the continuity equation and the boundary conditions. Chosen as the Lagrange multipliers for the solution of the following adjoint problem:

$$\frac{\partial DC^*}{\partial t} + L^*C^* = P(C_{me} - RC), \tag{7}$$

$$\Gamma : \frac{\partial C^*}{\partial n} = 0, \sigma = 0 : \frac{\partial C^*}{\partial \sigma} = 0, \sigma = -1 : \frac{\partial C^*}{\partial \sigma} = 0, \tag{8}$$

$$\mathrm{t} = \mathrm{T} : C^* = 0,$$

where L^* is the formally adjoint operator to the operator L.

Then we have from the stationarity of the functional and the definition of its gradation

$$\nabla_{C^0} I = C(\sigma, x, y, 0),$$

$$\nabla_{Q(x,y)} I = \int_0^T C^*(-1, x, y, t)dt.$$

The initial concentration distribution or power value of the source is searched iteratively

$$C_0^{n+1} = C_0^n + \tau \nabla_{C^0} I$$

$$Q^{n+1}(x, y) = Q^n(x, y) + \tau \nabla_{Q(x,y)} I.$$

where τ is an iterative parameter, which is determined (Kochergin and Kochergin 2010) taking into account the solution of the problem in variations, based on the minimum of the functional (5) by the following formula:

$$\tau = \frac{(P(RC - C_{me}), P(RC - C_{me}))}{(PR\delta C, PR\delta C)_{M_t}}$$

In (7) δC—the solution of the corresponding problem in variations.

4 Results and Discussions

Input information for calculations based on the impurity transport model was generated using a hydrodynamic model (Fomin 2002; Ivanov and Fomin 2008) for the Azov Sea. The dynamic model was integrated to a quasi-stable solution with a constant North-easterly wind influence 10 m/c. The obtained velocity fields and model coefficients were set as input parameters when integrating the passive impurity transport model for a period $T = 1$ day. The calculations used a step by time $\Delta t = 240$ s, and a step by space $\Delta x = 0.78$ km, $\Delta y = 1.125$ km. Vertically, the calculated grid is used σ-coordinates with 15 horizons.

Figure 1 shows MODIS AQUA data for which digital concentration values are obtained (Kremenchutsky et al. 2014). Data for October 5 were assimilated using the method (Kochergin et al. 2020) by identifying the initial concentration field for October 4. The atmospheric effect is characterized by a small North-easterly wind (http://dvs.net.ru/mp/data/201410vw.shtml). The resulting field is shown in Fig. 2.

Using this concentration distribution as the initial concentration field, as a result of assimilation of the information for October 6 (Fig. 3), by identifying the flow of matter on the sea floor, we minimize the functional of the forecast quality and obtain a model field (Fig. 4) consistent with the measurement data.

Note that the field shown in Figs. 3 and 4 was formed under the influence of intense wind action in the North-Eastern direction with values of about 10 m/s. The drawings clearly show the main features of the topography of the bottom of

Fig. 1 MODIS AQUA RGB composite, October 5, 2014

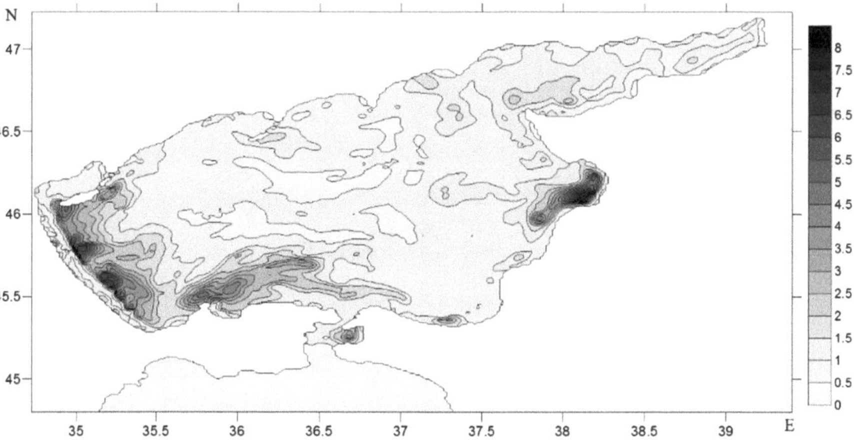

Fig. 2 Restored initial model field of concentration on the sea surface on October 5, 2014 ($\sigma = 0$ mg/m^3)

the Azov Sea—this is the Arabatskaya spit, and the shoals of the Northern spits. An increased concentration is observed in the area of Zhelezinskaya Bank and in the area of Kazantip Bay, as well as on the Azov coast of the Kerch Peninsula. The peculiarities of processing primary information made it impossible to obtain an increased concentration of digested digital information in the area of the Dolgaya spit and the Sandy Islands in the Taganrog Bay.

Additional research is needed to capture information in these areas. The use of a variational approach based on the solution of associated problems allowed us to determine the field of variable space-based flow of matter on the sea bottom. The identification result $Q(x, y)$ is shown in Fig. 5. In this way, when we set $Q(x, y)$ the result found by integrating the transfer model, we achieve a minimum functional

Fig. 3 RGB composite MODIS AQUA, October 6, 2014

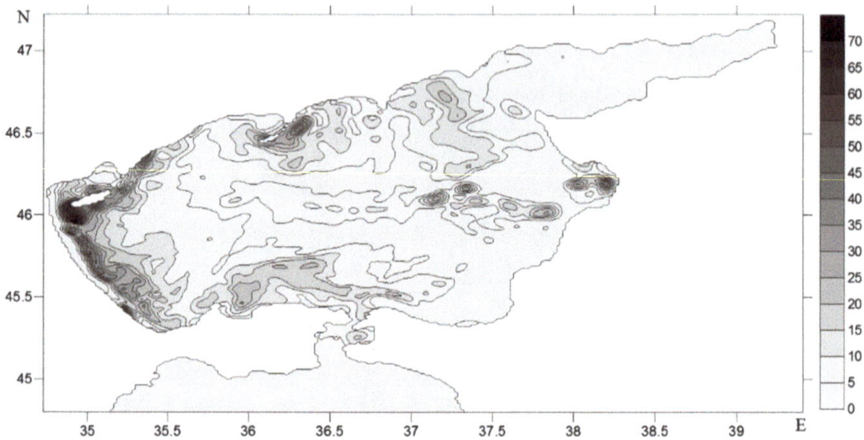

Fig. 4 Model field of concentration on the sea surface October 6, 2014 ($\sigma = 0$ mg/m^3)

with a certain accuracy. The found field of the substance flow at the sea bottom generally correlates well with the information used about the concentration of suspended matter in the upper sea layer (Fig. 4) and repeats not only dynamically active sea areas, but also the features of the bottom relief, i.e. the main known spits and banks where agitation occurs. Note that the alogrithm of identification $Q(x, y)$ is implemented iteratively. The process of falling of the normalized forecast quality functional depending on the iteration number is shown in Fig. 6. For 10 iterations, the error in the formation of the model concentration field at the final moment of time reaches values less than 15%.

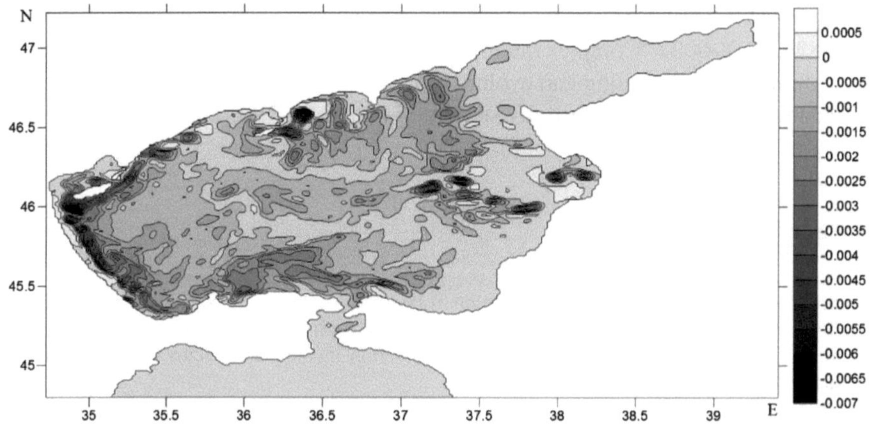

Fig. 5 Identified field of substance flow at the bottom of the sea (mg/m^{2}c)

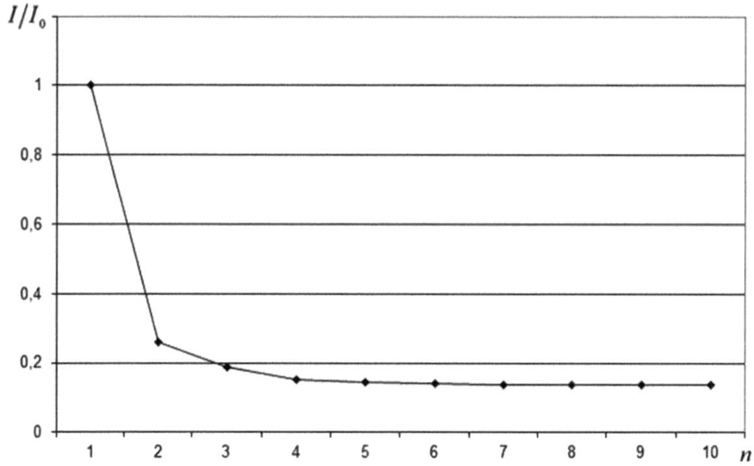

Fig. 6 Normalized values of the quality functional during iterations

5 Conclusions

Thus, as a result of numerical experiments, it is shown that when determining the spatial variable flow of matter, the variable method of assimilation of measurement data, based on the solution of adjoint problems, has good accuracy and convergence. The presented algorithm can be used for assimilation of satellite information about the concentration of suspended matter in the sea to solve various environmental problems in the aquatoires of the Azov and Black seas.

The work was carried out within the framework of the state task on the theme No. 0827-2018-0004 "Complex interdisciplinary studies of Oceanological processes determining the functioning and evolution of ecosystems in the coastal zones of the Black and Azov seas" and partially supported by the RFBR grant 18-45-920,035.

References

Blumberg AF, Mellor GL (1987) A description of the three-dimensional coastal ocean circulation model. Three-dimensional coastal ocean models. Am Geoph Union 4:1–16

Fomin VV (2002) Numerical model of water circulation in the sea of Azov. Nauchnye Trudy UkrNIGMI Iss 249:246–255

Harten A (1984) On a class of high resolution total-variation-stable finite-difference schemes SIAM J Numer Anal 21(1):1–23

Ivanov VA, Fomin VV (2008) Mathematical modeling of dynamic processes in the sea-land zone. In: ECOSI-hydrophysics, Sevastopol, 363 p

Kochergin VS, Kochergin SV (2010) Use of variational principles and solutions of the conjugate problem in identifying the input parameters of the passive impurity transfer model. Ecologicheskaya Bezopasnost' Pribrezhnoj i Shel'fovoj Zon Morya, MGI, Sevastopol, Iss 22:240–244

Kochergin VS, Kochergin SV (2015) Identification of a pollution source power in the Kazantip Bay applying the variation algorithm. Phys Oceanogr 2:69–76

Kochergin VS, Kochergin SV, Stanichny SV (2020) Variational assimilation of satellite data of surface concentration of suspended matter in the sea of Azov. Modern Probl Remote Sens Earth Space 17(2):40–48

Kremenchutsky DA, Kubryakov AA, Zavyalov PO, Konovalov BV, Stanichny SV, Aleskerova AA (2014) Determination of the concentration of suspended matter in the Black sea according to the MODIS satellite data. Environ Saf Coast Shelf Zones Integr Use Shelf Res 29:1–9

Marchuk GI (1982) Mathematical modeling in the environmental problem. M Nauka, 320 p

Marchuk GI, Penenko VV (1978) Application of optimization methods to the problem of mathematical simulation of atmospheric processes and environment. In: Modelling and optimization of complex systems, IFIP-TC7 working conference. Springer, NewYork, pp 240–252

Shutyaev VP (2019) Methods for observation data assimilation in problems of physics of atmosphere and ocean. Izv Atmos Ocean Phys 55:17–31

Shutyaev VP, Parmuzin EI (2019) Sensitivity of functionals to observation data in a variational assimilation problem for a sea thermodynamics model. Numer Anal Appl 12:191–201

Shutyaev VP, Le Dimet F-X, Parmuzin E (2018) Sensitivity analysis with respect to observations in variational data assimilation for parameter estimation. In: Nonlinear processes in geophysics, vol 25, Iss 2, pp 429–439

Zalesny VB, Agoshkov VI, Shutyaev VP, Le Dimet F, Ivchenko VO (2016) Numerical modeling of ocean hydrodynamics with variational assimilation of observational data. Izv Atmos Ocean Phys 52(4):431–442

Detached Flows at Rock Salt Exposition to Aqueous Solution

V. P. Malyukov

Abstract The results of experimental and field studies of breakaway currents when the solution affects rock salt are presented. Experiments are carried out on the samples of rock salt kern from the interval of underground shell-reservoirs for hydrocarbons storage, on the cubes of salt (models of vertical wells), large-scale horizontal models under field conditions. For the first time, it is found that during the mass transfer, when the solution affects the rock salt, a complex mechanism of various breaks of the boundary layer is formed: concentration detachment, surface detachment due to surface roughness (ledges of the trough), detachment at sedimentation of insoluble inclusions from salt—mechanical detachment, vortex detachment at formation of surface vortices, mechanodynamic break-off at the salt plate removal from the crystal, gas break-off at gas emission from rock salt (gas-mechanodynamic break-off at the salt plate removal and gas emission from the crystal, gas-metho-crystal break-off at gas emission from intercrystalline space).

Keywords Rock salt · Core · Solution · Boundary layer · Large-scale model · Vortex · Crater · Break-off currents

1 Introduction

If the area of detachment is small in comparison with the body, this type of flow is called "detached flow" and there occur "detachment of the flow in the boundary layer" or "detachment of the boundary layer" (Zhen 1972).

Prandtl found that a necessary condition for the flow to break-off from the wall is an increase in pressure in the flow direction, i.e., a positive (or inverse) pressure gradient in the flow direction (Prandtl 1939).

Construction of underground gas storage (UGS) facilities in rock salt is an urgent task to ensure reliable gas supplies to various regions of the country and for export. In January 2019 the President of the Russian Federation opened the Kaliningrad UGS in rock salt, specialists of "Gazprom geotechnologies" put into operation the Volgograd

V. P. Malyukov (✉)
Russian Peoples' Friendship University, 6 Miklukho-Maklaya str., 117198 Moscow, Russia

T. Chaplina (ed.), *Processes in GeoMedia—Volume IV*, Springer Geology,
https://doi.org/10.1007/978-3-030-76328-2_4

UGS in the horizontal production capacity (tank, cavern) (Khloptsov 2015). Shell-reservoir is an artificial hollow in the rock intended for reservation of gas or oil products.

The basic parameter that comprehensively characterizes the process of solution influence on rock salt (dissolution of soluble inclusions, destruction of insoluble inclusions and gas yield to the solution) is the mass transfer coefficient.

2 Mass Ejection at Influence of Aqueous Solution on Rock Salt

Mass ejection under the influence of solution on rock salt is combined of physical and chemical processes of mass ejection in combination with hydrodynamic and mechanical destruction of salt, insoluble inclusions and with periodic differentiated separation of boundary layer (Schlichting 1974). The mass efficiency depends on the intensity of detachment of the boundary layer from the rock salt surface under the action of solution.

The studied samples of rock salt differed from each other in the size of crystals, in the number and composition of soluble and insoluble inclusions, and different forms of gas inclusions. Under the influence of solution on the surface of the rock salt sample, vertical and horizontal models of shell-reservoirs and constructed shell-reservoirs a break-off concentration boundary layer is formed. There is a periodic detaching boundary layer from the surface of salt (vertical or at different angles to the vertical), as well as from the surface at the roof of the mine (concentration-gravity detach of the boundary layer at the roof—the upper part of the mine).

Construction of shell-reservoirs in rock salt by underground dissolution through drilling wells is characterized by the coefficient of mass output, the value of which determines the terms of construction and affects the reservoir configuration (Khloptsov 2015; Malyukov and Kazaryan 2007a; Malyukov and Vorobiev 2019a).

Studies are carried out and the values of mass efficiency coefficient are determined when the solution affects the vertical surface of cylindrical form (kern) rock salt samples from the interval of underground shell-reservoirs construction. The peculiarities of mass efficiency under the influence of solution on rock salt are considered for different deposits.

Different inclusions in the rock salt react differently to the solution. It has been experimentally established that the rate of gypsum dissolution increases with increasing NaCl concentration in the solution and with temperature increase (Chaplina 2019).

The laboratory facility for determination of rock salt mass transfer coefficient when the solution affects the vertical surface of the cylindrical salt sample is presented in Fig. 1. Mass efficiency of rock salt (K) under the influence of the solution is determined by the relation:

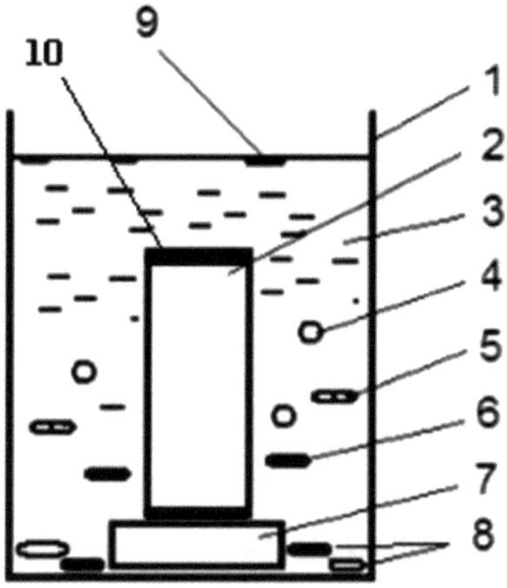

Fig. 1 Scheme of the laboratory facility for determination of rock salt mass transfer coefficient under the influence of solution: 1—glass container; 2—specimen; 3—solution; 4—gas bubbles; 5—salt plates; 6—particles of insoluble inclusions; 7—rack; 8—fallen insoluble particles and salt plates; 9—floating light admixtures; 10—insulation coating of core ends

$$K = \frac{\Delta G}{S_{av} t (C_n - C_s)} \tag{1}$$

where K is the coefficient of mass yield of rock salt, m/s; ΔG—sample weight loss, g; S_{av}—average value of rock salt surface under the influence of solution, cm^2; t—duration of solution influence, s; C_n and C_s—concentration of saturated and unsaturated solutions, g/cm^3.

Experimental studies of the mass transfer coefficient value under the influence of the solution are carried out on salt samples from the intervals of underground shell-reservoirs construction. Studies of mass efficiency and the form of vertical wells on salt cubes and horizontal workings in natural conditions of the salt deposit are carried out.

3 Detachment of the Concentration Boundary Layer

At exposure of the solution on the surface of the sample of rock salt and model specimens there is a local periodic detachment of the concentration boundary layer (manifestation of positive pressure gradient at flow, concentration detachment, Fig. 2).

At hydrodynamic influence of the solution (unsaturated aqueous solution) on the sample of rock salt at the flow of the vertical surface there are breaks concentration in the boundary layer (BL). The concentration boundary layer is formed when the solution affects the rock salt surface in the process of mass transfer. Observations

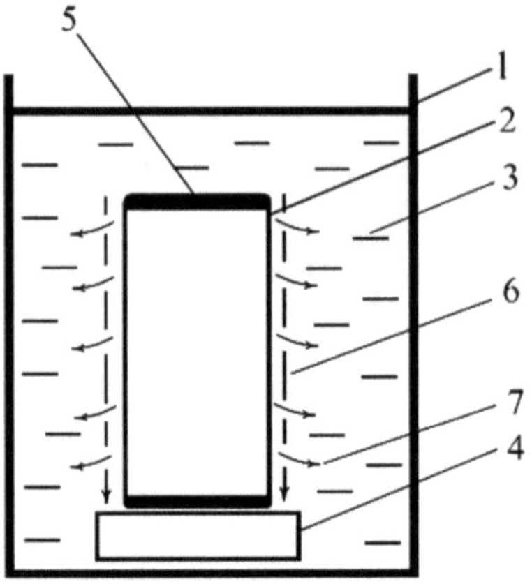

Fig. 2 Scheme of movement and detachment of the concentration boundary layer from the vertical surface of the core when the solution affects the salt surface: 1—glass container; 2—specimen; 3—solution; 4—rack; 5—insulating coating of specimen ends; 6—movement of the concentration boundary layer along the vertical surface of specimen; 7—detachment of the concentration boundary layer from the vertical specimen surface

show that BL is saturated with salt and a part of the BL is locally detached from the specimen surface. A small part of the BL moving down the rock salt surface thickening in the lower part can be seen as a "conical drop" with volume expansion at the bottom. The concentration "drop" of the BL is about 5 mm long and about 3 mm wide. The frequency of local detachment of the concentration BL for the vertical surface depends on the structure, material composition of the rock salt and concentration of the solution. For inclined surfaces of rock salt, the frequency of detachment of the concentration BL additionally depends on the sloping angle of the surface.

In the presented schematic diagram (Fig. 3) all designations are the same for left and right parts from the centerline.

In real conditions the surfaces of rock salt with different sloping angles can be located in different parts of the mine taking into account geological, technological factors and concentration of stratified solution. Columns 2 and 3 in Fig. 3 denote mine functioning by technology of counterflow, when water is supplied through column 3, located above column 2, with the help of which the solution is taken away.

In real conditions, the development of reservoirs out taking into account the specific situation is carried out using different technologies, not necessarily on the counterflow.

When the inclination angle to the horizon of the mine surface increases to α_1 (Malyukov and Vorobiev 2019a), the intensity of the concentration BL detachment increases and the mass transfer coefficient increases. With decrease in the sloping angle of the mine surface to the horizon α_2 (Trubetsky and Kaplunov 2016) the intensity of the concentration BL separation decreases and so is the value of the mass transfer coefficient. With increase of breaks number for BL from the surface of

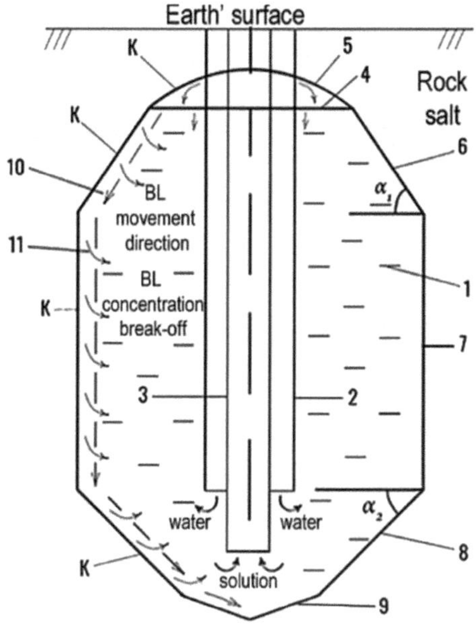

Fig. 3 Scheme of movement and detachment of concentration BL on different surfaces of vertical underground shell-reservoir contour in rock salt under the influence of solution: K—excavation contour; 1—solution; 2—water supply column; 3—solution lifting column; 4—horizontal surface of excavation roofing; 5—dome-shaped excavation roofing surface; 6—excavation surface inclined downwards; 7—vertical excavation surface; 8—excavation surface inclined upwards; 9—excavation hydraulic pipe; 10—movement of concentration BL along different excavation surfaces; 11—detachment of concentration BL from different excavation surfaces

rock salt, the value of mass efficiency increases (the highest mass efficiency under equal conditions, as a rule, is associated with the roof of the reservoir). At the roof (the upper part of the reservoir) is a concentration-gravity separation of the BL with greater intensity than on other surfaces.

The maximum intensity of detachment of the concentration BL from the surface of rock salt refers to the horizontal surface of the reservoir roof (Schlichting 1974).

In the actual conditions of reservoir development at the bottom of conical gidroprom is formed (Nadai 1969), and in the upper part may be a dome of the reservoir (Malyukov and Kazaryan 2007a) or horizontal surface of the roof of the reservoir (Schlichting 1974).

Under real conditions, when exposed to the surface of rock salt excavation, taking into account different values of mass transfer coefficient of salt inclusions and different destructibility of insoluble inclusions, all reservoirs have individual surface forms.

Experimental determination of mass efficiency value for the vertical surface of a sample of rock salt can be equated to the average value of the mass efficiency of different surfaces under the influence on the rock salt of the reservoir contour.

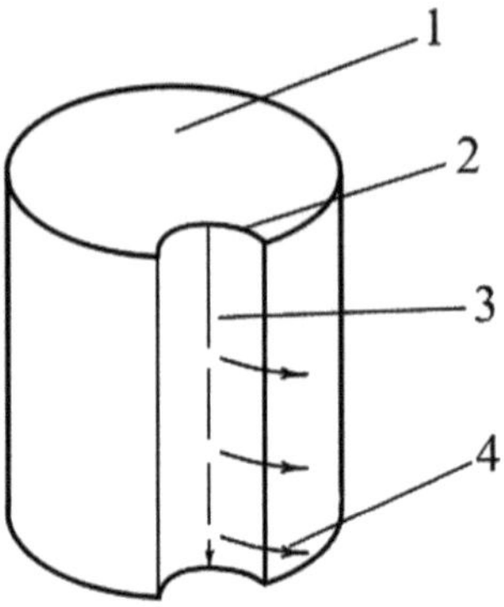

Fig. 4 Scheme of three-dimensional vertical formation on the surface of the kern of rock salt under the influence of a solution with local breaks of the BL from the surface of formation: 1—rock salt specimen; 2—three-dimensional vertical formation; 3—movement of the concentration BL along the vertical formation; 4—detachment of the concentration BL from the surface of vertical formation

When the vertical specimen surface is in the flow, three-dimensional vertical formations of different lengths with local breaks of the BL appear at the surface of formation (Fig. 4).

The velocity of BL movement on vertical formation boundary depends on geological, technological factors, concentration of stratified solution.

4 Mechanodynamic Detachment of the Boundary Layer

For conditions of the Far North (Republic of Sakha, Botuobinskoye deposit) on kern samples from a depth of 457–509 m the values of mass transfer coefficient of rock salt at low process temperatures (+3 °C) are determined. At the solution influence on the salt samples thin rock salt plates with thickness of ~1–2 mm and diameter of 2–3 mm (plates of ellipsoid form, sometimes of larger size ~5 mm) are torn from the crystals. During the whole time, the salt plates are detached from the specimen surface with a spacing of up to 5 cm. Under the influence of the solution, the detachment from the rock salt surface of lenticular plates with sharp edges is accompanied by a sound like a pencil strike against a wooden surface, "shooting" of rock salt (Trubetsky and Kaplunov 2016) under the hydrodynamic influence of the solution (Fig. 5).

It is found that in this case there is the destruction of the near-surface part of the salt crystal (Nadai 1969) with the formation of an indentation on the crystal surface (an increase in the mass transfer surface) and the rupture of the BL by the detachable salt plate with an additional local tear of the BL—the BL destruction on the salt surface.

The phenomenon of BL detachment by salt plates bouncing off crystals has been discovered—mechanodynamic detachment of the BL,—in addition to detachment

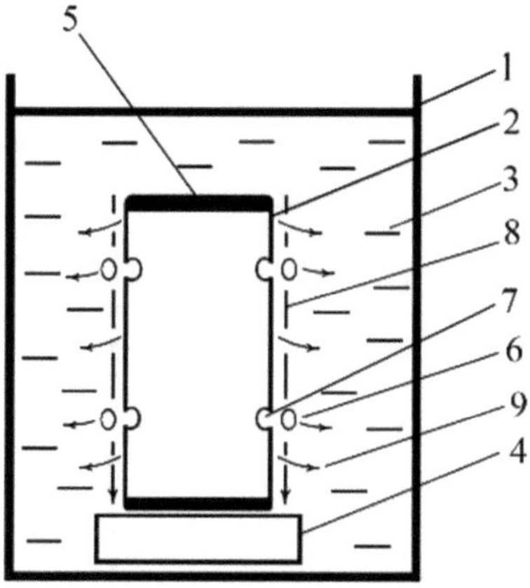

Fig. 5 Scheme of BL movement and complex concentration and mechanodynamic separation of the BL from the vertical core surface under the influence of the solution on the salt surface: 1—glass container; 2—specimen; 3—solution; 4—rack; 5—insulating coating of core ends; 6—bouncing salt plates with detachment of boundary layer; 7—salt surface trough; 8—movement of concentration BL along vertical core surface; 9—detachment of concentration BL from vertical specimen surface

of the concentration BL in the process of solution exposure (Fig. 5). Mass transfer coefficient of a sample with salt plates removal is 29% higher than that of a sample without salt plates removal.

5 Boundary Layer Surface Break-Off

Many papers consider "traditional" detachment of the BL in the flow near solid bodies of different shapes and sizes (cylinder, rectangle, square, cone, etc.). In real conditions, flow around various inclusions in rock salt of irregular shape, as well as a combination of inclusions (insoluble, soluble) of various shapes and sizes (protuberances) and hollows on the surface formed by the influence of the solution on the rock salt, occurs. In this case, the BL is detached from the surface due to surface irregularities (protrusions, hollows), there is a surface detachment of the BL when the surface is subjected to flow.

6 Mechanical Detachment of the Boundary Layer

When working out models of horizontal shell-reservoirs in rock salt at the Khodja-Mumyn deposit, self-organized regular spherical clay formations (Lizegang formations) are found, which under the influence of the solution disintegrated into the

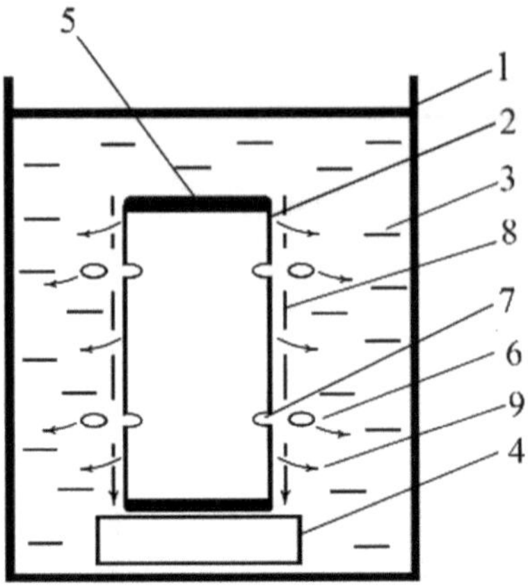

Fig. 6 Scheme of BL movement and complex concentration and mechanical detachment of BL from vertical core surface under the influence of solution on salt surface: 1—glass container; 2—specimen; 3—solution; 4—rack; 5—insulating coating of core ends; 6—deposition of insoluble inclusions with mechanical detachment of the boundary layer; 7—indentation; 8—movement of the concentration BL along the vertical surface of the specimen; 9—detachment of the concentration BL from the vertical surface of the core and BL detachment at the deposition of the insoluble inclusion

smallest particles (Malyukov 2016). This is an example of a possible flow of spherical formations with a BL under real conditions.

Kern samples from well 1P of the Kaliningrad UGS are characterized by a gradual increase in the feldspar structure (larger crystals) in the rock salt by depth and a decrease in the mass transfer coefficient. In the process of dissolution there is honeycomb-like structure elements of salt and anhydrite of different sizes with anhydrite frame. Salt crystals are ‘ringed’ with anhydrite layer. The anhydrite frames around the salt crystals are destroyed under the influence of the solution, and on the vertical surface of the sample protruding approximately 1–2 mm anhydrite frame is preserved around the crystal. The BL is locally bounded by anhydrite interlayers around the salt crystal and a mechanoanhydrite break-off of the BL occurs (Fig. 6).

7 Vortex Detachment of the Boundary Layer

It is found that hydrodynamic vortex structures are self-organized at rock salt surfaces under the influence of solution. Self-organized structures are formed at interconnected hydrodynamic and mass exchange processes due to interaction of self-organizing vortex structures and surfaces of salt of vertical sample and on the contour of underground vertical and horizontal reservoir are formed surface relief structures (imprints of self-organizing hydrodynamic vortex structures).

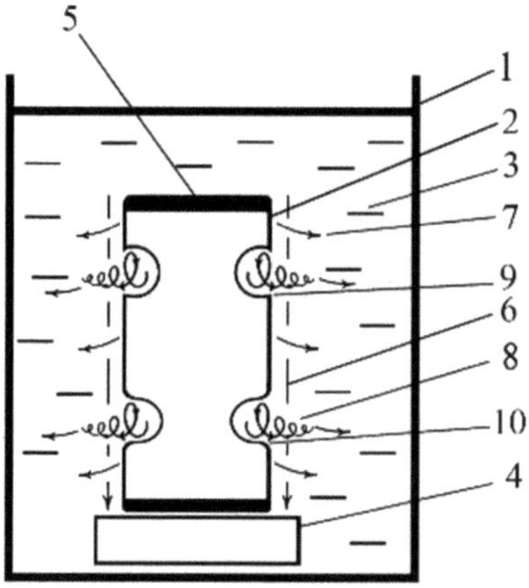

Fig. 7 Scheme of BL movement and complex concentration and vortex BL breakdown from vertical specimen surface: 1—glass container; 2—specimen; 3—solution; 4—rack; 5—insulating coating of core ends; 6—movement of concentration BL along vertical surface of specimen; 7—detachment of concentration BL; 8—vortices and vortex detachment of boundary layer; 9—three-dimensional crater with sharp edges; 10—three-dimensional crater with smooth contours

Emerging vortex structures intensify mass transfer efficiency. Vortices formed at the surface of rock salt in the process of solution influence are considered as concentrated vortices (Alexeyenko et al. 2005; Chaplina 2019).

Vortex structures formed when rock salt is exposed to the solution destroy the boundary layer, thus forming detachment of BL and increasing mixing of the solution in the near-circuit zone at the salt surface.

The phenomenon of the BL detachment by vortices formed at the surface of rock salt under the influence of solution—vortex detachment of the BL—in addition to the detachment of the BL concentration—has been revealed (Fig. 7).

As a result of experimental studies, the phenomena characterizing the break-off processes and rock salt mass ejection under the influence of the solution are found. The phenomenon of detachment of the BL by the salt plates bouncing off the crystals under the influence of the solution at low temperature has been revealed—mechanodynamic detachment of the boundary layer. The phenomenon of detachment of the BL by vortices formed at the surface of rock salt under the influence of solution—vortex detachment of the BL—is discovered.

Visual inspection shows that in the process of mass transfer on the kern surfaces, as well as on the surface of the cubic model of vertical salt excavation (reservoir) and large-scale horizontal models in the field conditions, a number of craters is formed—the surface is wrapped with a system of three-dimensional hollows.

It is found that vortices (Trubetsky and Kaplunov 2016) on the actual rock salt mass transfer surface form three-dimensional craters as from "large" vortices with smooth outlines and concave surface. On the kern sample three-dimensional craters with smooth contours and concave surface are also formed with additional craters with sharp edges (Fig. 8).

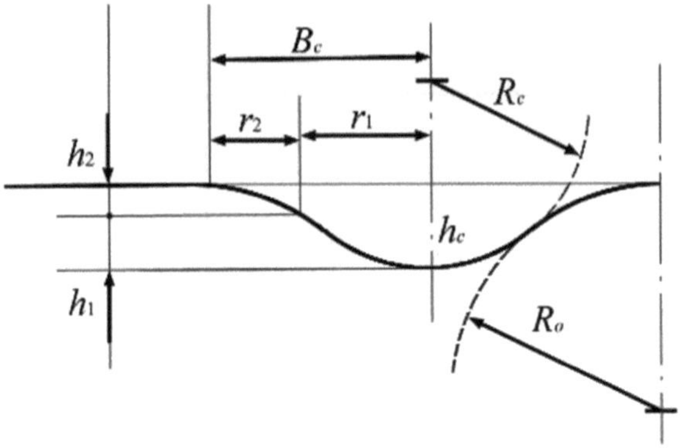

Fig. 8 Characteristic parameters of the crater: h_C—crater height; B_C—geometric transverse size of crater; R_C—crater radius; R_O—curvature radius of edge

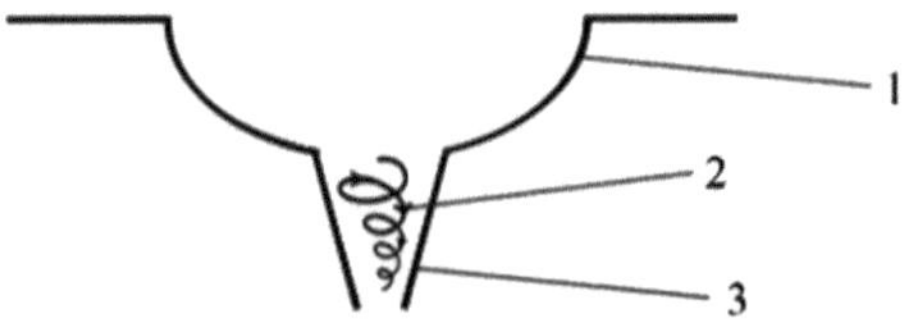

Fig. 9 Scheme of two-level crater formation on the surface of the rock salt specimen under the influence of a solution with consecutive impact of two different vortices and subsequent BL break-off: 1—larger crater size from the first "large" vortex; 2—the second "smaller" vortex spiral shape; 3 –crater of smaller size from the "smaller" vortex

The h_c/B_c ratio is important in terms of hydrodynamics and mass transfer, which is approximately 0.23 for the natural surface of rock salt. The resulting craters are relatively deep by this parameter.

During the visual inspection of the specimen it is found that a spiral-shaped print is formed on the crater surface formed by a "big" vortex under the influence of "smaller" vortex (Fig. 9).

8 Gas Detachment of the Boundary Layer

Gas-dynamic effects of the solution on rock salt with a significant amount of gas in the specimen and during the construction of shell-reservoirs are studied. In the process of salt mass transfer with gas inclusions in the crystal, double mechanism of influence on the BL is observed: salt particles and gas ejection from the crystal (gas-crystal separation). The mechanism of influence on the BL is also considered at salt mass transfer with inclusion of gas in intercrystalline space, and ejection of gas and particles of insoluble inclusions from intercrystalline space (gas intercrystalline separation) is noted.

When determining the mass transfer coefficient on the core samples of rock salt from the interval of mine working-out-capacities, increased gas content is found in the rock at Leikovskiy stock (Ukraine) and Tuz-Galu deposit (Turkey, Table 1).

Study of gas emission at mass ejection and its influence on the process of construction of shell-reservoirs is carried out at creation of underground storage at Leikovskiy salt stock. Gas emissions from rock salt are considered throughout the entire period of construction of shell-reservoirs for various technological operations.

Measurements of the coefficient of mass transfer of vertical surface of rock salt on core material from the intervals of underground shell-reservoirs in conditions of natural convection revealed gas emission, containing mainly methane (>99%).

Actual solubility of gas in the unit of brine volume when brought to the ground surface (C_v) according to the brine pipeline measurements was found to be 22.66 cm^3/l (analyses are performed by UKRNIIGAS after the 4T tank construction completion). Gas solubility in brine volume unit at phase equilibrium (temperature +21.5 °C, saturated brine) is equal to 533 cm^3/l. The maximum gas content at the kern mass transfer of the Leikovskiy salt stock is 136.26 ml/kg (13.63 ml/100 g).

Laws of underground reservoir formation are studied on model in laboratory conditions (salt block $100 \times 50 \times 45$ cm^3). Parameters of model processing are set on the basis of similarity theory provisions. Linear scale is equal to 250. Figure 10 shows the configuration of model reservoir.

In the process of gas emission from rock salt, the rock walls enclosing the gas are broken down towards the reservoir. The process of destruction of the near-border zone during gas emission affects the parameters of the BL and increases mass transfer. The value of mass ejection varies depending on the amount of gas and its pressure, as well as on the properties of the salt. In the process of the shell-reservoirs construction in rock salt gas manifestations (gas emission) are an integral part of mass transfer. At the Leikovskiy salt stock with a significant gas content in the intercrystalline space of rock salt revealed a factor of faster construction of shell-reservoirs (about 30%) compared to the calculated parameters.

The actual configuration of the shell-reservoirs built in rock salt with a significant gas content is shown in Fig. 11. At the Leikovskiy deposit with significant gas content 10 mine shell-reservoirs are created (valuable capacity of shell-reservoirs under design regulations is 50 or 75×10^3 m^3).

During mass transfer, gas and particles of insoluble inclusions are released from the rock salt. Depending on the stratification of the solution, insoluble heavy particles

Table 1 Mass transfer coefficient of vertical surface of rock salt under conditions of natural convection at 20 °C, determined on the kern samples of the UGS 2 well Tuz-Galu

Specimen no	Depth (m)	d (cm)	h (cm)	ρ (g/cm^3)	G_1 (g)	G_1 (g)	K × 10^{-5} (m/s)
27P	962.6–963.0	9.46	9.99	2.17	1523.45	1249.60	1.694

Fig. 10 Form and sizes of underground reservoir model with volume of 50×10^3 m^3

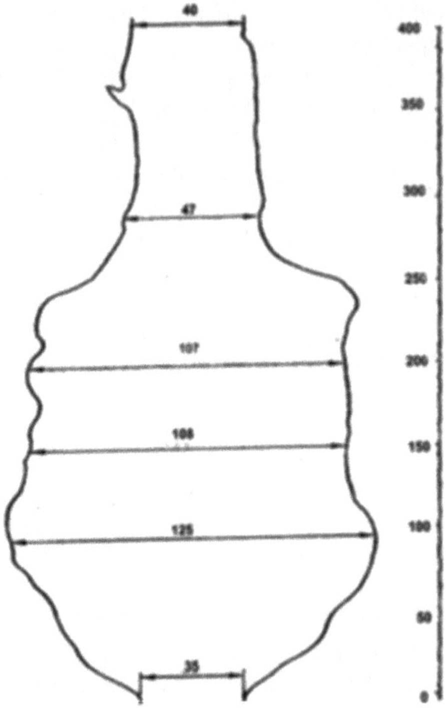

are lowered to the bottom of the mine, and light particles along with gas microbubbles rise up, resulting in a significant amount of particles get into the liquid solvent (diesel fuel, etc.), considerably affecting its properties.

For the first time, the phenomenon of salt rock destruction at mass transfer is observed as bouncing (microburst) of plates in the process of rock salt dissolution with intercrystalline gas content (Tuz-Galu deposit).

In determining the mass transfer coefficient of the vertical surface of the rock salt kern under conditions of natural convection, there is a uniform, almost of equal value, gas distribution in the salt crystals of the Tuz-Galu deposit, which creates a locally variable pressure in the rock salt (pressure increase in the cavities of the crystals filled with gas).

The process of mass transfer out of rock salt with increased gas content is characterized by the mass of rock salt transferred into the solution, the mass of bouncing rock salt plates and the volume of gas released.

This process of mass transfer can be considered as combined gas-dynamic phenomenon—gas emission with the bounce of salt plates (destruction of salt crystals accompanied by gas emission). Microburst of gas from the salt crystal with the bouncing of the salt plate in the process of mass transfer is a continuous effect (destruction of the contact zone of the rock) (Figs. 12 and 13).

Average values of mass transfer coefficient for vertical rock salt kern surface with significant gas content under conditions of natural convection at 20° C for Leikovskiy

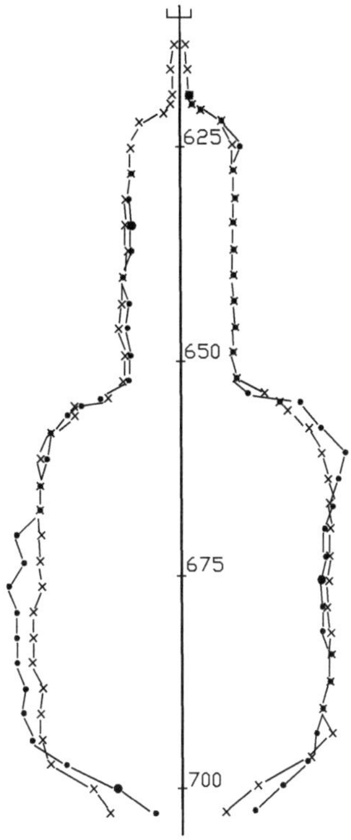

Fig. 11 Vertical cross-sections of underground 4T tank built in rock salt with significant gas content (according to acoustic surveys). Sections: •– – –•South-North; ×– – –×West–East

stock deposit are: from interval of well 3TK = 1.5972 $\cdot 10^{-5}$ m/s; 4TK = 1.5750 $\times 10^{-5}$ m/s; 5TK = 1.6667 $\times 10^{-5}$ m/s; 6TK = 1.6250 $\times 10^{-5}$ m/s; 7TK = 1.6611 $\times 10^{-5}$ m/s; 8TK = 1.6083 $\times 10^{-5}$ m/s; 9TK = 1.6139 $\times 10^{-5}$ m/s.

Average values of mass transfer coefficient of vertical surface of rock salt under conditions of natural convection at temperature 20 °C for core of Tuz-Galu deposit (with increased gas content inside salt crystals) are for well 1 K = 1.7500 $\times 10^{-5}$ m/s; well 2 K = 1.6167 $\times 10^{-5}$ m/s. The highest mass transfer coefficient (K = 2.1139 $\times 10^{-5}$ m/s) is registered for specimen 52P.

The value of mass transfer coefficient of rock salt with increased gas amount increases in comparison with the mass transfer coefficient of traditional salt, and therefore, in spite of different mechanisms of gas manifestations at dissolution of rock salt of Leikovskiy stock and Tuz-Galu, the values of their mass-effect ratio are quite close.

9 Conclusion

As a result of experimental studies, the phenomena characterizing break-off processes and mass output of rock salt under the influence of the solution are found. The

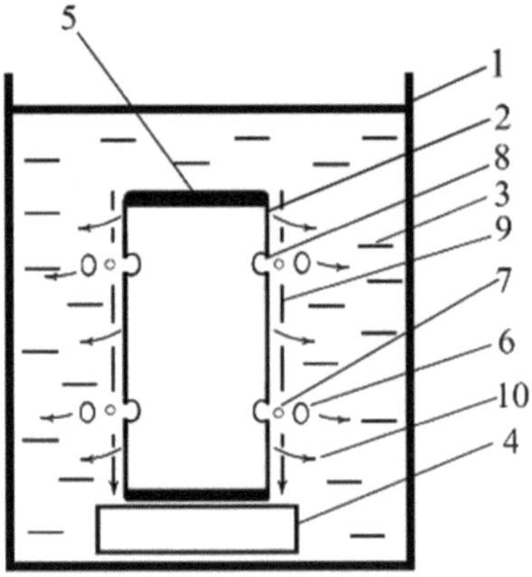

Fig. 12 Scheme of BL movement and complex concentration and gas-mechanodynamic separation of BL from vertical specimen surface under the influence of solution: 1—glass container; 2—specimen; 3—solution; 4—rack; 5—insulating coating of specimen ends; 6—bouncing salt plates with detachment of boundary layer; 7—gas bubbles, from rock salt crystal; 8—salt surface trough; 9—movement of concentration BL along vertical surface; 10—detachment of concentration BL from vertical kern surface

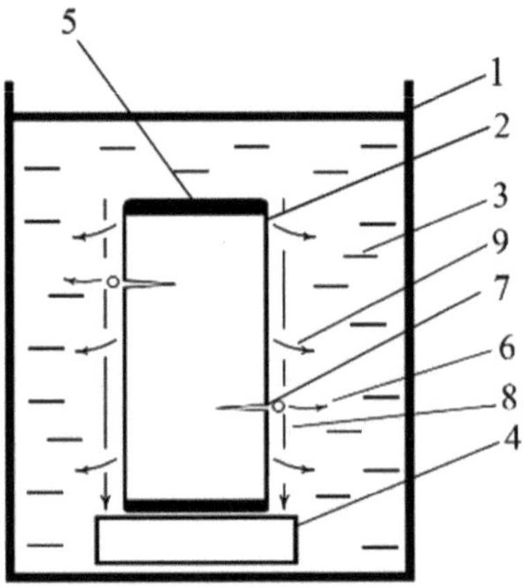

Fig. 13 Scheme of BL motion and combined concentration and gas-intercrystalline BL detachment from vertical specimen surface under the influence of solution on the salt kern surface: 1—glass container; 2—specimen; 3—solution; 4—rack; 5—insulating coating of core ends; 6—gas bubbles released from rock salt with BL detachment; 7—hollow space in rock salt; 8—movement of the concentration BL along the specimen vertical surface; 9—detachment of the concentration BL from the vertical surface and at gas emission

phenomenon of BL detachment at salt plates bouncing off the crystals under the influence of the solution at low temperature has been revealed—mechanodynamic detachment of the BL. The phenomenon of BL detachment by vortices formed at the surface of rock salt under the influence of solution—vortex BL detachment is discovered.

For the first time, it is found that in the process of mass transfer, when the solution affects the rock salt, complex mechanism of various BL detachment types is formed: concentration detachment, surface detachment due to surface roughness (hollow ledges), detachment at sedimentation of insoluble inclusions from salt—mechanical detachment, vortex detachment at formation of vortices on the surface, mechanodynamic break-off at the salt plate removal from the crystal, gas break-off at gas emission from rock salt (gas-mechanodynamic break-off at the salt plate removal and gas emission from the crystal, gas-intercrystalline break-off at gas emission from intercrystalline space).

The surface BL break-off occurs when moving BL meets an obstacle on the rock salt surface. Vortex detachment of the BL occurs when the moving solution produces vortices affecting the BL and rock salt. When the solution affects the rock salt, BL with constantly changing parameters at each point is formed at each point of the solution and salt contact and nearly constant local detachment of the BL occurs.

The intensity of BL detachment from the surface of rock salt under the influence of the solution determines the value of the mass transfer coefficient.

Change of value of rock salt mass ejection coefficient under the influence of solution is linked with geological parameters of rock, technological parameters of underground shell-reservoir construction, change of value of solution concentration and parameters of intensity of BL detachment (Malyukov and Kazaryan 2007b; Malyukov 2009a, b; Malyukov and Starovoitova 2017; Malyukov and Vorobiev 2019b).

Analogously to the Stanton's criterion, it is possible to present the ratio of mass efficiency coefficient K to the average velocity V of the solution between the BL breaks at the rock surface by following dependence:

$$M = \frac{K}{V} \tag{2}$$

The phenomena under consideration lay in the field of hydro and gas mechanics. The hydrogasodynamic phenomena characterize diffusion-convection-vortex process of mass transfer with different mechanisms of BL detachment at influence of water solution on rock salt.

References

Alexeyenko SV, Kuibin PA, Okulov VL (2005) Introduction to the theory of concentrated vortices. Institut Komputernyh Issledovany. Moscow-Izhevsk, 504 p [in Russian]

Chaplina TO (2019) Substance transfer in vortex and wave currents in single-component and multi-component media. Kim L. A. Publishing House, Moscow, 201 p [in Russian]

Khloptsov VG (2015) Underground natural gas storages in rock salt deposits. Gas Industry 9:28–31 [in Russian]

Lebedev AL (2015) Kinetics of plaster dissolution in water. Geochem 9:828–841 [in Russian]
Malyukov VP (2009a) Analysis of the self-organizing hydrodynamic structures and the topology of the rock salt dissolution relief. In: 9th international symposium on salt, vol 1. China, Beijing, pp 640–644
Malyukov VP (2009b) The effect of salt burst scaling in solution-mining. In: 9th international symposium on salt, vol 1. China, Beijing, pp 610–612
Malyukov VP (2016) Formation of Lysegang rings in rock salt with nanoparticles. Mining Inf Anal Bull Sci Tech J 10:242–248 [in Russian]
Malyukov VP, Kazaryan VA (2007a) Investigation of the mass transfer during the underground salt storage construction. Sci Tech Gas Industr 1(29):64–71 [in Russian]
Malyukov VP, Kazaryan VA (2007b) Study of mass transfer during construction of underground storage facilities in rock salt. Sci Tech Gas Industr 1(29):64–71 [in Russian]
Malyukov VP, Starovoitova YI (2017) Flotation processes in construction of mine shell-reservoirs by underground dissolution of rock salt with high gas content. Herald Rus Univ Peoples' Friendship Ser Eng Res 3:391–397 [in Russian]
Malyukov VP, Vorobiev KA (2019a) Complex use of oil and salt deposits. In: "Physical and mathematical modeling of processes in GeoMedia" 5th international science conference-School of Young Scientists. Moscow, Ishlinsky Institute for Problems in Mechanics RAS, pp 98–100
Malyukov VP, Vorobiev KA (2019b) Complex use of oil and salt deposits. In: "Physical and mathematical modeling of processes in geomedia" 5th international Science conference-School of Young Scientists. Moscow, Ishlinsky Institute for Problems in Mechanics RAS, pp 98–100 [in Russian]
Nadai A (1969) Plasticity and destruction of solids, vol 2. Mir, Moscow, 863 p [in Russian]
Prandtl L (1939) Mechanics of viscous fluids. In: Durend WF (ed) Coll. "Aerodynamics", vol 3. Oborongiz, Moscow-Leningrad
Schlichting G (1974) Theory of the boundary layer. Nauka, Moscow, 711 p [in Russian]
Trubetsky KN, Kaplunov DN (eds) (2016) Mining: terminological dictionary. Gornaya Kniga, Moscow. 635 p [in Russian]
Zhen P (1972) Breaking currents, vol 1. Mir, Moscow, 299 p [in Russian]

Innovative Technologies for Construction of Horizontal and Double-Deck Underground Tanks in Rock Salt

V. P. Malyukov and A. A. Shepilev

Abstract Innovative technologies of building underground shell-reservoirs in rock salt created by underground dissolution at the Volgograd underground gas storage (UGS) are analysed. The analysis of practical significance of scientific research on construction of tunnel (horizontal) underground shell-reservoirs built by the technology with the use of vertical-horizontal and vertical wells, as well as vertical double-deck shell-reservoirs built by underground dissolution of rock salt through one well developed for the first time in the world at Rossoshinskaya salt deposit are carried out. During the research it is revealed and confirmed the worldwide tendency to increase the number of UGS in the structures of rock salt, which can provide high rates of gas injection and pumping. Introduction of innovative technologies enables the most rational use of available mining and geological conditions—construction of horizontal reservoirs in formations of limited thickness and the most expedient use of salt structures, the thickness of which can reach hundreds of meters for the construction of double-deck tanks, while increasing the rate of maximum daily productivity of underground gas storage.

Keywords Shell-reservoir · Rock salt · Underground dissolution · Double-deck · Horizontal · Reservoir · Underground gas storage

1 Introduction

A record filling of underground gas storage facilities in Europe emerged in October 2019—as on October 27, according to AGSI+. the average occupancy in Europe was 97.84%: namely Germany—21.4 bln m^3 (98.6% filled of total capacity); Ukraine—20.3 bln m^3 (65.4%); Italy—18.3 bln m^3 (97.4%); France—12.4 bln m^3 (100%); Netherlands—12.3 bln m^3 (99.1%); Austria—8.91 bln m^3 (100%); Hungary—6.47 bln m^3 (97.1%); Slovakia—3.96 bln m^3 (97.2%); Czech Republic—3.38 bln m^3

V. P. Malyukov (✉) · A. A. Shepilev
Russian Peoples' Friendship University, 6 Miklukho-Maklaya str, 117198 Moscow, Russia

T. Chaplina (ed.), *Processes in GeoMedia—Volume IV*, Springer Geology,
https://doi.org/10.1007/978-3-030-76328-2_5

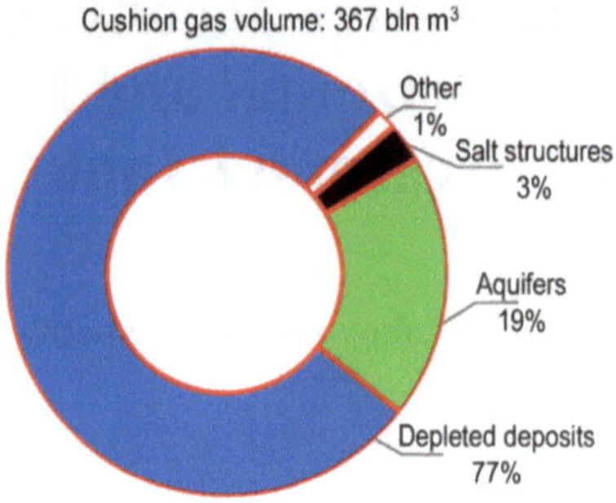

Fig. 1 Underground gas storage figures depending on the type of storage facility

(99.8%); Poland—3.27 bln m^3 (100%); Romania—2.89 bln m^3 (91.5%); Spain—2.82 bln m^3 (89.1%). Comparing these volumes as for the same date in 2018, all countries without exception increased their natural gas reserves.

There is a growing demand for storage and the role of storage as means of ensuring continuity and flexibility of supply, in addition to the function of smoothing out uneven consumption. Long-term investments in the construction of underground gas storage (UGS) facilities are thus justified.

When analyzing the distribution of underground gas storage facilities by storage type: in depleted (exhausted) deposits, in aquifers, in salt deposits,—there is a predominance of storage facilities built in depleted deposits, which allow storing large volumes of gas and mainly serve to regulate fluctuations in seasonal consumption. UGS facilities built in depleted deposits account for 73% of the total number of UGS facilities and concentrate 79% of the total active gas volume disposed in UGS facilities. Underground gas storage values are illustrated in Fig. 1 depending on the type of UGS.

However, important changes are taking place in the gas storage market and today flexibility is the key to maintaining market positions.

There is a trend in Europe and North America to increase the number of storage facilities in rock salt structures, which can provide high rates of gas injection and extraction. 99 UGS facilities in salt structures were in operation worldwide by the end of 2018, accounting for 15% of the total number of UGS facilities. Although such UGS facilities account for only 8% of the total active gas volume, they provide up to 26% of the global maximum gas withdrawal rate.

The operational gas reserve in UGS facilities was 72.232 bln m^3 by the autumn–winter period of 2019–2020, and the potential maximum daily capacity at the beginning of the withdrawal season in the Russian Federation reached a record level of 843.3 mln m^3. This is 30.8 mln m^3 higher than in the previous year. The dynamics of potential maximum daily capacity increase is shown in Table 1.

Application of innovative technologies of underground reservoirs construction in the rock-salt provides an opportunity to work in special mining and geological conditions—layers of limited thickness during the construction of horizontal tanks and for

Table 1 Dynamics of potential maximum daily capacity increase

Period	Maximum daily capacity (mln m^3)	Variance (mln m^3)
2013–2014	727.8	–
2014–2015	770.4	42.6
2015–2016	789.9	19.5
2016–2017	801.3	11.4
2017–2018	805.3	4.0
2018–2019	812.5	7.2
2019–2020	843.3	30.5

rational use of rock-salt layers of high thickness in the construction of double-deck vertical tanks, thereby opening up new prospects for the construction of underground tanks in previously inaccessible places and with greater production efficiency.

2 Results

Innovative underground storage construction technologies are used in the construction of the Volgograd underground gas storage facility. The Volgograd UGS is put into operation and became the second gas storage facility in the Russian Federation built in rock-salt structures. The Kaliningrad UGS facility was the first in Russia to be commissioned, while the Yerevan UGS facility in Armenia was commissioned earlier in the USSR.

Volgograd underground gas storage facilities are located in the salt bed of the Rossoshinskaya salt deposit, dedicated to the Kungur stage of the lower section of the Perm system, which thickness is 520 m, and the roof is located at a depth of 1143–1163 m (Litvinov 2002). The rocks forming the Kungur tier are represented by rock salt, anhydrides, potassium-magnesium salts and dolomites. It was also found out that the site for underground shell-reservoirs construction is represented by eight rhythmopacks within Rossoshinskaya salt deposit (Fig. 2).

In total it is planned to create 16 mine shell-reservoirs in V, VI and VIII rhythmopack units, the total geometric volume of which will reach 4.4 mln m^3 and will contain up to 1080 mln m^3 of gas, where 820 mln m^3 will be the share of active gas and 260 mln m^3 of cushion gas.

The project provides that out of the planned 16 workings out 14 will be built within the limits of V and VI rhythmopacks in the interval from 1350 to 1455 m. Tanks will have two tiers and vertical orientation with the volume of 284,000 m^3 each (Fig. 3).

In VIII rhythmopack, in the interval from 1160 to 1210 m, two reservoirs are put: the first—vertical, with the volume of 100,000 m^3, the second one—of tunnel type with use of two wells (vertical and inclined-horizontal), with the volume of 350,000 m^3 (Fig. 4) with application of non-solvent in the upper part of the reservoir to control

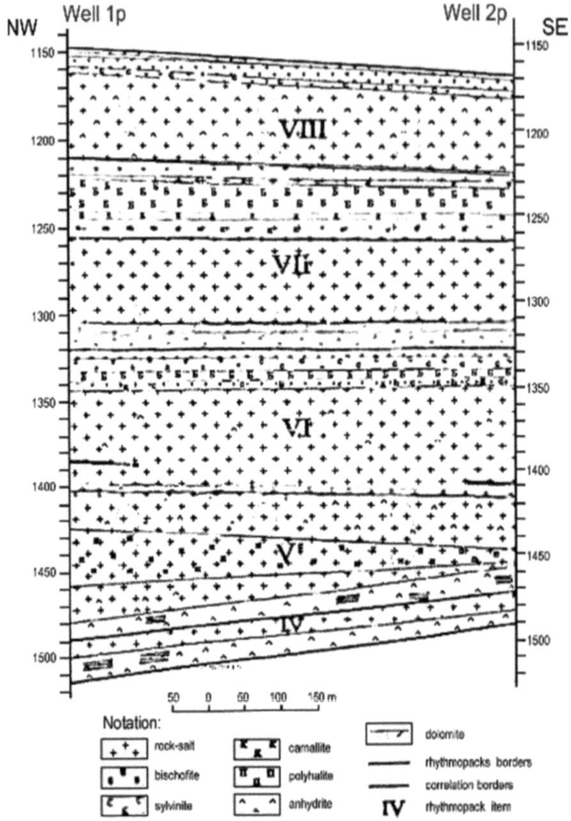

Fig. 2 Lithological section of the Rossoshinskaya salt deposit

the formation (Pozdnyakov et al. 1980).

Large-scale modeling of the tunnel-type reservoir construction at solvent supply with change of its input point in the construction process was first carried out at the Khoja-Mumyn rock salt deposit (Malyukov 2016). Based on the results of the performed physical modeling, the methodology is developed and analytical dependencies of horizontal reservoirs construction in real conditions of rock salt deposits (Malyukov 2016) are obtained.

In this case, modeling has shown that the applied technological scheme fully meets the requirements of the uniformity of rock salt dissolution in the extent of production and the optimality of its shape, including the dome-forming of the roof (Khloptsov 2015).

The creation process of horizontally located shell-reservoir, as well as a vertically located mine, can be reduced to control of the rock-salt mass transfer process under the influence of solvent. The main parameters of the mine workout construction process are: productivity of solvent supply; number of levels of column movement; amount of extracted salt at each level; numerical value of extracted brine concentration, issued in the process of mine workout formation.

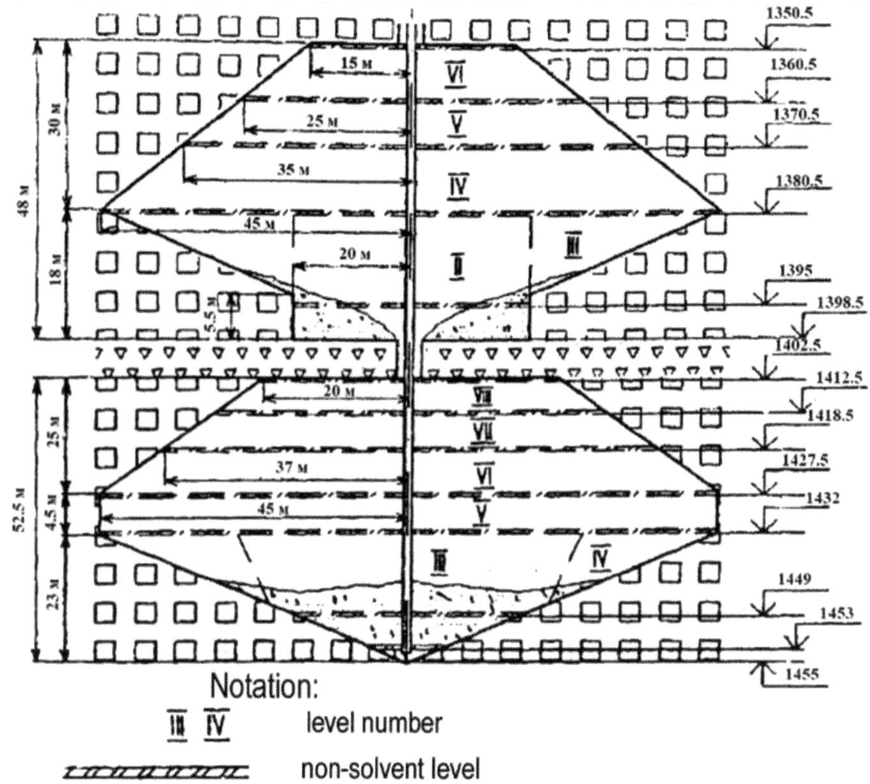

Fig. 3 Scheme of formation of a double-deck tank at the V-th and VI-th rhythmopacks of Rossoshinskaya square

The process of rock salt mass ejection depends on its parameters and on hydro-gasodynamic structures formed in mine (including vortex structures at the rock-salt surface) under hydrodynamic influence of solution.

Construction of the horizontal tank is carried out through two wells: vertical-horizontal and vertical. The shoe of the vertical-horizontal well is located at the designed location at the beginning of the first level of reservoir creation. The tier of vertical well shoe placement should not exceed the level of the vertical-horizontal well. Solvent is supplied in levels through the vertical-horizontal well. During the design formation of the shell-reservoir first phase, the shoe of the string is moved to the next design mark. The brine produced during construction is pumped out through vertical well.

The construction of a double-deck shell-reservoir begins with the formation of the lower part of the storage using traditional technologies. According to the results of laboratory modeling of the construction process of double-deck reservoirs, the method is developed and analytical dependencies for double-deck reservoirs

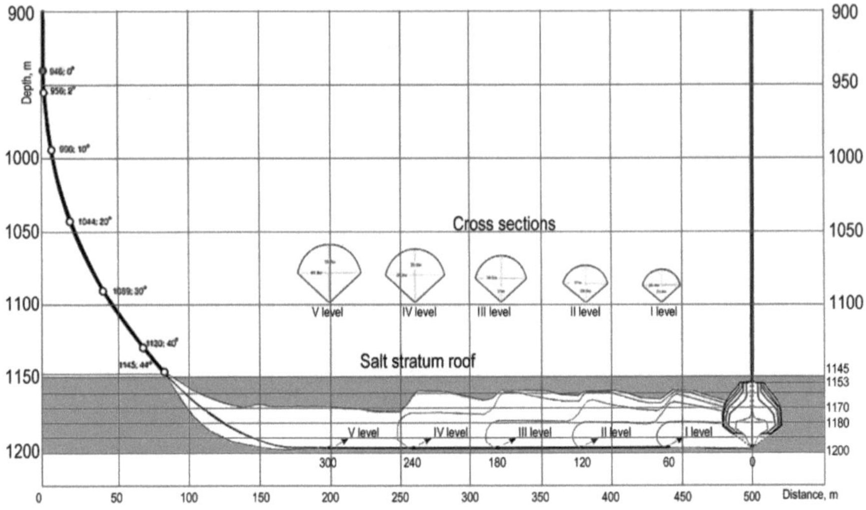

Fig. 4 Scheme of the vertical tank formation at the 8th rhythmopack of Rossoshinskaya square

construction are obtained parameters in real conditions of rock salt deposits (Pozdnyakov and Malyukov 1982). Subsequent construction of the upper part of the underground storage is carried out through the holes in the principal column at the level where shell-reservoir will be placed. Perforation of the column is carried out before its launching into the well or immediately before the upper reservoir construction (Malyukov and Kazaryan 2007). A double-deck reservoir was built for first time in the USSR at Yerevan UGS in Armenia.

Two shell-reservoirs tanks in the 8th rhythmopack are already commissioned at Volgograd UGS—vertical reservoir and the only one tunnel-type reservoir in the world built using the new technology (diameter of about 60 m and a length of about 350 m). Reservoirs of double-deck type are under construction.

At present, the gas storage facility in the Volgograd Region operates with maximum daily capacity of 10.0 mln m^3 at the beginning of the pumping season; the operational reserve is 135.0 mln m^3.

3 Conclusions

Underground storage of gas in rock-salt is an effective method of ensuring reliability of gas supply, which allows to solve issues in management of gas consumption irregularity and gas supply stockpiles. Application of innovative technologies allows to expand the sphere of construction of underground gas storage in rock salt, increase their production efficiency and increase the rate of maximum daily productivity (Liu et al. 2019; Li et al. 2019; Bérest 2017; Zhang et al. 2019).

References

Bérest P (2017) Cases causes and classifications of craters above salt caverns. Int J Rock Mech Mining Sci 100:318–329

Khloptsov VG (2015) Underground storage of natural gas in rock salt deposits. Gas Ind 9:28–31 [in Russian]

Li J, Tang Y, Shi X, Xu W, Yang C (2019) Modeling the construction of energy storage salt caverns in bedded salt. Appl Energy 255

Litvinov SA (2002) Geoecological aspects of underground gas storage in rock salt: on example of the Volgograd UGS under construction. Thes Cand Geol Mineral Sci Volgograd [in Russian]

Liu W, Zhang X, Fan J, Li Y, Wang L Potential evaluation of salt cavern gas storage in the Yangtze River Delta region and integration of brine extraction-cavern utilization. Nat Resour Res

Liu W, Zhang Z, Chen J, Fan J, Jiang D, Jjk D et al (2019) Physical simulation of construction and control of two butted-well horizontal cavern energy storage using large molded rock salt specimens. Energy 185:682–694

Malyukov VP (2016) Formation of Lysegang rings in stone salt with nanoparticles. Mining Inf Anal Bulletin 10:242–248 [in Russian]

Malyukov VP, Kazaryan VA (2007) Physico-technical processes in construction of underground storage of rock salt. In: UGS: reliability and efficiency international conference, October 11–13, 2006, vol 2. VNIIGAZ, Moscow, pp 32–49 [in Russian]

Pozdnyakov AG, Malyukov VP (1982) Mass exchange during creation of two-storied underground storages in rock salt through perforated column. In: Transport and storage of oil products and hydrocarbon raw materials, vol 1, pp 6–10 [in Russian]

Pozdnyakov AG, Sidorov IA, Malyukov VP, Mazurov VA (1980) Method of underground reservoir creation. Author's certificate of USSR No 972893 [in Russian]

UGS in rock salt deposits as an effective way to solve the problem of peak gas consumption. Gas Industr 11:66–67 (2019) [in Russian]

Zhang Z, Jiang D, Liu W, Chen J, Xie K (2019) Study on the mechanism of roof collapse and leakage of horizontal cavern in thinly bedded salt rocks. Environ Earth Sci 78(10):292

Coastal Geology and Geomorphology of the Kasatka Bay (Iturup Island, South Kuril Islands, SE Russia)

M. A. Kuznetsov and D. E. Edemsky

Abstract Iturup is the largest and most explored island of the Kuril Archipelago. It is known for the activity of tectonic processes, active volcanoes and geothermal sources, and is rich in minerals. However, much less attention is paid to the study of the geological and geomorphological structure of the coast. As a result of our studies, we identified 3 morpholithogenetic types of coasts in Kasatka Bay, of which 49% are in the accumulative coasts with sand-pebbles beaches, and 51% are on the erosion and erosion-denudation coasts (bay flanks and Chertovka rock). Using the joint use of geomorphological and geophysical methods, the geological and geomorphological structure of marine terraces in the central part of the bay has been clarified. Geological and geomorphological sections of low sea terraces were obtained on the basis of GPR sounding using the common midpoint method. It has been established that a 5–15 m accumulative terrace level lies on top of an ancient bench, developed, most likely, in proluvial pre-Holocene deposits. The measurements showed that the pyroclastic cover with buried soils at terrace levels of 25–40 m and 45–60 m has a subhorizontal flat-layered structure from 3 to 5–6 m thick. with a thickness of 3 to 5–6 m. Studies of terrace 45–60 m structure above Chertovka rock showed the presence of ancient lahar deposits under pyroclastic cover with buried soils.

Keywords Kuril Islands · Coastal classification · GPR · Marine sediments · Marine terraces

1 Introduction

Iturup Island is located in the southern part of the Great Kuril Ridge. Its area is 3175 km^2, with a length of about 200 km, a width from 7 to 27 km. Iturup is part of the Pacific seismic belt, which is characterized by the intensity of volcanic and tectonic processes, a high level of seismicity and may be affected by tsunamis

M. A. Kuznetsov (✉) · D. E. Edemsky
Faculty of Geography, Lomonosov Moscow State University, Moscow, Russia

Pushkov Institute of Terrestrial Magnetism, Ionosphere and Radiowave Propagation of the Russian Academy of Sciences, Troitsk, Moscow, Russia

T. Chaplina (ed.), *Processes in GeoMedia—Volume IV*, Springer Geology,
https://doi.org/10.1007/978-3-030-76328-2_6

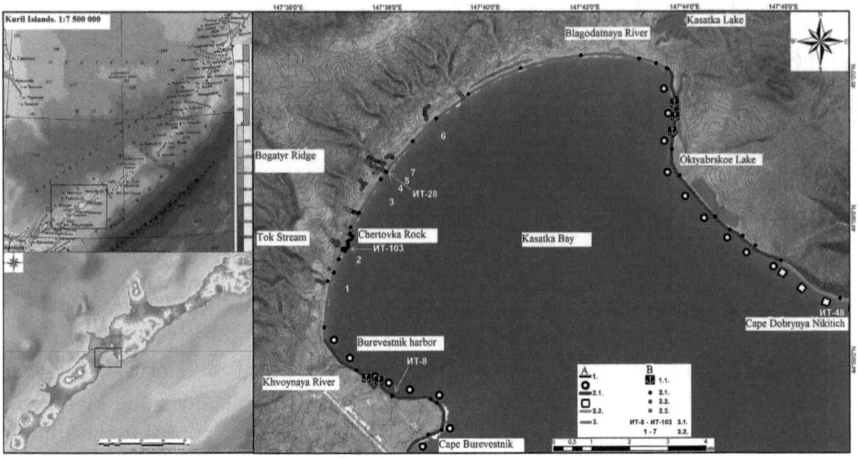

Fig. 1 Map of the study area and morpholithogenetic types of coasts of Kasatka Bay. **a**–types of coasts. Indices see Table 1. **b**—other symbols: 1.1—coastal technogenic landforms; 2.1—waypoints, 2.2—pickets of georadar profiling, 2.3—points of geomorphological profiling; 3.1—points with the most characteristic coastal structure, 3.2—numbers of georadar profiles

(Sergeyev 1976). There are 9 active volcanoes on the island (Atlas of the Kuril Islands 2009), the last phreatic eruption of Ivan Groznyy volcano occurred in 2013 (Zharkov and Kozlov 2013). Iturup is the most studied among all the islands of the Kuril archipelago. Main researcher's attention was paid to study its geological structure, minerals and relief (Geology of the USSR. T. XXXI 1964; State geological map of the Russian Federation 1980; Grabkov and Ishchenko 1982; Fedorchenko et al. 1989). However, a much smaller number of works is devoted to the study of the geological and geomorphological structure of the coast (Bulgakov 1994; Razjigaeva et al. 2004; Afanas'ev et al. 2018; Dunaev et al. 2019).

In 2019, during a expedition of the Russian Geographical Society "East Bastion—Kuril Ridge", field studies were conducted in Kasatka Bay, located on the Pacific side of Iturup Island (Fig. 1). The article considers the typification of the coast of the bay and clarifies the geological and geomorphological structure of marine terraces in the central part of the island.

This study has not only theoretical but also applied significance in connection with the planned expansion of the economic development of shore of central part of the Iturup island.

2 Materials and Research Methods

Route studies were planned based on a preliminary analysis of topographic, geological and geomorphological maps of different scales and open-source satellite imagery from DigitalGlobe [QuickBird, WorldView, GeoEye with meter spatial resolution in the GoogleEarth program (https://www.google.com/earth/)]. During the routes, geomorphological descriptions and coastal profiling were performed and morpholithogenetic types of coasts were identified. The leading relief-forming processes on the coasts and adjacent marine terraces are identified.

Study of the geological and geomorphological structure of marine terraces in the central part of Kasatka Bay was carried out by the method of pitting and deposition description, as well as using the geophysical method of georadiolocation—an effective method of engineering and geotechnical inspection of soils at depths from a few centimeters to tens of meters. The work on the research site was carried out by the Loza—V GPR and radar antenna systems with central frequencies of 300 and 50 MHz. To correctly interpret the obtained radar profiles and reconstruct the geological section from them in characteristic areas, sounding was carried out using the common midpoint (CMP), which allows one to determine the speed of electromagnetic waves in the medium and to convert the GPR section from the time scale to the depth scale without involving a priori information (Vladov and Starovoitov 2004; Edemskiy et al. 2010).

3 Research Results and Discussion

The marine relief of Iturup Island is divided into surface and underwater, which in turn are divided into modern and ancient. The surface marine relief is most fully represented in the southwestern and central parts of the island, including Kasatka Bay. Within the bay, we distinguished three morpholithogenetic types of coasts: erosion (stable) coast in volcanic rocks with steep cliff, erosion-denudation (stable) coast in volcanic rocks and lithified pyroclastics and accumulative coast with a wide sandy beach with rare pebbles. The identification of coast types is based on morphogenetic classification (Ionin et al. 1961). The distribution by their length is shown in Fig. 1 and Table 1.

1. **Erosion (stable) coast in volcanic rocks with steep cliff**, located in the area of the Chertovka Rock. They are confined to places where erosion-resistant effusives enter the water area. A section of this coast in the plan has the form of

Table 1 Length of morpholithogenetic types of coasts of Kasatka Bay

Type (subtype) of coast	Length (km)	%
1. Erosion (stable) coast in volcanic rocks with steep cliff	0.9	3
2. Erosion-denudation (stable) coast in volcanic rocks and lithified pyroclastics	12.6	48
2.1. With a boulder-block pavement	10.5	40
2.2. With rockfall cones	2.1	8
3. Accumulative coast with a wide sandy beach with rare pebbles	12.6	49
Total	26.1	100

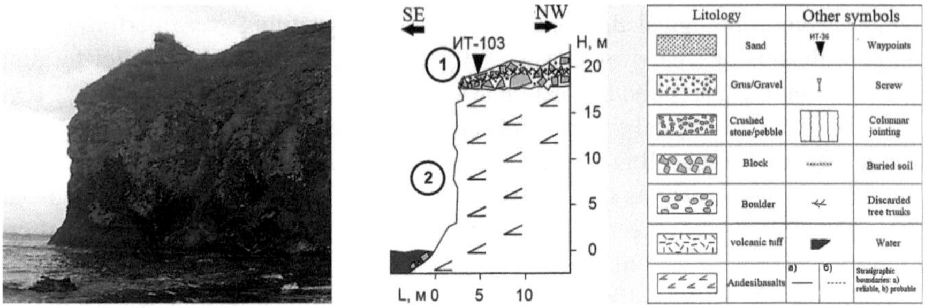

Fig. 2 Geological and geomorphological structure of erosion (stable) coast in volcanic rocks with steep cliff (point ИТ-103). Symbols for profiles 2–5

heavily indented capes of complex shapes. At point ИТ-103 (Fig. 1), the most characteristic structure of this type of coast is observed. In a steep abrasion ledge (Fig. 2) 18–20 m high, the following are discovered: (1)—unsorted, compacted gravel-block deposits with a dark brown sand aggregate and buried soils. The maximum thickness of the stratum is up to 3 m, the sole of the stratum is quite even and clear. Judging by the nature of the occurrence and the composition of the deposits, this is a mixture of pyroclastic and proluvial material. Massive gray andesitobasalts are revealed below (2), the apparent thickness of the formation is about 15 m. In some places, the overhang of the roof of effusive rocks over shallow bays is observed. According to State geological map of the Russian Federation (2002), the age of these volcanic rocks is Miocene-Pliocene.

At the base of the cliff, there is a bench with a visible underwater width of up to 10–15 m. It consists of boulder-block material lying on top of the highly fragmented surface of the erosion terrace. Between the headlands at the base of the cliff leaning pebble-gravel "pocket" beaches with a width of up to 20 m are observed. They have pebble drying, pebble festoons with gravel in the splash zone. In some places at the base of the cliff, shallow (up to 2–3 m), uneven (from 2–3 to 8–9 m) wave-breaking niches were formed.

2. **Erosion-denudation (stable) coast in volcanic rocks and lithified pyroclastics** are located on the flanks of the bay (Fig. 1), and they can be divided into 2 subtypes.

2.1. Erosion-denudation (stable) coast in volcanic rocks and lithified pyroclastics with a boulder-block pavement. Developed on the bay flanks and in the vicinity of Cape Burevestnik, point ИТ-8 (Fig. 1). In a steep (from 35–40° to 60°) erosion-denudation ledge 13–15 m high (Fig. 3), the upper 2–3 m are represented by dark brown hummed loams (1). Under them, 4–5 m (visible thickness) of sand with gravel and pebbles were revealed, while the number and average size of pebbles increases downward (2). According to State geological map of the Russian Federation (2002), these are deposits of marine genesis, late Pleistocene age. Marine sediments lie on the Miocene-Pliocene effusives of andesitobasaltic composition, which do not come

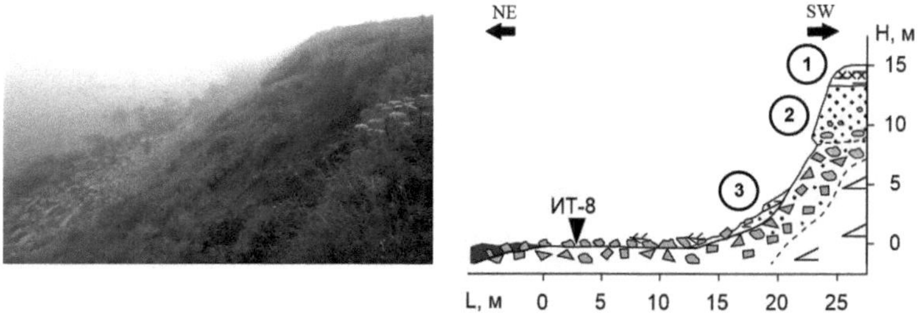

Fig. 3 Geological and geomorphological structure of erosion-denudation (stable) coast with a boulder-block pavement (point ИТ-8)

to the surface, but their fragments of boulder-block size form the bench and the lower part of the cliff (3).

The surface of the bench is a boulder block area up to 20 m wide, taking into account its visible underwater part, up to 40–50 m, block sizes up to 2–3 m. As a rule, in the river mouths (the Khvoynaya River) and streams, flowing into Kasatka Bay, there are remnants of eroded alluvial-marine terraces. In these places, the blind area is the widest, up to 100 m.

In the boulder-block blind area on the flanks of the bay are anthropogenic landforms, technogenic passages and berthing facilities of the port on Cape Burevestnik. One of such passages on the eastern flank of the bay (north of Oktyabrskoe Lake) was used during the landing of the expedition in 2019. It should be noted that most of the berthing facilities at Cape Burevestnik are located correctly, however, significant investments are needed for their stable and safe use.

2.2. Erosion-denudation (stable) coast in volcanic rocks and lithified pyroclastics with rockfall cones, located at Cape Dobrynya Nikitich in the southeast of the bay. This type of coast is formed at the foot of high (more than 150 m) and steep ledges composed of effusives and lithified pyroclastic. In the steep slopes (point ИТ-48, Fig. 1), intercalation of gray andesite-basalt composition effusives (1) (with a noticeable coarse separation) and red-brown tuffs (2) is revealed. According to State geological map of the Russian Federation (2002), the age of these rocks is Pliocene—Mid Quaternary (Fig. 4).

The foot of the ledge is a rockfall cone, composed of large block fragments. The main feature of such deposits is their unsorted nature. A feature of the coast is that its formation is not determined by the activity of wave processes. Deformation of the shore topography has gravitational and seismic causes.

3. **Accumulative coast with a wide sandy beach with rare pebbles.** This coast type is most pronounced in the center of the inner part of Kasatka Bay (Fig. 1), where we laid several geological, geomorphological, and georadar profiles, the main of which is located in the alignment of point ИТ-28 (Fig. 5).

In the relief, a full profile beach is represented by a coastal ridge up to 30 m wide, 0.5–1 m high, piled with different-grained sand with rare pebbles in the lower part, and exclusively sand in the upper part. The external appearance of the beach

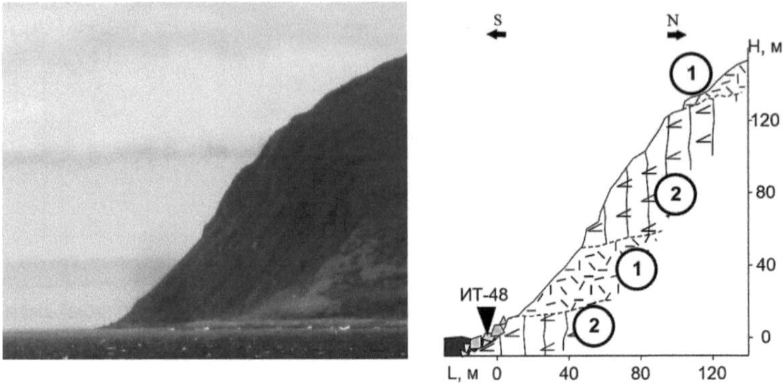

Fig. 4 Geological and geomorphological structure of erosion-denudation coast with rockfall cones (point ИТ-48)

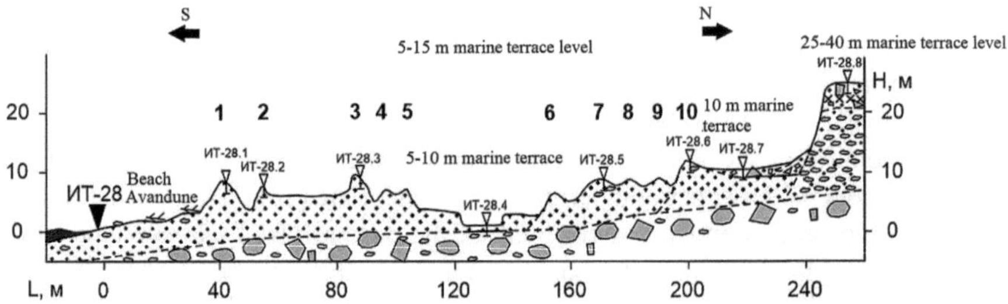

Fig. 5 Geological and geomorphological structure of accumulative coast and low sea terraces in the central part of Kasatka Bay, point ИТ-28. 1–10—serial numbers of coastal ridges

in the central part of Kasatka Bay is shown in Fig. 6. The surface of the beach

Fig. 6 Accumulative coast with a wide sandy beach with rare pebbles in Kasatka Bay, point ИТ-28

is technogenically disturbed. The avandune adjoins the beach (Fig. 5), up to 5 m wide, up to 0.5 m high. In the rear part it is partially sodded, sometimes it passes into a modern terrace of 2–3 m, most pronounced near the stream mouths. Further towards land on the profile line, a system of coastal ridges of a 5–15-m terrace level is developed.

On the coast of Iturup Island, different authors distinguish a different number of marine terraces from 200–300 m to 2–3 m: Chemekov Yu.F. allocates 9 terraces (Chemekov 1961); Grabkov V.K.]—7 terraces (Grabkov and Ishchenko 1982); Kulakov A.P.—6 terraces (Kulakov 1973); Razzhigaeva N.G. et al.—4 terraces for the central part of the island (Razjigaeva et al. 2003). The authors of this work observed marine terrace levels at Kasatka Bay at absolute heights of 100–120 m, 45–60 m, 25–40 m, 5–15 m, and 2–3 m.

A georadar survey of the coast of the central part of Kasatka Bay was carried out according to seven georadar profiles on a stretch of more than 5 km (Fig. 1). The lengths of the profiles ranged from 160 to 360 m; they were laid at terrace levels of 5–15 m, 25–40 m, and 45–60 m. Figure 7 shows the main geomorphological profile of ИТ-28, combined with the results of georadar survey. On this profile: beach with avandune—0–35 m; terrace level of 5–15 m with preserved coastal ramparts—35–240 m; terrace level of 25–40 m, partially covered with bamboo, shrubs and cedar dwarf—240–700 m.

A marine terrace level of 5–15 m is everywhere observed in Kasatka Bay, with a maximum width of up to 500 m (near Lake Blagodatnoe). It is composed mainly of sandy material. On the accumulative terrace of 5–10 m, coastal ridges stand out—up to 13 in the western part of the bay, and up to 3 in the eastern part. In the alignment of the ИТ-28 profile (Fig. 8), 10 ridges with a width of up to 30 m, a relative height of up to 3–4 m and up to 10 m in the eastern part of the bay (due to aeolian accumulation and technogenic interference) are observed. The length of the ramparts reaches 8.5 km, they are divided by numerous valleys of streams and the valley of the Blagodatnaya River.

In the east of the inner part of Kasatka Bay, in the areas of large lagoon lakes Blagodatnoe and Kasatka, at heights of 5–7 m, there are small sections of flat marshy

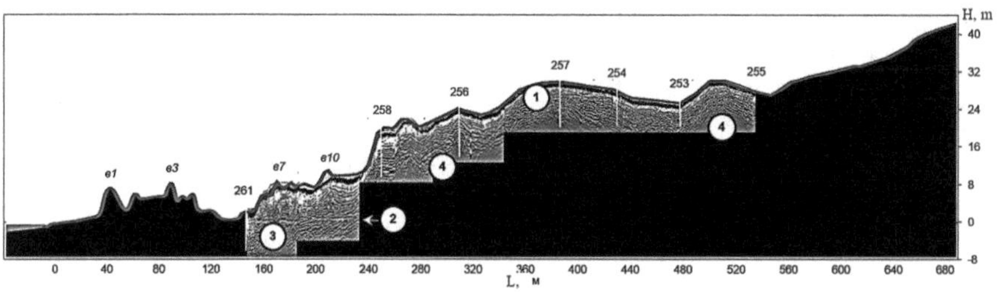

Fig. 7 Combined geomorphological and georadar profiles along the ИТ-28: 1—pyroclastic cover with buried soils; 2—groundwater level; 3—pebble-boulder deposits; 4—bedrock, basement of the marine terrace

Fig. 8 Coastal ridges in the alignment of profile ИТ-28 in the central part of Kasatka Bay

alluvial-lake and lagoon-sea plains composed of loam, sandy loam and clay. The age of the level is Holocene (Razjigaeva et al. 2002).

On the ИТ-28 profile (Fig. 5), eight pits are laid: five on the ridges, one in the hollow between ridges, one on the 10-m marine terrace adjacent to the ledge 25–40 m of the terrace level, and also one in the near-shore part of this level. The structure of the most characteristic ridges is shown in Fig. 9.

Description of the ИТ-28.3 and ИТ-28.5 pit deposits allows us to conclude the presence of genetically different layers: the upper 100–150 cm are represented by aeolian deposits (silty sand of different grains) with buried soil horizons (sand is moist, unwashed, humified), below are well-washed marine sands. This indicates the activity of the aeolian transfer of material from the beach.

The coastal ridges are located on a rather wide (up to 250 m) subhorizontal surface (Figs. 5 and 8), on which the cone of removal from gully valleys and the sea terrace rest, adjacent to a dying cliff about 25 m high. By pitting this subhorizontal surface in several places, including at the bottom of a wide hollow between the 5th and 6th ridges (point ИТ-28.4, Fig. 5), it was found that under sea sands with a thickness of up to 80 cm pebble-boulder sediments with a visible fragment size of up to 20 cm are revealed. Judging by the nature of the sediments composing this surface and the presence of a dead paleocliff, we can assume that this is a bench developed in proluvial pre-Holocene sediments that entered the coastal zone, presumably from the Bogatyr Ridge (Fig. 1). The coastal ridges are adjoined by a lined 10-m terrace with a width of about 30 m (Figs. 5, 7 and 9), partially covered by a pyroclastic cover or slope deposits, adjacent to a paleo-erosion ledge of 25–40 m of a marine terrace level.

Using the georadar survey, the structure of the upper coastal ramparts was clarified (*6–10*, Fig. 5), which confirmed the presence of a bench *(3)* on the radarogram (Fig. 7), which is recorded as unstructured multiple reflections in the form of hyperbolas from local objects. The presence of regular structures on the back of the ridges *(3)*

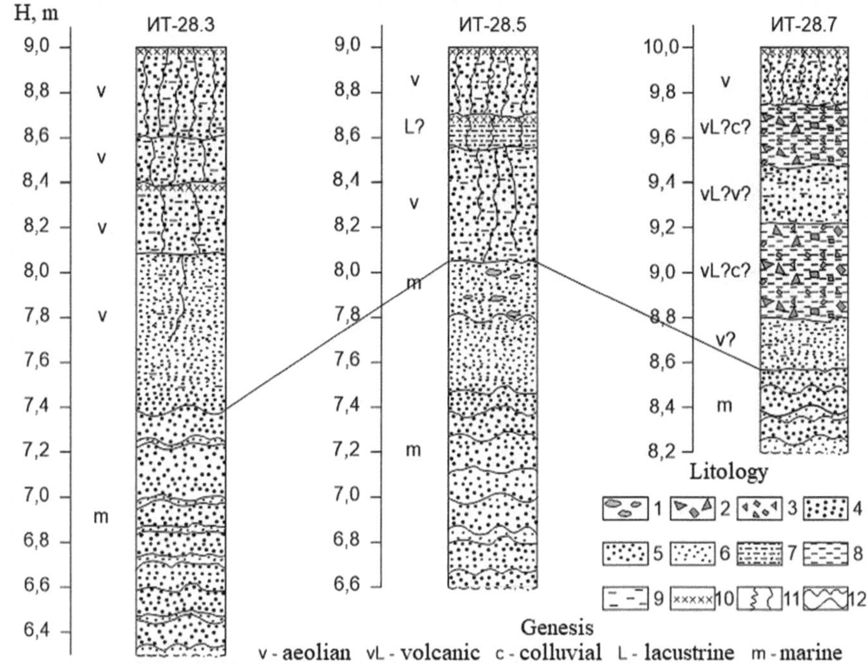

Fig. 9 The geological structure of coastal ridges and a 10-m marine terrace in the center part of Kasatka Bay (profile ИТ-28): 1—pebbles, 2—crushed stone, 3—grus, 4—coarse sand, 5—medium sand, 6—fine sand, 7—sandy loam, 8—loam, 9—silt, 10—soil (including buried), 11—roots of vegetation, 12—wavy layering

(Fig. 10a, b) confirms that the main tendency for the development of such shores is rare episodes of active, sometimes catastrophic, tsunamigenic erosion against the background of prolonged gradual accumulation of beach sediments, accompanied by aeolian processing of the ridges. On these profiles, according to the nature of the wave pattern, it is possible to distinguish the boundary of pebble-boulder deposits *(2)* and the intensity of the common-mode line *(1)* to trace the level of groundwater.

The terrace level of 25–40 m is hollow (3–5°) towards the coast and consists of 2 terraces: 25–30 m and 30–40 m. The rear seams and edge of the terraces are clearly defined, the total width of about 400 m (Fig. 7) Above the terraces are overlain by alternating layers of brown humus loam and a woody-gravelly stratum with a sand filler of the same color. Judging by the composition of the deposits and the nature of their occurrence, this is a pyroclastic cover with buried soils. The thickness of the loose cover of this terrace level, according to State geological map of the Russian Federation (2002), is 5–13 m. The terrace of 25–30 m is of marine origin, as can be judged by the pebbles opening in the valley of the Tok Stream at a depth of 4.5 m from the edge. The age of this terrace is defined as the Middle Pleistocene (Razjigaeva et al. 2003). The terrace is heavily technogenically disturbed, on its surface are located the Shumi-gorodok village and Burevestnik village. The terrace of 30–40 m in its

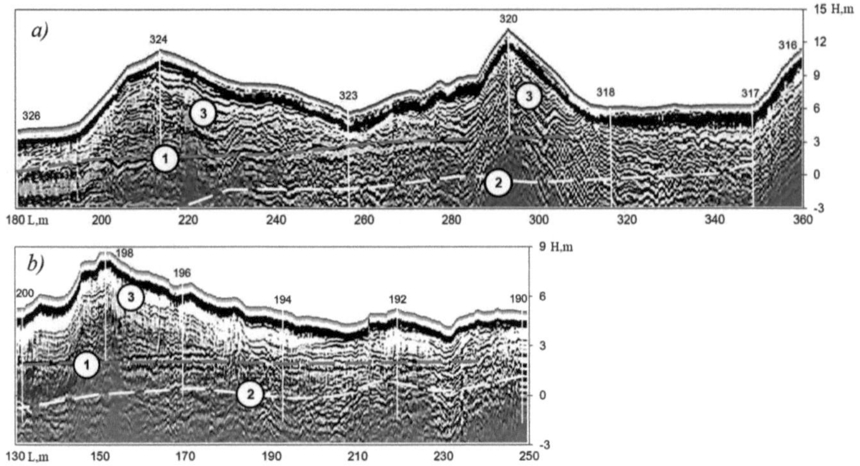

Fig. 10 Marine terrace level 5–15 m: **a**—GPR profile, central frequency 300 MHz, profile *(6)* (Fig. 1); **b**—GPR profile, center frequency 300 MHz, profile *(7)* (Fig. 1); 1—groundwater level; 2—pebble-boulder deposits; 3—sea sands homogeneous, layered

morphological appearance practically does not differ from the 25–30 m of the terrace; we can assume their similar structure. In the alignment of profile ИТ-28, the rear seam of the terrace 30–40 m is located on abs. 35 m high (in the southwest of the inner part it is located at an abs. height of 40 m).

In Fig. 11 shows the GPR profile of the terrace 25–30 m. The analysis of the wave pattern (Fig. 11a), the georadar profile *(5)* (Fig. 1), made it possible to distinguish two georadar complexes in its structure (Vladov and Starovoitov 2004), the interpretation of the georadar data is shown in Fig. 11b. The GPR complex *(1)* is presented in the form of extended subhorizontal in-phase axes and is a pyroclastic cover with buried soils, with alternating ashes and sands with a thickness of 3 m in the picket area *(244)* to 10–12 m in the picket area (370). The sole of the complex *(1)* is the roof of bedrock (terrace basement) *(2)*, the upper part of which *(3)* is heterogeneous, partially destroyed. Its increased fracturing is emphasized by characteristic radio images, which are marked on the profile by subvertical lines. Similar results were obtained on the terrace of 30–40 m, profile *(1)*, *(6)* (Fig. 1), except that the power of the PCP along its entire profile is almost unchanged and averages 4–5 m.

A **terrace of 45–60 m** is slightly inclined towards the coast (2–4°), width is up to 200 m. The rear seam is quite clear; the edge is smoothed. According to State geological map of the Russian Federation (2002), the thickness of the loose cover is 5–30 m. According to Grabkov and Ishchenko (1982), the age of the terrace from is estimated as mid-Quaternary. The terrace can be diagnosed by vegetation—a cedar dwarf tree is actively growing on it, while on a ledge below and above the terrace there is a forest of stone birch with bamboo.

Analysis of the wave pattern of the georadar profile of the terrace (Fig. 12a), profile *(2)* (Fig. 1), allowed us to distinguish three georadar complexes. The interpretation of

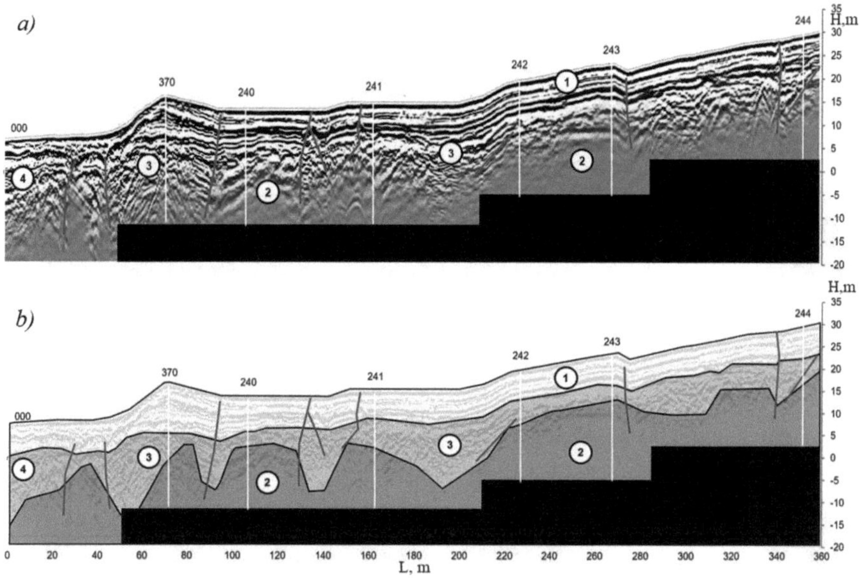

Fig. 11 Marine terrace 25–30 m, profile *(5)* (Fig. 1): **a**—GPR profile, central sounding frequency of 50 MHz; **b**—a profile with the allocation of georadar complexes; 1—pyroclastic cover with buried soils; 2—bedrock, basement of the marine terrace; 3—destroyed, highly fractured basement of marine terrace; 4—pebble-boulder deposits

georadar data is shown in Fig. 11b, the boundaries between the complexes are drawn along the line of changing the morphology of the in-phase axes corresponding to the boundaries of disagreement (Vladov and Starovoitov 2004). The georadar complex *(1)*, 2.5–4 m thick, whose pattern is presented in the form of extended subhorizontal in-phase axes, the intensity of which is stable along the profile, is a pyroclastic cover with buried soils. Starting from the bottom of this complex, the GPR section acquires a characteristic irregular wave pattern *(2)* with a thickness of 0–6 m, which indicates the presence of local unsorted gravel-block deposits. The sole of this georadar complex, the high-intensity phase axis *(3)*, is the roof of bedrock *(4)*, the basement of the terrace, covered with loams with a thickness of about 1 m.

A **terrace of 100–120 m** is represented in the area of Cape Burevestnik, as well as on the site of Chertovka rock—Lake Blagodatnoe. Its surface is flat, slightly inclined towards the coast, dissected by river and stream valleys from U-shaped to canyon-shaped profiles. On the sides of the valleys, effusive openings are revealed, over which a cover of loose deposits lies with a visible thickness of 20–30 m. It is composed of layered loams, tephra, sand and pebbles (State geological map of the Russian Federation 2002). The rear seam is indistinct, the edge of the terrace is flattened and most pronounced in places where it is bordered by a steep (30–40°) ledge up to 50 m high. In the coastal part of the terrace, we laid ИТ-28.9 pit, 1.5 m deep. The upper meter is represented by hummed loams of dark brown color, overlapping with sandy loam of brown color. 1–1.5 m are represented by highly compacted loam

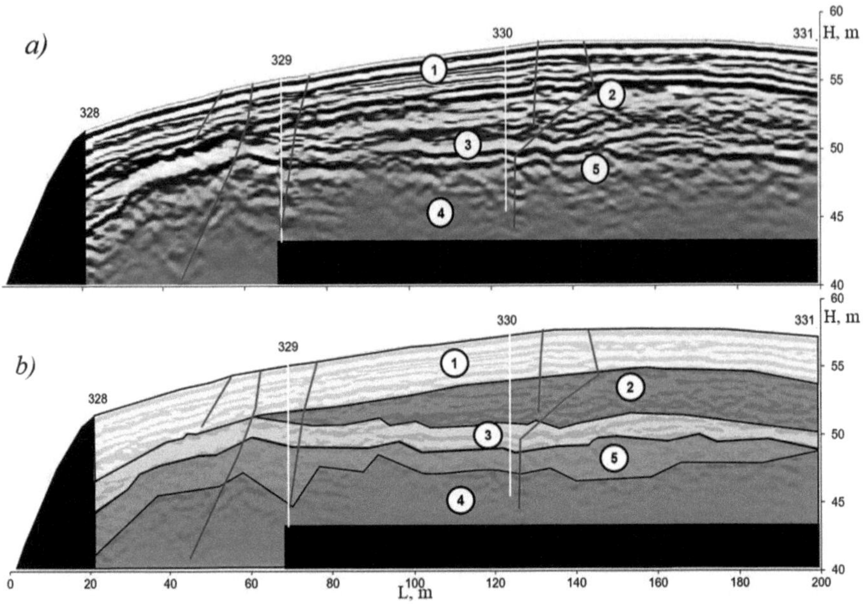

Fig. 12 Marine terrace 45–60 m, profile (2) (Fig. 1): **a**—GPR profile, center frequency 50 MHz; **b**—GPR profile with the allocation of GPR complexes; 1—pyroclastic cover with buried soils; 2—unsorted gravel-block deposits; 3—loam; 4—bedrock, basement of the marine terrace; 5—destroyed, highly fractured basement of marine terrace

with numerous fragments of wood-grained dimension. This is a pyroclastic case with buried soils. The age of the terrace is estimated as early Quaternary (Grabkov and Ishchenko 1982).

4 Conclusion

Three morpholithogenetic types of shores are distinguished in Kasatka Bay, of which 49% are in Accumulative coast with a wide sandy beach with rare pebbles (inner parts of the gulf), and 51% are in erosion (stable) and erosion-denudation (stable) coast (gulf flanks and Chertovka rock).

Using the joint use of geomorphological and geophysical methods, the geological and geomorphological structure of marine terraces in the central part of the bay has been clarified. Description of pit deposits in the coastal ridges allows us to conclude that the aeolian transport of material from the beach is active (the upper 100–150 cm are represented by aeolian sediments). It was established that a 5–15 m accumulative terrace level lies on top of an ancient bench, developed, most likely, in proluvial pre-Holocene sediments that entered the coastal zone, presumably from the Bogatyr ridge.

Marine terraces of 25–30 m and 30–40 m are almost identical in morphological appearance. The results of the analysis of georadar data obtained in the study of the erosion-accumulative terrace of 25–30 m made it possible to distinguish a pyroclastic cover with buried soils from 3–4 m thick (at the rear of the terrace) to 6–12 m (at the edge), and for the terrace 30– 40 m—up to 4–5 m. According to georadar sounding, an increase in the thickness of the cover in the region of the edge of the terrace occurs due to an increase in the number and thickness of individual interlayers. The sole of this complex is adjacent to the base of the fractured bedrock, their upper part is heterogeneous, partially destroyed.

A georadar survey of 45–60 m of the marine terrace revealed the presence of ancient, presumably, lahar deposits under a pyroclastic cover with buried soils. This complex on the GPR section has a characteristic irregular wave pattern and was observed by us only in the area of Chertovka rock. The maximum power of this complex on the profile is ~6 m.

The data obtained during the work on the geological and geomorphological structure of the coasts and marine terraces are necessary for the rational use of natural resources both in the coastal zone and on the shore and can be used for the economic development of this territory.

Acknowledgements The authors are grateful to the Russian Geographical Society, and personally Binyukov E.A. for the organization and assistance in conducting research, as well as Ph.D. Dunaev N.N. and Ph.D. Rybin A.V. for valuable tips and tricks for improving the article.

References

Afanas'ev VV, Dunaev NN, Gorbunov AO, Uba AV (2018) Manifestation of Caldera-forming volcanism in the formation of the sea coast (on the example of Iturup island of the Large Kuril Arc). Processes Geoenviron 3(16), pp 990–998

Atlas of the Kuril Islands (2009) IPTS "DIK", 516 p. [in Russian]

Bulgakov RF (1994) History of the development of the southern islands of the Great Kuril arc in the Pleistocene. Avtoref. dis. kand. geogr. nauk. MSU. 20 p. [in Russian]

Chemekov YuF (1961) Quaternary transgressions of the Far Eastern Seas and the North Pacific. Tr Inst geol AN SSSR, vyp. 8. [in Russian]

Dunaev NN, Repkina TYu. Baranskaya AV, Afanasiev VV (2019) Modern dynamics of an accumulative coast composed by pyroclastics of an underwater volcanic eruption. Geosistemy perekhodnykh zon 3(2):237–244. [in Russian]

Edemskiy DY, Edemskiy FD, Morozov PA (2010) Profiling and determination of environmental parameters during georadar surveys. Elektromagnitnyye Volny i Elektronnyye Sistemy, T. 15(9):57–63 [in Russian]

Fedorchenko VI, Abdurakhmanov AI, Rodionova RI (1989) Volcanism of the Kuril island arc: geology and petrogenesis. Nauka, 239 p. [in Russian]

Geology of the USSR. T. XXXI (1964) Kamchatka, Kuril and Commander Islands. Part 1. Geological description. Nedra, 733 p. [in Russian]

Grabkov VK, Ishchenko AA (1982) Morphological relief types of the Kuril Islands. Rel'yef i vulkanizm Kuril'skoy ostrovoduzhnoy sistemy. DVNTS AN SSSR, pp 13–24 [in Russian]

Ionin AS, Kaplin PA, Medvedev VS (1961) Classification of the world coasts (applied for Physiographical atlas of the world). Tr. okeano-grafich. komissii. v. 12:94–108. [in Russian]
Kulakov AP (1973) Quaternary coastlines of the Okhotsk Sea and the Japan Sea. Nauka. 187 p. [in Russian]
Razjigaeva NG, Grebennikova TA, Ganzey LA et al (2004) The role of global and local factors in determining the middle to late Holocene environmental history of the South Kurile and Komandar Islands, northwestern Pacific. Palaeogeogr Palaeoclimatol Palaeoecol 209:313–333
Razjigaeva NG, Korotky AM, Grebennikova TA et al (2002) Holocene climatic changes and environmental history of Iturup Island, Kurile Islands, northwestern Pacific. The Holocene 12
Razjigaeva NG, Grebennikova TA. Mokhova LM (2003) Middle Pleistocene coastal deposits about. Iturup, Kuril Islands. Tikhookean. geol. T. 22(3):48–58 [in Russian]
Sergeyev KF (1976) Tectonics of the Kuril island system. Nauka, 320 p. [in Russian]
State geological map of the Russian Federation (2002) Ser. Kuril. Sheet L-55-XXIII, XXIX: Expl. app. Izd. 2-ye. FGUGP SakhGRE, 117 p. [in Russian]
State geological map of the Russian Federation (1980) Ser. Kuril. Sheet L–55–XXVIII: Expl. app. Izd. 2-ye. FGUGP SakhGRE, 96 p. [in Russian]
Vladov ML, Starovoitov AV (2004) Introduction to GPR. Tutorial. MGU, 153 p. [in Russian]
Zharkov RV, Kozlov DN (2013) Explosive eruption of Ivan Groznyy volcano in 2012–2013 (Iturup Island, Kuril Islands). Vestnik DVO RAN. №3, pp 39–44. [in Russian]

Movement of the Particles Around Particle in a Shear Flow

A. I. Fedyushkin

Abstract The problem of particle motion around a particle in a shear flow is considered by numerically for conditions of weightlessness and normal earth gravity. n. The minimum distance between liquid small drops and a streamlined drop is shown.

Keywords Numerical simulation · Motion of liquid particles · Shear flow · Particle closest approach

1 Introduction

The flow of liquid and solid particles in the shear velocity field is widely studied in connection with many technical applications for the separation of liquid and solid inclusions in a fluid. In addition, the problem of small particles flowing around a small cylindrical obstacle (thread) is relevant today for studying the protective properties of individual masks in the fight against the spread of viruses.

Cases with particle densities greater or less than the density of the main buffer moving fluid (oil particles in water and water particles in oil) are presented. Based on numerical modeling, the nature of the influence of gravitational, viscous, lifting, and drag forces on the trajectory of liquid particles in a shear fluid flow is show. The minimum distance between liquid small drops and a streamlined drop is shown.

2 Mathematical Model

It is assumed that the number of particles is too small and they do not affect the main flow, and the shear flow of the buffer fluid affects the movement of particles. In Fig. 1 a simulation region and the velocity vector field (left) and the particle tracks

A. I. Fedyushkin (✉)
Ishlinsky Institute for Problems in Mechanics RAS, Moscow, Russia
e-mail: fai@ipmnet.ru

T. Chaplina (ed.), *Processes in GeoMedia—Volume IV*, Springer Geology,
https://doi.org/10.1007/978-3-030-76328-2_7

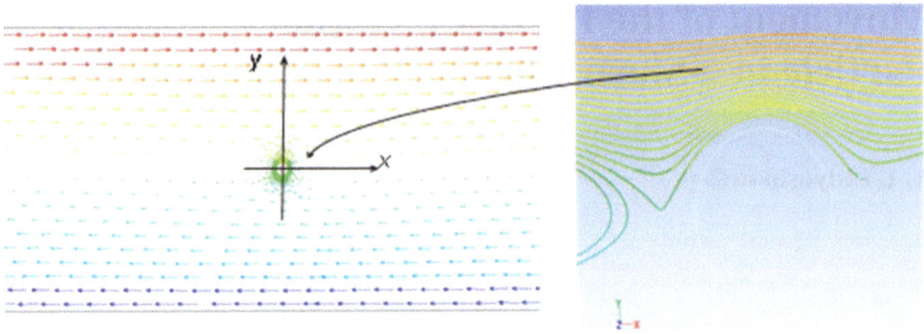

Fig. 1 The scheme of simulation region, field of vector of velocity (at the left) and tracks of particles around a cylindrical particle which there is in the beginning of co-ordinates (on the right)

around the particle with center located in begin of coordinates (right) are shown. On the right, in Fig. 1 a streamlined particle shows at an enlarged scale.

The particle in the center of calculation region is simulated by not deformable cylindrical surface and rotatable on account of a shear flow. Cases of the task for surface shear stress equal to zero and tasks of a surface tension for water and oil are considered. The calculations have shown that velocity on a cylindrical surface coincides with angular speed ω = 1/2 rotV (ω—a vector of angular rotation of an element of environment in a point).

Mathematical simulation is performed on the basic of numerical solutions of unsteady 2D Navier–Stokes equations for incompressible laminar fluid flow (Landau and Lifshitz 1987; Batchelor 2002).

$$\frac{\partial u}{\partial t} + u\frac{\partial u}{\partial x} + v\frac{\partial u}{\partial y} = -\frac{1}{\rho}\frac{\partial P}{\partial x} + \nu\left(\frac{\partial^2 u}{\partial x^2} + \frac{\partial^2 u}{\partial y^2}\right), \tag{1}$$

$$\frac{\partial v}{\partial t} + u\frac{\partial v}{\partial x} + v\frac{\partial v}{\partial y} = -\frac{1}{\rho}\frac{\partial P}{\partial y} + \nu\left(\frac{\partial^2 v}{\partial x^2} + \frac{\partial^2 v}{\partial y^2}\right) + F_g, \tag{2}$$

$$\frac{\partial u}{\partial x} + \frac{\partial v}{\partial y} = 0, \tag{3}$$

where x, y—horizontal and vertical Cartesian coordinates; u, v—components of the velocity vector; t—time; P—pressure; ρ—density and $F_g = -g$—gravity force; g—gravitational acceleration of the earth's free fall; ν—coefficients of kinematic viscosity. Equations (1–3) were solved numerically by the control volume method.

A movement of the particles is calculated in Lagrangian variables.

$$\frac{d\vec{u}_p}{dt} = \mathbf{F}_D + \mathbf{F}_M + \mathbf{F}_L + \mathbf{F}_{pg} \tag{4}$$

Equation (4) was solving by the Runge–Kutta method of the 4th order of accuracy. In the model of interaction of particles with the fluid flow, the friction forces $\mathbf{F}_D$, the virtual masses $\mathbf{F}_M$, and the Saffman forces $\mathbf{F}_L$ and gravity forces $\mathbf{F}_{pg}$ were taken into account. The $\mathbf{F}_D = (\vec{u} - \vec{u}_p)/\tau_r$ is the drag force per unit particle mass. Here $\tau_r = \frac{\rho_p d_p^2}{18\nu\rho}\frac{24}{C_d \mathrm{Re}}$ is the droplet or particle relaxation time, where C_d is the drag factor, $\vec{u}$ is the fluid phase velocity, $\vec{u}_p$ is the particle velocity of the particle. Re is the relative Reynolds number, which is defined as $\mathrm{Re} = \frac{|\vec{u} - \vec{u}_p|}{\nu}$. The p index for velocity, density and diameter refers to particles. The "virtual mass" force $\mathbf{F}_M$ can be written as $\mathbf{F}_M = C_{vm}\frac{\rho}{\rho_p}\left(\vec{u}_p\nabla\vec{u} - \frac{d\vec{u}_p}{dt}\right)$, where C_{vm} is the virtual mass factor in this work was equal to 0.5. The virtual mass force is important when the density ratio ρ/ρ_p is greater than 0.1. The Saffman's lift force (Saffman PG 1965) for small particle Reynolds numbers is $\mathbf{F}_L = \frac{2K\nu^{1/2}\rho d_{ij}}{\rho_p d_p (d_{lk} d_{kl})^{1/4}}(\vec{u} - \vec{u}_p)$ where $K = 2.594$ and d_{ij} is the deformation tensor (Li and Ahmadi 1992). Particle gravity force is $\mathbf{F}_{pg} = \vec{g}(\rho_p - \rho)/\rho_p$.

Initial distances of particles were 2d from the center of region and they regular intervals located on a vertical on a range from y = 0 to y = 0.5d. Initial speeds of particles corresponded to velocities of environment in the given points.

In given paper results of simulation for following values of parameters: a diameter of particles d = 0.0001 m, X = 0.0025 m, Y = 0.001 m, $\alpha = 300\ \mathrm{s}^{-1}$ are presented.

Two variants of particles and a buffer liquids are considered with following properties:

(1) the drops of diesel oil (a density is 730 $\mathrm{kg/m^3}$) in water (a density is 875 $\mathrm{kg/m^3}$, a viscosity is 0.000589 kg/m sec),
(2) the drops of water in diesel oil (a density is 730 $\mathrm{kg/m^3}$, a viscosity is 0.0024 kg/m sec).

3 Results

The problem of particle motion around a particle in a shear flow is considered for conditions of weightlessness and normal earth gravity. Cases with particle densities greater or less than the densities of the main buffer moving fluid (oil particles in water and water particles in oil) are presented. The time dependences of position (y co-ordinate) of ten oil particles in water shear flow are shown in Fig. 2 for normal and weightlessness gravity conditions. The time dependences of position (y co-ordinate) of ten water particles in oil shear flow are shown in Fig. 3 for normal and weightlessness gravity conditions. In all cases, at the initial moment, the particles were equidistant on the x axis at a distance (0.5 d) from center of the streamlined particle and were located at the heights shown in Figs. 2 and 3 on the ordinate axis at t = 0 s. Despite the fact that the initial vertical location of the particles differed by a small amount, the trajectories of the particles changed significantly over time. Some particles did not reach the streamlined particle and turned in the opposite direction.

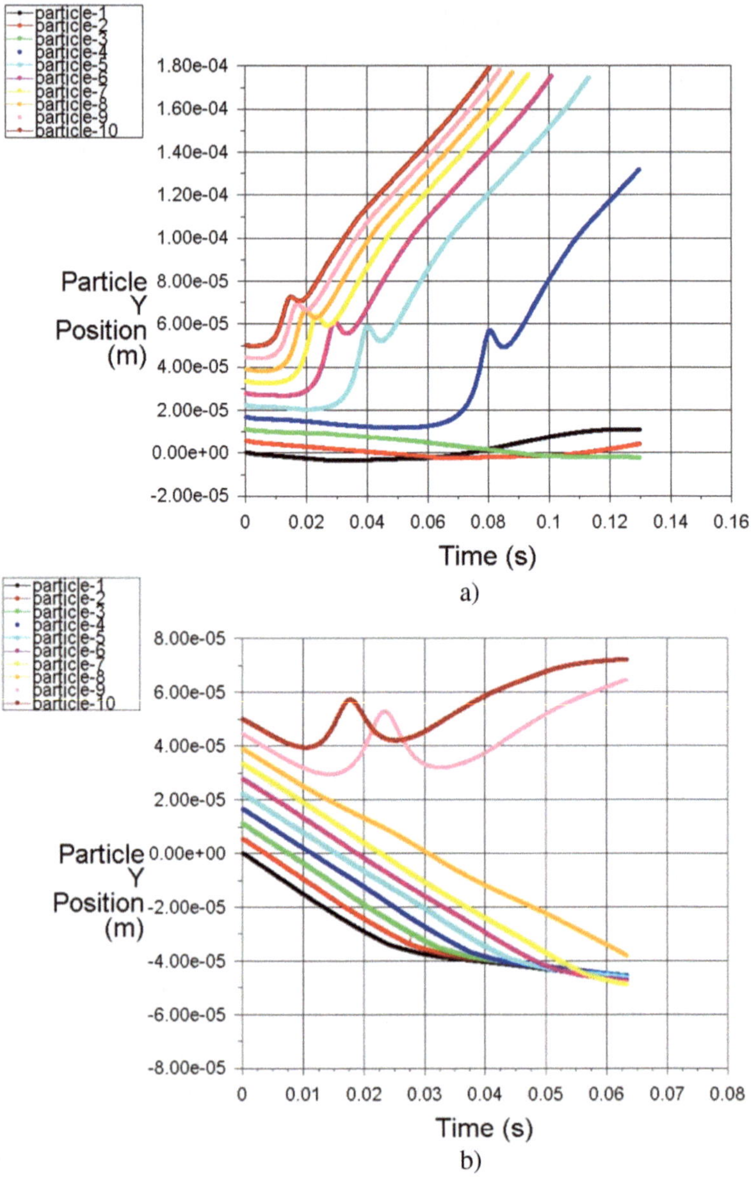

Fig. 2 Y co-ordinate position of the oil particles in water from time for Ground **a**—$g = 9.81$ m/s^2 and weightlessness conditions **b**—$g = 0$ m/s^2)

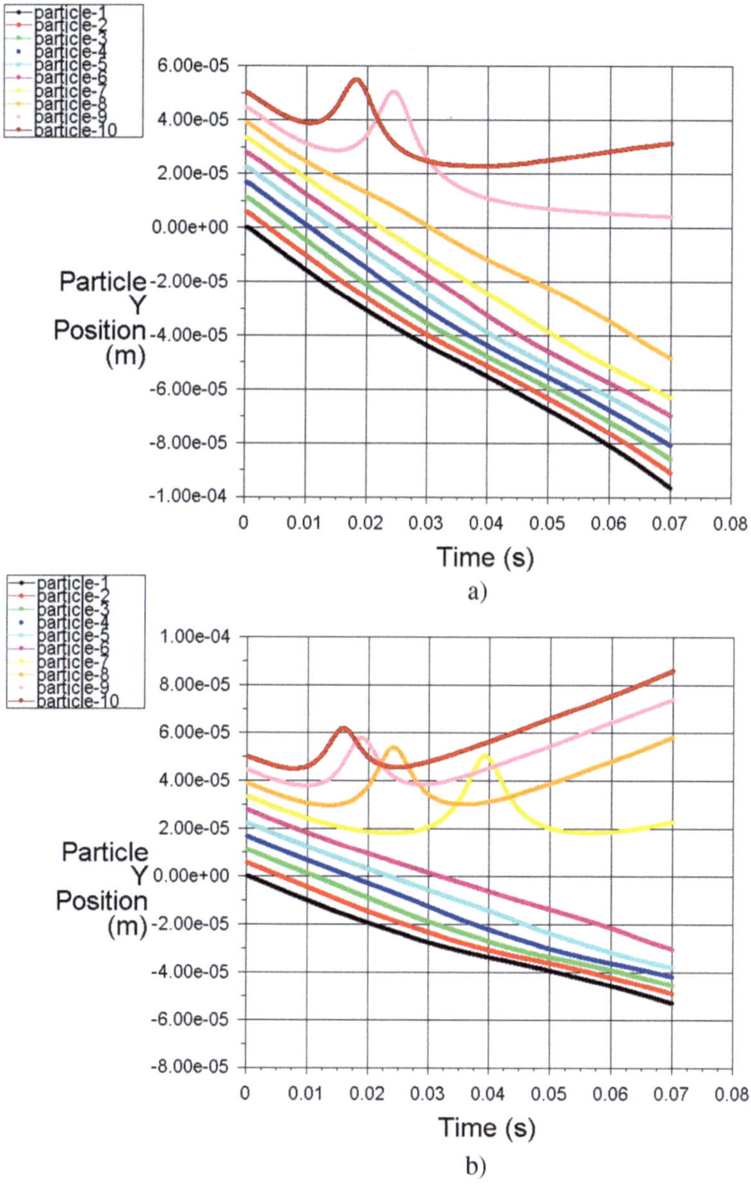

Fig. 3 Y co-ordinate position of the *water particles in oil* from time for Ground **a**-$g = 9.81$ m/s^2 and weightlessness conditions **b**—$g = 0$ m/s^2)

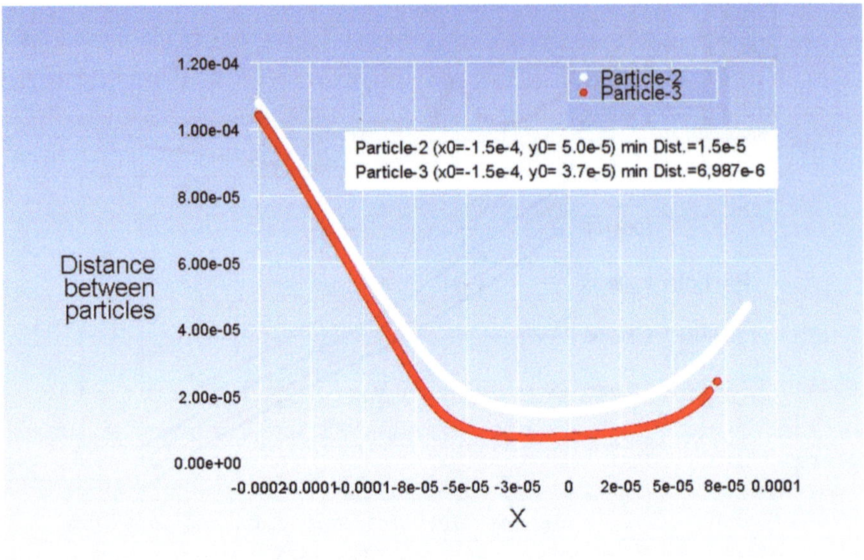

Fig. 4 The dependences of distance of two oil running particles to surfaces of the particle (which is located in the center of region) from x co-ordinate

On the trajectories of the particles, you can see the areas of their approach to the streamlined particle and their departure from it.

Time dependences of distances of two oil running particles to a surface of particle from x co-ordinate are presented in Fig. 4. The given two oil particles (in Fig. 4) are chosen, as the coming most nearer to the cylindrical particle for the given series of calculations. The minimum distance of closeness with the cylindrical particle approximately is 0.07d.

4 Conclusion

The results of numerical simulation have shown character of influence of gravity, viscosity, lift and drag forces on trajectories of a movement of water and oil particles in a shear flow. The minimum distance between liquid small drops and a streamlined drop is shown.

Acknowledgements The study was supported by the Government program (contract # AAAA-A20-120011690131-7) and was funded by RFBR, project number 20-04-60128.

References

Batchelor GK (2002) An introduction to fluid dynamics. Cambridge University Press, 615 p
Landau LD, Lifshitz EM (1987) Fluid mechanics: Volume 6 (Course of Theoretical Physics S) 2nd edn. Pergamon, 554 p
Li A, Ahmadi G (1992) Dispersion and deposition of spherical particles from point sources in a turbulent channel flow. Aerosol Sci Technol 16:209–226
Saffman PG (1965) The lift on a small sphere in a slow shear flow J Fluid Mech 22:385–400

Internal Gravity Waves Fields Dynamics in Vertically Stratified Horizontally Inhomogeneous Medium

V. V. Bulatov and Yu. V. Vladimirov

Abstract The paper considers the problem of the harmonic internal gravity waves dynamics in a vertically stratified horizontally inhomogeneous medium. It is assumed that the scale of the horizontal variability of the medium density exceeds the characteristic lengths of internal gravity waves. The solution is constructed using a modified version of the spatio-temporal ray method. An asymptotic solution is constructed that allows one to calculate the amplitude and phase characteristics of wave fields. The solutions obtained make it possible to calculate the internal gravity waves fields for arbitrary disturbances sources.

Keywords Stratified Medium · Internal Gravity Waves · Asymptotics · Ray Method

Horizontal heterogeneity and non-stationarity have a significant impact on the propagation of internal gravity waves (IGW) in stratified natural environments (Ocean, Atmosphere) (Fabrikant and Stepanyants 1998; Miropol'skii and Shishkina 2001; Mei et al. 2017; Morozov 2018). If the ocean depth and its density are changing slowly as compared to the characteristic length (period) of IGW, which is well done in the real ocean, then for solving the mathematical modeling of IGW dynamics we may use the space–time ray-optics method (geometrical optics method) and its generalizations (Bulatov and Vladimirov 2012, 2018, 2019; Kravtsov and Orlov 1999). The requirement of slowness as compared with characteristic lengths (periods) of internal gravity waves, changing horizontal (time) parameters of the medium is a mandatory condition for using the geometrical optics method. This condition, however, may not be used to formulate sufficient applicability conditions for this method. It's evident that making an error estimation of the geometrical optics method requires more accurate analytical approaches than the space–time ray-path method, but because of significant computational difficulties it is not realistic. The analytical methods available at this time for researching the IGW dynamics in stratified, non-uniform and non-stationary mediums are severely limited and do not allow us to estimate a faulty

V. V. Bulatov (✉) · Yu. V. Vladimirov
Ishlinsky Institute for Problems in Mechanics RAS, Vernadskogo ave. 101-1, 119526 Moscow, Russia

T. Chaplina (ed.), *Processes in GeoMedia—Volume IV*, Springer Geology,
https://doi.org/10.1007/978-3-030-76328-2_8

proportion of the geometrical optics method for real-world ambients. In most cases there are no accurate analytical solutions, and the well-known rigorous solutions can only instruct about a possible error rate for typical situations. The same instructions about a possible error rate for the space–time ray-path method can be obtained by comparing the resulting asymptotics with approximated, but more general than the ray-path, solutions of basic wave problems. Thus, the correctness of using the space–time ray-path method and its results should be considered in the frame of consistency of theoretical results with the field measurement data (Mei et al. 2017; Morozov 2018; Kravtsov and Orlov 1999; Froman and Froman 2002; Broutman and Rottman 2004).

In this paper we consider harmonic internal gravity waves. For determination of a small perturbations of pressure $p = \exp(i\omega t)P(z, x, y)$ and velocity components $U_j = \exp(i\omega t)u_j(z, x, y)$, $j = 1, 2, 3$ we obtain (Bulatov and Vladimirov 2012, 2019; Bulatov 2014; Matyushin 2019; Bulatov et al. 2019)

$$-\omega^2\left(\frac{\partial^2 u_3}{\partial z^2} + \delta^{-2}\Delta u_3\right) + \delta^{-2}\frac{g}{\rho_0}\left(\delta^{-1}u_1\frac{\partial \rho_0}{\partial x} + \delta^{-1}u_2\frac{\partial \rho_0}{\partial y} + u_3\frac{\partial \rho_0}{\partial z}\right) = 0,$$

$$\delta^{-1}\Delta u_1 + \frac{\partial^2 u_3}{\partial z \partial x} = 0, \delta^{-1}\Delta u_2 + \frac{\partial^2 u_3}{\partial z \partial y} = 0,$$

$$\delta^2\frac{\partial}{\partial z}(\frac{\partial D}{\partial z}\alpha^{-1}) - \Delta D = -\frac{\partial}{\partial z}(\alpha^{-1}\nabla D \nabla \chi), \tag{1}$$

$$\delta^2\frac{\partial D}{\partial z} + \nabla D \nabla \chi = 0, \text{ for } z = 0, -H$$

$$\chi(z, x, y) = g\omega^{-2}\ln \rho_0, \alpha(z, x, y) = \omega^{-2}N^2(z, x, y) - 1, D = P/\rho_0$$

$$\Delta = \partial^2/\partial x^2 + \partial^2/\partial y, \nabla = (\frac{\partial}{\partial x}, \frac{\partial}{\partial y}), N^2(z, x, y) = g\frac{\partial \ln \rho_0}{\partial z}$$

where $\rho_0(z, x, y)$—undisturbed density, $N^2(z, x, y)$—buoyancy frequency, g—gravity acceleration, $D(z, x, y) = P(z, x, y/\rho_0(z, x, y)$, $\delta >> 1$. The large parameter δ characterizes the ratio of the horizontal and vertical scales of density variability in real ocean (Miropol'skii and Shishkina 2001; Mei et al. 2017; Morozov 2018). The asymptotic solution of systems (1) shall be found in the form (Miropol'skii and Shishkina 2001; Kravtsov and Orlov 1999; Froman and Froman 2002; Bulatov 2014)

$$u_i(z, x, y) = \sum_{m=0}^{\infty}(i\delta)^{-m}u_{mj}(z, x, y)\exp(i\delta L(x, y)), \quad j = 1, 2, 3$$

$$P(z, x, y) = \sum_{m=0}^{\infty}(i\delta)^{-m}P_m(z, x, y)\exp(i\delta L(x, y))$$

Functions $L(x, y)$, P_m, u_{jm}, $j = 1, 2, 3$, $m = 0, 1, \ldots$ are subject to definition. Using relations: $u_{10} = -i|\nabla L|^{-2} L'_x \frac{\partial u_{30}}{\partial z}$, $u_{20} = -i|\nabla L|^{-2} L'_y \frac{\partial u_{30}}{\partial z}$ we can get:

$$\Pi u_{30} = 0, \Pi = \frac{\partial^2}{\partial z^2} + |\nabla L|^2\big((N(z, x, y)/\omega)^2 - 1\big)u_{30}(0, x, y) = u_{30}(H, x, y) = 0,$$

This spectral problem has a calculation setup of eigenfunctions ϕ_{0n} and eigenvalues $\Omega_n(x, y, \omega) \equiv |\nabla L_n|$, which are assumed to be known (Miropol'skii and Shishkina 2001; Bulatov and Vladimirov 2012, 2019; Bulatov 2014) From here on the index n will be omitted while assuming that further calculations belong to an individually taken mode. For determination of the function $L(x, y)$ we have the eikonal equation: $\nabla^2 L = \Omega^2(x, y, \omega)$. For solving this equation we have the rays systems:$x'_\sigma = e\,\Omega(x, y, \omega)^{-1}$, $y'_\sigma = b\,\Omega(x, y, \omega)^{-1}$, $b'_\sigma = \Omega(x, y, \omega)'_y$, $e'_\sigma = \Omega(x, y, \omega)'_x$, where $e = \partial L/\partial x$, $b = \partial L/\partial y$, $d\sigma$ is the length element of the ray, σ- ray coordinate. The function $u_{30}(z, x, y)$ is defined to the accuracy of multiplication by the some arbitrary function. The solution to problem (1) can be sought in the form:$D(z, x, y) = \sum_{n=1}^{\infty} A_n(z, x, y)\ \Phi_n(z, x, y) \exp(i\delta L_n(x, y))$. Functions $\Phi_n(z, x, y)$ are determined from the vertical spectral problem

$$\frac{\partial}{\partial z}\left(\frac{\partial \Phi_n(z, x, y)}{\partial z}\alpha^{-1}\right) + \Omega_n^2(x, y, \omega)\Phi_n(z, x, y) = 0 \qquad (2)$$

$$\frac{\partial \Phi_n(z, x, y)}{\partial z} = 0, \text{ for } z = 0, -H$$

Functions $A_n(z, x, y)$ can be represented as (index n is omitted hereafter):$A(z, x, y) = A_0(x, y) + \sum_{m=1}^{\infty} (i\delta)^{-m} A_m(z, x, y)$. Then, to determine the function $A_0(x, y)$, one can obtain the transfer equation

$$2\Lambda\nabla S\nabla A_0 + \Lambda A_0 \Delta S + A_0(\nabla\Lambda + \boldsymbol{I}) = 0$$

$$\Lambda = \int_{-H}^{0} \Phi^2(z, x, y)dz, \boldsymbol{I} = \frac{1}{2}\int_{-H}^{0} \frac{\nabla\chi}{\alpha}\frac{\partial\,\Phi^2}{\partial z}dz.$$

Along rays this equation has a solution

$$A_0^2(\sigma)\Lambda(\sigma)\Omega(\sigma)da(\sigma) = A_0^2(\sigma_0)\Lambda(\sigma_0)\Omega(\sigma_0)da(\sigma_0)T$$

$$T = \exp(-\int_{-\sigma_0}^{\sigma} \boldsymbol{I}\nabla S(\Lambda\Omega)^{-1}d\sigma$$

where $da(\sigma) = R(\sigma)d\sigma$—ray tube width, $R(\sigma)$—geometric ray divergence (Kravtsov and Orlov 1999; Froman and Froman 2002; Bulatov 2014). It can be shown that the exponential factor T is $\Omega(\sigma)/\Omega(\sigma_0)$. Therefore, the solution for the $A_0(\sigma)$ is simplified: $A_0^2(\sigma)\Lambda(\sigma)da(\sigma) = A_0^2(\sigma_0)\Lambda(\sigma_0)da(\sigma_0)$. Then the expression for $\boldsymbol{I}$ can be represented as

$$\boldsymbol{I} = \int\limits_{-H}^{0} \frac{\nabla\chi}{\alpha}\Phi\frac{\partial\,\Phi}{\partial z}dz = -\frac{1}{2\,\Omega^2}\int\limits_{-H}^{0}\frac{\partial}{\partial z}((\alpha^{-1}\frac{\partial\Phi}{\partial z})^2)\nabla\chi dz$$

$$= \frac{1}{2\,\Omega^2}\int\limits_{-H}^{0}\frac{\partial}{\partial z}(\nabla\chi)\left(\left(\alpha^{-1}\frac{\partial\Phi}{\partial z}s\right)^2\right)\nabla\chi dz$$

$$= \frac{1}{2\,\omega^2\,\Omega^2}\int\limits_{-H}^{0}\nabla N^2(z,x,y)\left((\alpha^{-1}\frac{\partial\Phi}{\partial z})^2\right)dz$$

It is possible to obtained

$$-\Lambda\nabla\ln\Omega = \frac{1}{2\,\omega^2\,\Omega^2}\int\limits_{-H}^{0}\nabla N^2(z,x,y)((\alpha^{-1}\frac{\partial\Phi}{\partial z})^2)dz \tag{3}$$

Apply to Eq. (2) the operator ∇

$$\frac{\partial}{\partial z}(\frac{\partial\boldsymbol{F}}{\partial z}\alpha^{-1}) + \Omega^2\,\boldsymbol{F} = -\frac{\partial}{\partial z}(\frac{\partial\Phi}{\partial z}\nabla(\alpha^{-1})) - \Phi\nabla\,\Omega^2,\ \boldsymbol{F} = \nabla\Phi \tag{4}$$

$$\frac{\partial\boldsymbol{F}}{\partial z} = 0,\ \text{for}\ z = 0, -H.$$

From (3)–(4) we get

$$\nabla\Omega^2\int\limits_{-H}^{0}\Phi^2(z,x,y)dz = \int\limits_{-H}^{0}\frac{\partial}{\partial z}(\omega^{-2}\alpha^{-2}\frac{\partial\Phi}{\partial z}\nabla N^2)\Phi dz$$

As a result we can get

$$\boldsymbol{I} = -\Lambda\nabla\ln\Omega$$

The conservation law can be written as well in the form suitable for finding the function A_0:

$$\frac{A_0^2(x(\sigma,\alpha), y(\sigma,\alpha))}{\Omega^2(x(\sigma,\alpha), y(\sigma,\alpha))} d\,a(x(\sigma,\alpha), y(\sigma,\alpha)) = \frac{A_0^2(x_0(\alpha), y_0(\alpha))}{\Omega^2(x_0(\alpha), y_0(\alpha))} d\,a(x_0(\alpha), y_0(\alpha))$$

where $d\,a(x(\sigma,\alpha), y(\sigma,\alpha)) = J(x(\sigma,\alpha), y(\sigma,\alpha))d\alpha$ is the unit ray tube width, J– Jacobian from ray coordinates to Cartesian coordinates. Note that the wave energy flash is proportional to $A_0^2\Omega^{-1}da$, thus, in this case, there survives the value equal to the wave energy flash divided by the wave vector modulus (Miropol'skii and Shishkina 2001; Bulatov and Vladimirov 2012, 2019; Bulatov 2014). In real oceanic conditions, an arbitrary source of disturbances can be represented as a superposition of harmonic waves

$$(U_i, p) = \int \omega \sum_{m=0}^{\infty} (i\delta)^{-m}(u_{jm}, P_m)\ \exp(i\omega t - i\delta L_m(\omega, x, y))d\omega, \quad j = 1, 2, 3$$

In this paper we have built asymptotics notations of the solutions on IGW propagation in the non-uniform medium in the horizontal direction. Numerical calculations for typical oceanic parameters indicate essential influence of the factors of horizontal non-uniformity of natural parameters on real IGW dynamics. All obtained analytical results of this paper were generally obtained for arbitrary density distributions and other parameters of non-uniform stratified natural mediums, and moreover, the basic physical mechanisms of forming the investigated phenomena of IGW dynamics in non-uniform stratified mediums were examined in the frame of the available field data. The popularity of suggested approaches to analyze the IGW dynamics could be supported by the fact of existing a wide scope of interesting physical problems which are quite adequately defined by these approaches, because the multiplicity of issues of the kind is related to the variety of non-uniform stratified mediums (Morozov 2018; Matyushin 2019; Lecoanet et al. 2015; Voelker et al. 2019; Wang et al. 2017). The importance of such methods for analyzing the wave field is attributed to not only their visualization, universal applicability and effectiveness in solving various problems, but also to a possibility that they become a semi-empirical basis for other method of approximations in the propagation theory for wave packets of other physical nature. Therefore, to understand the physics of many linear and non-linear wave processes within non-uniform media we have to study the full complex of hydrodynamic effects going with the propagation of IGW in a stratified medium, moreover, it so happened that the analytical constructions of the sort can be easily observed under the real ocean conditions (Mei et al. 2017; Morozov 2018).

Acknowledgements The work is carried out with financial support from the Russian Foundation for Basic Research, project 20-01-00111A.

References

Broutman D, Rottman J (2004) A simplified Fourier method for computing the internal wave field generated by an oscillating source in a horizontally moving, depth-dependent background. Phys Fluids 16:3682–3689

Bulatov VV, Vladimirov YuV (2012) Wave dynamics of stratified mediums. Nauka, Moscow

Bulatov VV, Vladimirov YuV (2014) Asymptotical analysis of internal gravity wave dynamics in stratified medium. Appl Math Sci 8:217–240

Bulatov VV, Vladimirov YuV (2018) Internal gravity waves in horizontally inhomogeneous ocean. In Velarde MG et al (eds) The ocean in motion, Springer Oceanography. Springer International Publishing AG, Berlin pp 109–126

Bulatov VV, Vladimirov YuV (2019) A general approach to ocean wave dynamics research: modelling, asymptotics, measurements. OntoPrint Publishers, Moscow

Bulatov VV, Vladimirov YuV, Vladimirov IYu (2019) Far fields of internal gravity waves from a source moving in the ocean with an arbitrary buoyancy frequency distribution. Russian J Earth Sci 19:ES5003

Fabrikant AL, Stepanyants YuA (1998) Propagation of waves in shear flows. World Scientific Publishing, London

Froman N, Froman P (2002) Physical problems solved by the phase-integral method. Cambridge University Press, Cambridge

Kravtsov Yu, Orlov Yu (1999) Caustics, catastrophes and wave fields. Springer, Berlin

Lecoanet D, Le Bars M, Burns KJ, Vasil GM, Brown BP, Quataert E, Oishi JS (2015) Numerical simulations of internal wave generation by convection in water. Phys Rev E Stat Nonlinear Soft Matter Phys 9:1–10

Matyushin PV (2019) Process of the formation of internal waves initiated by the start of motion of a body in a stratified viscous fluid. Fluid Dyn 54:374–388

Mei CC, Stiassnie M, Yue DK-P (2017) Theory and applications of ocean surface waves. Advanced series of ocean engineering. V. 42. World Scientific Publishing

Miropol'skii YuZ, Shishkina OV (2001) Dynamics of internal gravity waves in the ocean. Kluwer Academic Publishers, Boston

Morozov TG (2018) Oceanic internal tides. Observations, analysis and modeling. Springer

Voelker GS, Myers PG, Walter M, Sutherland BR (2019) Generation of oceanic internal gravity waves by a cyclonic surface stress disturbance. Dyn Atm Oceans 86:16–133

Wang H, Chen K, You Y (2017) An investigation on internal waves generated by towed models under a strong halocline. Phys Fluids 29:065104

Reconstruction of Spacious Stress Fields in a Heavy Elastic Layer from Discrete Data on Stress Orientations

A. N. Galybin

Abstract The problem considered in this paper is aimed at modification of earlier developed algorithms for the stress field reconstructions from discrete data on stress orientations in the earth's crust. The modification is made for an elastic layer and takes into account the weight of the rockmasses and the vertical loads acting on the interfaces of the layer. We present the general solutions for the 3D layer using the assumption that the vertical stress is one of the principal stresses. Then the modification of the algorithm is described and followed by examples of stress reconstruction for the region near Taiwan.

Keywords 3D elastic layer · Body forces · Stresses functions · Complex potentials · Stress trajectories · Maximum shear stresses · Earth's crust

1 Introduction

This report deals with further development of the stress field reconstruction methods that use as input the discrete data on stress orientations in the earth's crust. Such data on azimuths of the maximum compressive horizontal stresses can be found in the World Stress Map Project (WSMP) (Heidbach et al. 2016). The previously developed methods (Galybin and Mukhamediev 2004; Irša and Galybin 2010) addressing 2D cases have been aimed at the areas whose horizontal sizes are much greater than the thickness of the lithosphere. Therefore the depth of the data have not been taken into account but referred to the whole crust. For more compact areas the question regarding necessity to include the depth is much more pronounced. In this paper a modification of the earlier developed techniques is suggested in order to analyse how strongly the results may be affected if gravity is taken into account.

The work consists of three parts. Firstly, a general solution for a spacious elastic layer subjected to lateral and normal loading and gravity is obtained. Then, the algorithms developed earlier are modified to include extra terms that appear in the expression for the stress function: mean stresses and stress deviator. The third part

A. N. Galybin (✉)
The Schmidt Institute of Physics of the Earth, RAS, Moscow, Russia

T. Chaplina (ed.), *Processes in GeoMedia—Volume IV*, Springer Geology,
https://doi.org/10.1007/978-3-030-76328-2_9

presents the results of stress reconstruction in the earth's crust near Taiwan by using the data on stress orientations from the WSMP database. (overall 42,870 data and near 32,000 data of quality A–C, with the errors of A $\pm$ 15°, B $\pm$ 20°, C $\pm$ 25°).

The first part is a generalisation of the results reported in Timoshenko and Goodier (1970) where one can find a discussion on the approximate character of the plane stress solutions problem. In particular an exact solution for that the Airy stress function has been found for an elastic homogeneous layer of finite thickness for the case $\sigma_{33} = \sigma_{13} = \sigma_{23} = 0$, where the x_3-axis is vertical and perpendicular to the plane of the layer and the x_1- and x_2-axes lie in the plane. It has been shown that the spacious Airy function, $\mathrm{A} = \mathrm{A}(x_1, x_2, x_3)$, is quadratic with respect to the variable x_3, i.e. through the thickness of the layer. It has the following form

$$\mathrm{A}(x_1, x_2, x_3) = -\frac{\nu}{1+\nu} P(x_1, x_2) x_3^2 + \mathrm{A}_1(x_1, x_2) x_3 + \mathrm{A}_0(x_1, x_2). \quad (1)$$

Here ν is Poisson's ratio, P and A_1 are harmonic functions and A_0 is biharmonic function that coincides with the Airy stress function for the plane problems. The function A_1 is independent while P is expressed as the Laplace operator applied to A_0. Thus

$$2P(x_1, x_2) = \left(\frac{\partial^2}{\partial x_1^2} + \frac{\partial^2}{\partial x_2^2}\right) \mathrm{A}_0(x_1, x_2), \quad \left(\frac{\partial^2}{\partial x_1^2} + \frac{\partial^2}{\partial x_2^2}\right) \mathrm{A}_1(x_1, x_2) = 0. \quad (2)$$

Non-zero stress components are expressed through the Airy function as follows (Timoshenko and Goodier 1970)

$$\sigma_{11} = \frac{\partial^2 \mathrm{A}}{\partial x_2^2}, \quad \sigma_{22} = \frac{\partial^2 \mathrm{A}}{\partial x_1^2}, \quad \sigma_{12} = -\frac{\partial^2 \mathrm{A}}{\partial x_1 \partial x_2}. \quad (3)$$

Therefore the stress field of an elastic plate is defined by two independent functions A_1 and A_0. Meanwhile, for all problems in which the stress field is symmetrical about the middle plane, $x_3 = 0$, the second term in the right hand side of (1) vanishes and all non-zero stress components are determined through the Airy function for the plane problems, A_0, only.

In this paper we generalised the solution obtained in Timoshenko and Goodier (1970) for elastic layers by taking into account the weight of the layer introducing specific gravity $\gamma = \gamma(x_1, x_2, x_3) > 0$ and by assuming that $\sigma_{33} = \sigma_{33}(x_1, x_2, x_3) \neq 0$. Then we use this generalisation to modify the methods of stress reconstruction reported in Galybin and Mukhamediev (2004), Irša and Galybin 2010).

2 Solution for Heavy Layer

Let us consider an elastic layers located between other (not usually elastic and/or homogeneous) layers and introduce a Cartesian coordinate frame as shown in Fig. 1. It is assumed that the top layer is loaded by normal and tangential forces of intensity

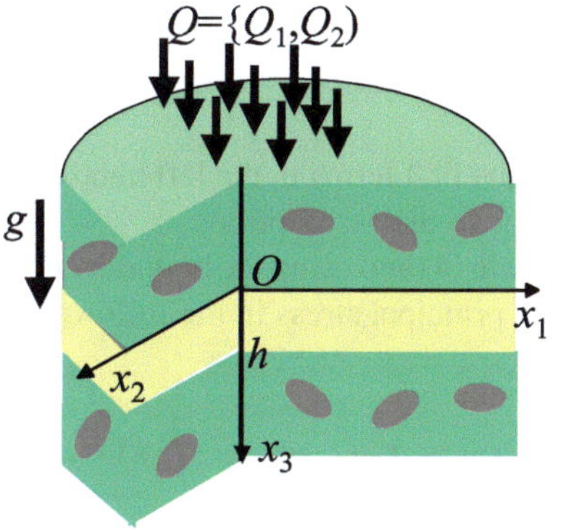

Fig. 1 An elastic homogeneous layer $0 < x_3 < h$ surrounded by other layers

$Q = \{Q_1, Q_2\}$ and as a result the layer $0 < x_3 < h$ is loaded on $x_3 = 0$ by known stress vector $\sigma^0 = \{\sigma_1^0, \sigma_2^0\}$. It is further assumed that the shear component $\sigma_2^0 = 0$, which mechanically corresponds to the case of low friction on the interfaces between the layer and surrounding rocks.

When using the WSMP data it is usually accepted that the vertical stress is one of the principal stresses. Finding the general solution employing this assumption is the focus of this section.

2.1 *The Governing System of Differential Equations*

The governing system consists of three differential equations of equilibrium, DDE, and six Beltrami-Michell compatibility equations imposed on the stress components, BME. We further use the following fundamental assumption.

Assumption. Inside the layer the vertical stresses σ_{33} is one of the principal stresses and weight is the only body force. Therefore $\sigma_{13} = \sigma_{23} = 0$ at every point inside the layer including its boundaries and DDE and BME assume the following form.

DDE:

$$\frac{\partial \sigma_{kj}}{\partial x_j} = 0, \quad k, j = 1, 2; \quad \frac{\partial \sigma_{33}}{\partial x_3} + \gamma = 0. \tag{4}$$

BME:

$$\Delta \sigma_{kj} + \frac{1}{1+\nu} \frac{\partial^2 \sigma}{\partial x_k \partial x_j} = X_{kj}, \quad k, j = 1, 2, 3; \quad \sigma = \sigma_{11} + \sigma_{22} + \sigma_{33}. \tag{5}$$

Here $\Delta = \frac{\partial^2}{\partial x_1^2} + \frac{\partial^2}{\partial x_2^2} + \frac{\partial^2}{\partial x_3^2}$ is the Laplace operator and the right hand sides are

$$X_{11} = X_{22} = \frac{-\nu}{1-\nu}\frac{\partial \gamma}{\partial x_3}, \quad X_{33} = \frac{\nu-2}{1-\nu}\frac{\partial \gamma}{\partial x_3}, \quad X_{12} = 0, \quad X_{13} = -\frac{\partial \gamma}{\partial x_1}, \quad X_{23} = -\frac{\partial \gamma}{\partial x_2}. \tag{6}$$

It should be noted that the first terms in the left hand sides of (5) corresponding $k = 1, 2$ and $j = 3$ vanish due to the assumption.

Integrating the 3rd DDE in (4) and using the boundary condition one finds spacious distribution of the vertical principal stress in the following form

$$\sigma_{33} = \sigma_{33}(x_1, x_2, x_3) = \sigma_{33}^0(x_1, x_2) - \int_0^{x_3} \gamma(x_1, x_2, t)\, dt, \quad x_3 > 0. \tag{7}$$

Here $\sigma_{33}^0(x_1, x_2)$ is the vertical stress component caused by applied normal loads on the top and bottom sides of the layer. Besides the load acting on the bottom side $\sigma_{33}^0(x_1, x_2, h)$ should compensate both the load $\sigma_{33}^0(x_1, x_2)$ and the weight in order to keep the layer in global equilibrium.

2.2 *Complex Variables*

It is convenient to introduce a complex variable $z = x_1 + ix_2$ and its complex conjugate. The differential operators of the first order take the form

$$\frac{\partial}{\partial x_1} = \frac{\partial}{\partial z} + \frac{\partial}{\partial \bar{z}}, \quad \frac{\partial}{\partial x_2} = i\frac{\partial}{\partial z} - i\frac{\partial}{\partial \bar{z}}, \quad 2\frac{\partial}{\partial z} = \frac{\partial}{\partial x_1} - i\frac{\partial}{\partial x_2}, \quad 2\frac{\partial}{\partial \bar{z}} = \frac{\partial}{\partial x_1} + i\frac{\partial}{\partial x_2}. \tag{8}$$

The second order derivative are obtained by successive application of the first order derivatives, in particular, the Laplace operators for plane and spacious problems become

$$\frac{\partial^2}{\partial x_1^2} + \frac{\partial^2}{\partial x_2^2} = 4\frac{\partial^2}{\partial z \partial \bar{z}}, \quad \Delta = \frac{\partial^2}{\partial x_1^2} + \frac{\partial^2}{\partial x_2^2} + \frac{\partial^2}{\partial x_3^2} = 4\frac{\partial^2}{\partial z \partial \bar{z}} + \frac{\partial^2}{\partial x_3^2} \tag{9}$$

With the use of (8), (9) one can present the first two DDE in (4) in the following complex form

$$\frac{\partial P}{\partial z} = \frac{\partial D}{\partial \bar{z}}, \tag{10}$$

where P and D are the stress functions: mean stresses (real-valued function) and complex stress deviator respectively. They are defined as follows

$$P(z, \bar{z}, x_3) = \frac{\sigma_{11} + \sigma_{22}}{2}, \quad D(z, \bar{z}, x_3) = \frac{\sigma_{22} - \sigma_{11}}{2} + i\sigma_{12} \tag{11}$$

Hereafter we use the following convention $F(x_1, x_2, x_3) = F(z, \overline{z}, x_3)$, which assumes that one should substitute $x_1 = \frac{1}{2}(z + \overline{z})$, $x_2 = \frac{1}{2i}(z - \overline{z})$ but keep the same notation for the function as the one used in the Cartesian coordinates.

The stress functions P and D are expressed via the Airy function as follows

$$P(z, \overline{z}, x_3) = 2\frac{\partial^2 A(z,\overline{z},x_3)}{\partial z \partial \overline{z}}, \quad D(z, \overline{z}, x_3) = 2\frac{\partial^2 A(z,\overline{z},x_3)}{\partial z^2}. \tag{12}$$

The arguments are omitted further on for briefness unless in those cases where we want to emphasize the dependence.

The BME system assume the following form

$$\frac{1}{1+\nu}\Delta\sigma = \frac{1}{1-\nu}\frac{\partial^2 \sigma_{33}}{\partial x_3^2} \tag{13}$$

$$\Delta D - \frac{2}{1+\nu}\frac{\partial^2 \sigma}{\partial z^2} = 0 \tag{14}$$

$$\frac{\partial^2 \sigma}{\partial z \partial \overline{z}}\left(\frac{1}{1+\nu}\sigma - \sigma_{33}\right) = 0 \tag{15}$$

$$\frac{\partial^2}{\partial z \partial x_3}\left(\frac{1}{1+\nu}\sigma - \sigma_{33}\right) = 0 \tag{16}$$

Here Eq. (13) is obtained by summing three Eq. (5) corresponding $k = j = 1, 2, 3$, divided by $(2 + \nu)$; Eq. (14) is found by subtracting (5) for $k = j = 1$ from (5) for $k = j = 2$ followed by summing with Eq. (5) for $k = 1, j = 2$ multiplied by $2i$; Eq. (15) is the difference of (13) and Eq. (5) corresponding $k = j = 3$, and (16) is the complex composition of the last two Eq. (5) for $k = 1, 2$ and $j = 3$. It is worth to mention that (13), (15) are real-valued while (14), (16) are complex-valued; therefore the total number of real BME equations remains the same as (5) and these BME systems are equivalent.

It follows from (16) that the term in brackets can be presented as follows.

$$\sigma - (1 + \nu)\sigma_{33} = T(x_3) + \lambda(z, \overline{z}), \tag{17}$$

which is due to the fact that the derivative of the left-hand side of (16) with respect to x_3 depends on x_3 only.

Substitution of (17) into (15) shows that the function λ is harmonic in plane, therefore it can be considered as a real part of arbitrary holomorphic function. Thus (17) assumes the following form

$$\sigma = 2\Big(T(x_3) + \Phi(z) + \overline{\Phi(z)}\Big) + (1 + \nu)\sigma_{33}, \tag{18}$$

where the coefficient 2 is introduced for convenience.

As $\sigma = 2P + \sigma_{33}$, one finds the following expression for the mean stresses

$$P(z,\overline{z},x_3) = P_0(z,\overline{z}) + T(x_3) + \tfrac{\nu}{2}\sigma_{33}(z,\overline{z},x_3), \quad P_0(z,\overline{z}) = \Phi(z) + \overline{\Phi(z)}, \tag{19}$$

where the term P_0 is separated to emphasize its relevance to the plane stress (strain) conditions.

Substitution of (19) into (13) results in

$$\tfrac{2}{1+\nu}T''(x_3) = \tfrac{\nu}{1-\nu}\tfrac{\partial^2\sigma_{33}}{\partial x_3^2} - 4\tfrac{\partial^2\sigma_{33}}{\partial z\partial\overline{z}}. \tag{20}$$

It is evident from (20) that the following relationship can be imposed on the vertical component of the spacious stress tensor

$$\frac{\partial}{\partial z}\left(\tfrac{\nu}{1-\nu}\tfrac{\partial^2\sigma_{33}}{\partial x_3^2} - 4\tfrac{\partial^2\sigma_{33}}{\partial z\partial\overline{z}}\right) = 0. \tag{21}$$

This should be considered as a criterion that can be used for validation of the fundamental assumption accepted in this paper, i.e. the vertical stress component σ_{33} is one of principal stresses if (21) is satisfied.

The general solution for the stress deviator can be obtained by integration of DEE (10), which gives

$$D(z,\overline{z},x_3) = \overline{z}\Phi'(z) + \tfrac{\nu}{2}\int \tfrac{\partial\sigma_{33}(z,\overline{z},x_3)}{\partial z}\,d\overline{z} + F(z,x_3), \tag{22}$$

where F is an arbitrary function.

Substitution of (22) into (14) followed by simplifications employing (21) leads to the following differential equation imposed on F

$$\tfrac{\partial^2 F(z,x_3)}{\partial x_3^2} = -\tfrac{4\nu}{1+\nu}\Phi''(z). \tag{23}$$

Integrating (23) one finds the expression for the stress deviator

$$\begin{gathered} D(z,\overline{z},x_3) = D_0(z,\overline{z}) + x_3\mathrm{E}(z) - \tfrac{2\nu}{1+\nu}x_3^2\Phi''(z) + \tfrac{\nu}{2}\int \tfrac{\partial\sigma_{33}(z,\overline{z},x_3)}{\partial z}\,d\overline{z}, \\ D_0(z,\overline{z}) = \overline{z}\Phi'(z) + \Psi(z). \end{gathered} \tag{24}$$

Here Φ(z) and Ψ(z) are complex potentials (holomorphic functions) as in the plane problem and E(z) is an additional sought complex potentials. Note that we keep the notations usually used for the Kolosov-Muskhelishvili formulas (Muskhelishvili 1963), the bi-holomorphic function D_0, and complement these formulas by adding the third potential E(z) and other terms expressed through σ_{33} including $T(x_3)$.

It is evident that the result reported in Timoshenko and Goodier (1970) is a particular case considered here. Indeed, if $\sigma_{33}^0 = 0$ and $\gamma = 0$ then $\sigma_{33} = 0$, the criterion given by (21) is satisfied and $T(x_3) = 0$. Therefore the expressions for P and D coincide with those obtained from (1) by using (12) and employing the Kolosov-Muskhelishvili formulas.

It is understood that the criterion given by (21) theoretically could be suitable for quite wide class of loads. However the situations where specific gravity γ varies within the homogeneous layer are limited. We further focus on the case that is suitable for applications to rock mechanics.

2.3 Linear Load

Let us consider the case $\sigma_{33}^0 = \sigma_0 = \text{const}$, $\gamma = \text{const}$, which will be used in further calculations

$$\sigma_{33} = \sigma_0 - \gamma_0 x_3, \quad 0 \le x_3 \le h \tag{25}$$

It is evident that (25) satisfies the 3rd DDE in (4) and the criterion in (21), also $\mathrm{T}(x_3) = 0$. Therefore, the stress functions are

$$P(z, \bar{z}, x_3) = \Phi(z) + \overline{\Phi(z)} + \tfrac{\nu}{2}(\sigma_0 - \gamma_0 x_3), \tag{26}$$

$$D(z, \bar{z}, x_3) = \bar{z}\Phi'(z) + \Psi(z) + x_3 \mathrm{E}(z) - \tfrac{2\nu}{1+\nu} x_3^2 \Phi''(z). \tag{27}$$

It follows from the formulas above that the mean stress is harmonic and the complex stress deviator is bi-holomorphic as in the case of plane stress (strain) conditions but contain extra terms. These are known for the mean stress and unknown for the stress deviator.

It is remarkable that the stress deviator does not depend directly on the vertical stress. Moreover, one can introduce the average of the stress deviator by using integration over the thickness of the layer.

Firstly, let us introduce a new holomorphic function in the form

$$\Psi_1(z) = \Psi(z) + \tfrac{1}{2} h \mathrm{E}(z) - \tfrac{2\nu}{1+\nu} \tfrac{1}{3} h^2 \Phi''(z). \tag{28}$$

This allows one to present (27) as shown below

$$D(z, \bar{z}, x_3) = \bar{z}\Phi'(z) + \Psi_1(z) + (x_3 - \tfrac{1}{2}h)\mathrm{E}(z) - \tfrac{2\nu}{1+\nu}(x_3^2 - \tfrac{1}{3}h^2)\Phi''(z). \tag{29}$$

Secondly we integrate (29) with respect to x_3 from 0 to h to obtain

$$\langle D(z, \bar{z})\rangle = \bar{z}\Phi'(z) + \Psi_1(z), \quad \langle D(z, \bar{z})\rangle = \tfrac{1}{h}\int_0^h D(z, \bar{z}, x_3)dx_3. \tag{30}$$

This formula demonstrates that the averaged stress deviator has the form similar to the stress deviator for the plane problems.

In the same way one can average the mean stress (26) to obtain

$$\langle P(z,\overline{z})\rangle = \Phi_1(z) + \overline{\Phi_1(z)},$$

$$\langle P(z,\overline{z})\rangle = \frac{1}{h}\int_0^h P(z,\overline{z},x_3)dx_3, \quad \Phi_1(z) = \Phi(z) + \frac{\nu}{4}\sigma_0 - \frac{\nu}{8}\gamma_0 h. \tag{31}$$

It is seem that both averaged stress functions have the form of the Kolosov-Muskhelishvili solution for the plane problems.

3 Algorithm

It is evident from (26)–(27) that in contrast to the plane problems the exact solution for the 3D layer (subjected to the assumption regarding σ_{33}) requires finding one extra harmonic function, E(z), which should be taken into account in turn with $\Phi''(z)$ when considering the cases where the weight is important. It is the case when the thickness of the layer is not small enough (as compare to other dimensions) in order to neglect the influence of the additional terms in (26)–(27).

The complex stress deviator (27) can be presented in the form.

$$D(z,\overline{z},x_3) = \tau_{\max}(z,\overline{z},x_3)e^{-2i\theta(z,\overline{z},x_3)}, \tag{32}$$

where $\tau_{\max} = (\sigma_H - \sigma_h)/2$ is the maximum shear stress acting in the horizontal plane with the principal stresses $\sigma_H > \sigma_h$, θ is the principal direction of the in-plane stress tensor (the angle between σ_H and the x-axis). Further under the term "trajectory field" we understood a set of curves tangent to which at any point is inclined at the angle θ at this point. The maps of $\tau_{\max}$ are also analysed as they provide information regarding stress regimes.

The stress field reconstruction problem is formulated as follows. Let principal directions θ_j be known at given points ζ_j ($j = 1\ldots N$), located within the domain or on its boundary Γ. The objective is to determine the stress field including stress trajectories, maximum shear stress and means stresses that provide the best fit to the data and satisfy Eqs. (26)–(27).

The data are taken from the WSMP database release 2016, see (Heidbach et al. 2016).

In contrast to the original approach developed in Galybin and Mukhamediev (2004) we take the depth of the data and need to reconstruct one extra holomorphic function E(ζ).

The proposed algorithm assumes the following steps.

- The holomorphic functions $\Phi(\zeta)$, $\Psi(\zeta)$ and E(ζ) are sought as polynomials of n-th degree with unknown complex coefficients.

- These coefficients are found by minimizing the functional

$$\sum_{j=1}^{N}\left\{\mathrm{Im}\left[\exp(2i\theta_j)\left(\overline{\zeta}_j\Phi'(\zeta_j)+\Psi(\zeta_j)+x_{3j}\mathrm{E}(\zeta_j)-\frac{2\nu}{1+\nu}(x_{3j})^2\Phi''(\zeta_j)\right)\right]\right\}^2\to\min \tag{33}$$

complemented by the condition that the mean of $\tau_{\max}$ is equal to unity over the domain considered, where N is the number of data.

- The optimisation problem is reduced to solving an overspecified system of linear algebraic equations, SLAE, of the type **AC** = **b**, where **A** is $(N+1)$ by $(6n+3)$ matrix, **C** is $(6n+3)$ vector of unknown coefficients, **b** is $(N+1)$ vector, all components of which are zero except of the last one equal to N. Details of system forming can be found in Galybin and Mukhamediev (2004); Irša and Galybin 2010).
- The SLAE is solved by the least squares method provided that the condition number of the system is not large; otherwise the SVD regularisation is applied to obtain stable solutions of the system.
- The number of terms in the approximating polynomials n is selected by analysing both the residual and the condition number; for real data n is not big due to scattering in experimentally determined azimuths of the stress orientations.
- As soon as the solution for the coefficients is found the holomorphic functions becomes known and it is possible to determine the complex stress deviator function normalized by a positive real constant, which means that the field of the stress trajectories (angles θ) is found uniquely while the field of $\tau_{\max}$ within a real parameter.

Some results of calculations using the real data form the WSMP database for a particular geographic region are reported in the next section.

4 Examples

4.1 WSMP Data Near Taiwan

The WSMP data on azimuths of the maximum horizontal compressive principal stresses are shown in Fig. 2 for the depths in the range of 0–60 km. The data quality shown are with the errors of ±15° (quality A), ±20° (quality B), ±25° quality C. The stress regimes are by presented different colours as shown in the legend of the figure. They are: Strike-Slip, SS, Normal Faulting, NF, Thrust Faulting, TF. The SS regime can produce sliding in horizontal plane; the NF and TF regimes out of plane.

Data distribution with the depth of quality A is 0, B is 1, C is 356. About 60 data are on shallow depth or for the cases where the depth was not defined. The depth of

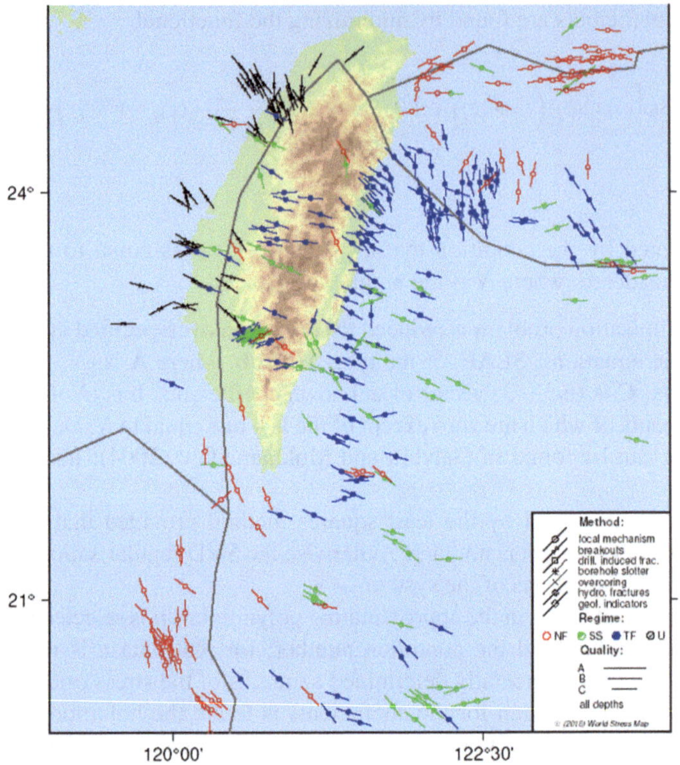

Fig. 2 Orientations of the maximum principal horizontal stress from WSMP of quality A-C at depth 0–60 km

data distribution is show in Fig. 3. It is evident from the figure that majority of the data are located on the depth of 10–40 km.

4.2 Stress Field Reconstruction

The results of calculations of the stress trajectory patterns and maximum shear stress maps for different depth are shown in Figs. 4 and 5 respectively. It is worth to mention that the maps on the free surface are the same as the averaged maps because at $x_3 = 0$ the stress deviator given by (27) coincides with that given by (30). It is evident that with increasing depth the stress field is changing. Thus, there are no isotropic points at shallow depth while they appear at the depth of 25 and 50 km. Therefore the account for the depth for the investigated region is meaningful. It has been pointed out in Mukhamediev et al. (2006); Mukhamediev and Galybin 2006) that the existence of such points is important to tsunami prediction.

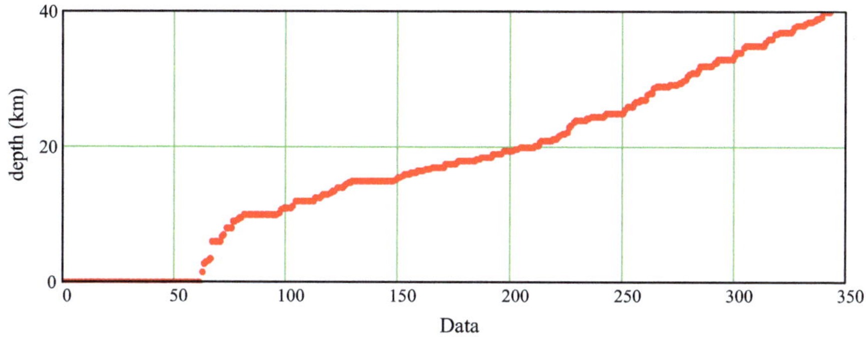

Fig. 3 Data distribution with depth

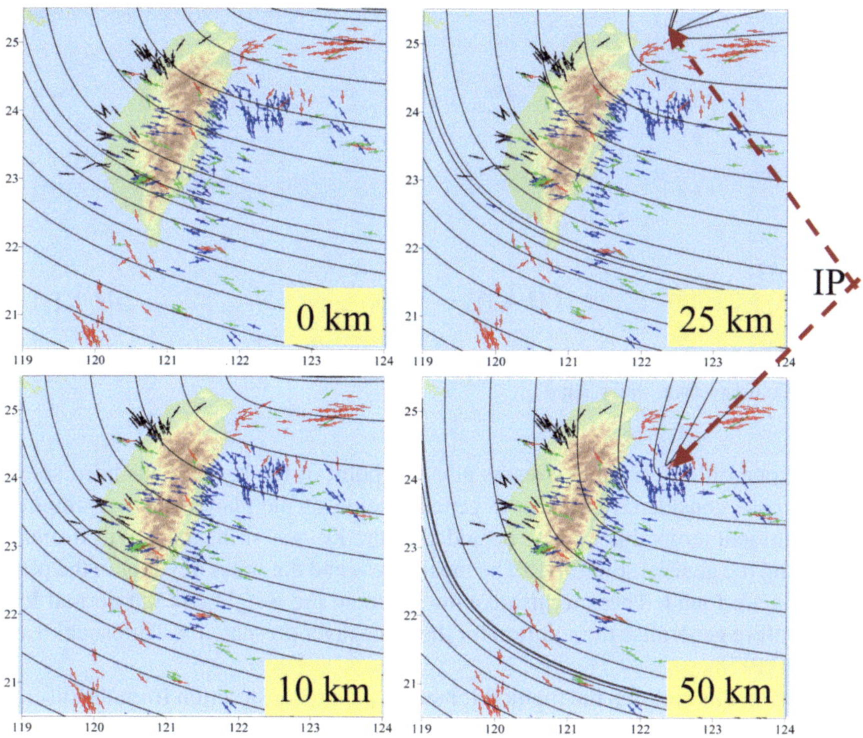

Fig. 4 Trajectory maps at different depth (IP = isotropic point)

5 Conclusions

This paper presents the general solution for the case of a heavy elastic layer loaded by normal and longitudinal stresses that is valid for the assumption that the vertical

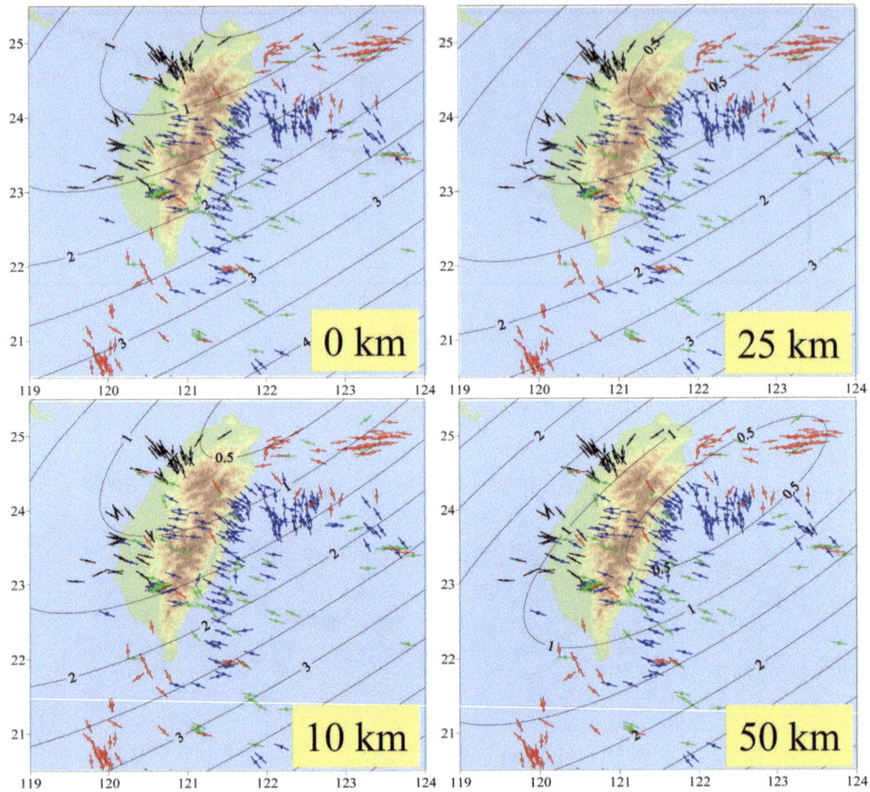

Fig. 5 Maximum shear stress maps

stress is one of the principal stresses at every point of the layer. It has been shown that the case considered here is the generalisation of the plane stress assumption. The additional terms that have to be added to the Kolosov-Muskhelishvili formulas presenting the general solution for the mean stress and the stress deviator in the plane problems are found. These terms take into account the weight and the normal load and therefore generalise the solution for the finite layer presented in Timoshenko and Goodier (1970).

Based on this solution the method for stress field reconstruction from the discrete WSMP data on stress orientations reported in Galybin and Mukhamediev (2004) has been updated to take into account the gravitational forces acting in the earth's crust and the depth of the WSMP data. It is shown that one extra holomorphic function and the second derivative of one of the Kolosov-Muskhelishvili potentials should be determined to complement the stress deviator for the plane problems. This does not affect much stability of calculations.

The examples of the stress field reconstruction in the region near Taiwan have been reported. They demonstrate that gravity is essential and can change qualitatively the stress field at deeper levels of the earth's crust.

Acknowledgements The work is supported by the Government Program 0144-2019-0005 and partially by the RSF grant N20-49-08002 (stress field reconstruction near Taiwan).

References

Galybin AN, Mukhamediev ShA (2004) Determination of elastic stresses from discrete data on stress orientations. Int J Solids Struct 41:5125–5142

Heidbach O., Rajabi M., Reiter K.; Ziegler M.; WSM Team (2016) World stress map database release 2016. GFZ Data Serv. https://doi.org/10.5880/WSM.2016.001

Irša J, Galybin AN (2010) Stress trajectories element method for stress determination from discrete data on principal directions. Eng Anal Boundary Elem 34:423–432

Mukhamediev ShA, Galybin AN (2006) Where and how did the ruptures of December 26 2004 and March 28, 2005 earthquakes near Sumatra originate? Dokl Earth Sci 406(1):52–55

Mukhamediev ShA, Galybin AN, Brady BHG (2006) Determination of stress fields in elastic lithosphere by methods based on stress orientations. Int J Rock Mech Min Sci 43(1):66–88

Muskhelishvili NI (1963) Some basic problems of the mathematical theory of elasticity. P. Noordhoff, Groningen, the Netherlands

Timoshenko SP, Goodier JN (1970) Theory of elasticity. McCraw-Hill, Singapore

Impact of Warm Winters on the White Sea: In Silico Experiment

I. Chernov and A. Tolstikov

Abstract We use the numerical model JASMINE for the in silico experiment on the White Sea by estimating the influence of two warm winters on water temperature, salinity, and pelagic ecosystem. The reference run consisted of 11 identical years, in the experiment two years in a row had positive air temperature, after that the conditions were the same. Another experiment added changes in wind speed and precipitation. The simulation showed that all variables return to the reference values within one year.

Keywords White Sea · In silico experiment · Global warming · Pelagic ecology

1 Introduction

Sea ice plays an important role in the functioning of the White Sea pelagic ecosystem. In winter, biogenic elements necessary for the development of phytoplankton in spring accumulate directly in the ice mass and under the ice cover. A combination of closely related factors in the late spring period (an increase in solar radiation and air and sea-surface water temperature, ice melting, growth of autochthonous matter concentration with increasing river runoff, etc.) allows phytoplankton to rapidly increase production. For the White Sea, this time at the end of May and beginning of June.

On the one hand, the absence of ice in the White Sea may seem to be positive for the economy: longer navigation, better navigating and fishing conditions, easier mining of natural resources. However, on the other hand, it is quite clear that if the tendency of the ice-cover reduction from year to year remains the same, this may significantly change the ecological systems of the White Sea; the consequences of such changes are by no means obvious.

I. Chernov (✉) · A. Tolstikov
Institute of Applied Math Research of Karelian Research Centre of RAS, Petrozavodsk, Russia
e-mail: chernov@krc.karelia.ru

Northern Water Problems Institute of Karelian Research Centre of RAS, Petrozavodsk, Russia

T. Chaplina (ed.), *Processes in GeoMedia—Volume IV*, Springer Geology,
https://doi.org/10.1007/978-3-030-76328-2_10

The *aim of this work* is to consider a model scenario in which the ice does not form in the White Sea in winter for two years in a row and to show possible consequences of such "warm winters" for the marine ecosystems.

Sea ice in the White Sea appears in the late autumn, when the surface water becomes cold enough, near the freezing point. Depending on the salinity (which is low in the White Sea compared to the neighbour Barents Sea and can be even lower in the bays influenced by rivers), this freezing temperature may vary from −1.9 °C in the core area to −0.3 °C in bays (−1.5 °C on the average throughout the sea) (Hydrometeorology and hydrochemistry of seas of the USSR 1991). During half a year (from November to May), there is floating ice in the White Sea and fast ice in the bays; but there is no continuous ice cover nor multiyear ice.

The earliest ice appears in October at river mouths, usually starting with Mezen river (the last decade of October) and on the drainage of the eastern coast of the Mezen Bay. The average thickness of fast ice, as a rule, is 40–70 cm, sometimes it can reach up to one and a half meters in severe winters (Hydrometeorology and hydrochemistry of seas of the USSR 1991). However, in recent years, such thick ice has not been observed at all due to warming. The more than 60-year observations of the near-surface air temperature in the White Sea reveal the significant positive trend (Serykh and Tolstikov 2020), i.e., he warming. This work shows that the observed increase in the near-surface air temperature is probably associated with the Global Atmospheric Oscillation (GAO); therefore, a large part of the atmosphere is involved in this process, and the same trends are revealed, for example, for the Barents Sea. Periods from two to seven years and from seven to ten years are distinguished in the interannual variability of the near-surface air temperature of the White Sea; they are associated, respectively, with El Niño (the Southern Oscillation) and NAO (the North Atlantic Oscillation), which are parts of GAO.

The largest landfast ice in the area is formed in the Kandalaksha Bay (up to 5 km from the coast) (Elisov 1997). The lowest area of the fast ice is near the Kaninsky, Konushinsky, Tersky coasts, in Voronka, Gorlo, and Bassein, along the Zimny coast of the Gorlo. Almost everywhere, except some protected bays, ridged ice exists; the local name of stranded ridges is *stamukha*. Ridging can reach 4–5 points (Hydrometeorology and hydrochemistry of seas of the USSR 1991).

The water temperature in most of the Sea in May and June is positive (about 1–2 °C). The White Sea becomes usually ice-free in May, but occasional floating ice may persist until June.

The wind is the most important for the ice drift. From September to March, most winds are southern, southwestern, and western (the autumn–winter monsoon trend); from April to August northern, northeastern, and eastern winds are more often (the spring–summer monsoon trend) (Sailinig directions for the White Sea 1995).

However, the presented numbers are for the past years, because now the ice area and mean thickness decrease not only in the White Sea, but in other Arctic sea as well (Stroeve and Notz 2018), in the context of the global warming (The Bulletin of the World Meteorological Organization 2018). The ice-free period is also increasing.

The abnormally warm winter of 2020 showed that an ice-free winter is quite possible; just one or two months of fast-ice presence in the bays of the White Sea with very thin and scarce floating ice are even more likely. Therefore, it is interesting to study in silico the influence of one or two ice-free winters on the pelagic ecosystems.

2 Numerical Experiment: Set-Up

The simulation was carried out using the JASMINE numerical model of the White Sea (Chernov et al. 2018). The spatial resolution of the model is about 3 km (200 × 200 points of the horizontal grid), along the vertical 46 z-horizons are taken with a step of 5 m to a depth of 150 m and further with a step of 10 m. The time step was 3 min; the strict limitation on the step is associated with the requirements of stability and high speeds of currents in the sea. For ice, the distribution of the volume of ice and snow on it, as well as the concentration of ice for 15 categories of thickness, corresponding to the classification of the USSR (Yakovlev 1948) is described. A harmonic semi-daily tide is set at the border with the Barents Sea. The line of 69°N is the model-area boundary; thus, a part of the Barents Sea is included. Atmospheric forcing is given: air pressure, temperature, and relative humidity, precipitation, cloudiness, wind speed.

The simulation set-up was the following. The reference simulation was 11 identical (2010) years. This year was chosen because it was relatively warm and extremely cloudless; so there are many satellite data for the White Sea (Zimin and Tolstikov 2010). In the in silico experiment, the first two years were identical (2010, a spin-up); then, starting from July of the third year up to July of the fifth, the air temperature near the sea surface was at least 1 °C. After that, several identical years were simulated in order to see the adaptation/relaxation of the components of the sea system. In the second experiment, there were additional changes: 20% stronger wind (with the same direction) and 15% stronger precipitation. The boundary values of sea-water temperature (the Flather boundary condition was used) was the same for all runs: the monthly-average piecewise constant values. The same for the rivers, see Table 1. The boundary condition for floating tracers (sea-ice components) is "the same concentration on the other side": during outflow, the tracer is taken out the sea, during inflow the concentration does not change. This is an assumption that ice conditions in the Barents Sea near the boundary are approximately the same as in the White Sea.

Inter-sea boundary −1.5, −1.7, −1.8, 0.0, 2.5, 3.0, 7.0, 8.5, 8.0, 5.0, 0.0, −1.5

Rivers 0.1, 0.1, 0.1, 0.2, 5.5, 15.0, 19.6, 17.1, 11.1, 4.0, 0.3, 0.1

The boundary salinity was constant: 35 per mille; the river water is fresh.

3 Discussion

Since the air temperature near the sea surface is positive during a "warm year," sea ice does not appear, though the temperature is negative at the inter-sea boundary. Due to the boundary condition, no ice is taken to the sea from the Barents Sea if there is no ice, and any new ice is taken out by the outflow (note than the net amount of water outflowing the White Sea is about 200km3/year, almost equal to the total river discharge). In the next year after the "warm" year, ice appears later compared to the reference run and observations (at the end of December) and the average thickness is lower since the water has accumulated more heat compared to the average climatic year.

The simulation has shown that the average values both for the surface and the bulk return to the reference values in one year. "No memory" property of the White Sea has been noted by other researchers (Semenov 2004). In Fig. 1 we compare the sea-surface temperature and the sea-ice concentration for the reference run and the experiment. The accumulated heat is seen in hte next year, but within one year the difference becomes negligible. Very little ice appears during the warm winter near the boundary where the incoming water has negative temperature.

In the experiment with the changed wind, the concentrations of nutrients and salinity remain slightly lower than the reference values. This is due to wind because with the unchanged wind these values restore completely. The reason is the water exchange at the sea boundary. From September to March, southern winds are more often: they prevent taking the matter in the White Sea from the Barents Sea. Stronger wind of the same direction reduces this even more.

The main observed effect for the simulated ice-free White Sea is a change in the chlorophyll-a concentration (Figs. 2 and 3). Figure 2 shows chlorophyll distribution over sea surface for two days (May 10 and May 15) for the reference run and the experiment. The bloom flash is clearly visible in a reference run when the bloom is

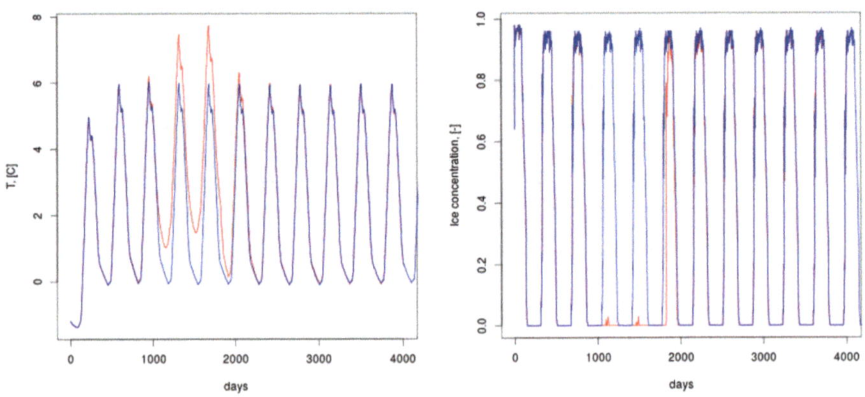

Fig. 1 The bulk-average daily water temperature (left) and sea-ice concentration (right); the blue line stands for the reference run, the red one is for the "warm winters" experiment

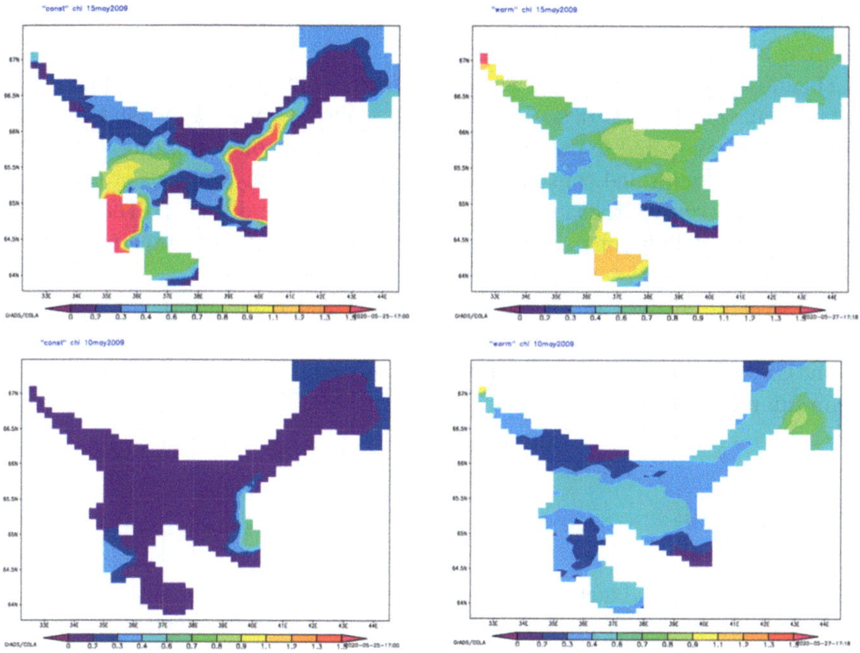

Fig. 2 Chlorophyll-a on the sea surface; the reference year on the left, the "warm" year on the right; the top row is for May 15, the second row is for May 10

Fig. 3 Chlorophyll-a surface concentration

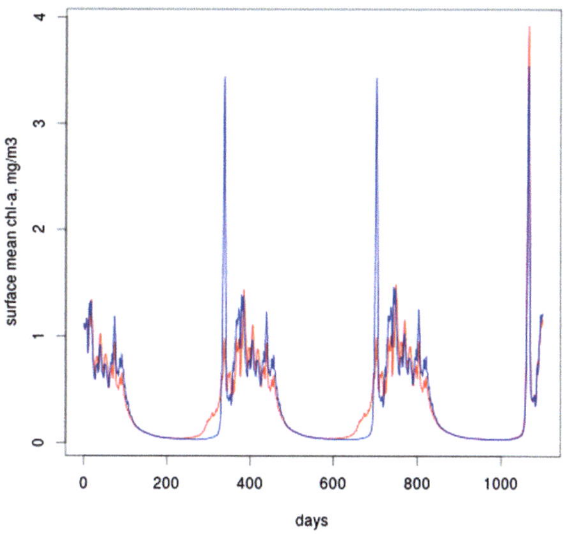

blocked by non-transparent ice up to ice melt; the bloom in a warm year starts earlier (no light-blocking ice) and is not so active.

The most important factor for phytoplankton growth is light, so the bloom starts earlier in an ice-free year (a soon as the polar night is over), but the peak is significantly lower compared to the reference run, because of other limiting factors: the presence of nutrients and consumption by zooplankton. Phytophages are playing a more important role in the change in phytoplankton biomass dynamics. Their biomass grows together with biomass of the phytoplankton and so it starts growing earlier in the ice-free warm year.

So, atypical changes in sea-ice concentration and reduction of the ice period change the dynamics of the marine ecosystems. The total biomass remains the same, but the chlorophyll peak is lower compared to the reference year (which agrees with the in situ measurements).

4 Conclusion

The simulation showed that two no-ice winters with positive air temperature has no long-term influence over the White Sea. Both bulk and surface variables return to the non-disturbed values within one year.

Acknowledgements The work was supported by the Russian Foundation for Basic Research (18-05-60184).

References

Chernov I, Lazzari P, Tolstikov A, Kravchishina M, Iakovlev N (2018) Hydrodynamical and biogeochemical spatiotemporal variability in the White Sea: a modeling study. J Mar Syst 187:23–35. https://doi.org/10.1016/j.jmarsys.2018.06.006

Elisov VV (1997) Estimation of water, thermal, and haline balance of the White sea. Meteorol Hydrol 9:83–93. In Russian

Hydrometeorology and hydrochemistry of seas of the USSR (1991) The White Sea. Reference book «Seas of the USSR».— L. : Gidrometeoizdat—T. II.—V.1. 196 p. In Russian

Sailinig directions for the White Sea (1995) Saint-Petersburg: Defence Ministry of the USSR. № 1110. 336 pp. In Russian

Semenov EV (2004) Numerical modelling of White Sea dynamics and monitoring problem. Izvestia Atmos Oceanic Phys 40(1):114–126

Serykh IV, Tolstikov AV (2020) On the climatic changes of the surface air temperature in the White Sea region. IOP Conf Ser Earth Environ Sci 606:EESE6061054

Stroeve J, Notz D (2018) Changing state of Arctic sea ice across all seasons. Environ Res Lett IOP Sci 13:103001. https://iopscience.iop.org/article/https://doi.org/10.1088/1748-9326/aade56

The Bulletin of the World Meteorological Organization (2018) Climate change: science and solutions 67(2). https://library.wmo.int/doc_num.php?explnum_id=5439

Yakovlev NG (2009) Reproduction of the large-scale state of water and sea ice in the Arctic ocean from 1948 to 2002: part II. The state of ice and snow cover. Izvestia Atmos Oceanic Phys 45(4):478–494

Zimin AV, Tolstikov AV (2019) Structure and variability of the main frontal zones in the White Sea during the warm season of 2010. In: Proceedings of Karelian Research Centre of RAS. № 3. Ser. “Limnology and oceanology”, pp 5–15. https://doi.org/10.17076/lim891

On the Possibility of Use the Himalayan-Tien Shan Mountain Ring Relief to Assess the State of the Substance at the Boundary of Intense Vortex in the Underlying Mantle

S. Yu. Kasyanov

Abstract Within the hypothesis of the ideal compressible fluid movement along the flat circular flow line in the Earth's mantle, where the pressure in the fluid is equal to the hydrostatic pressure in the mantle, taking into account the elevation of the lithosphere relief over the flow line, qualitative one-dimensional model of the substance's movement on the intense intra-mantle vortex boundary under the Himalayan-Tien Shan mountain ring is proposed. Due to the fact that the spatial changes in pressure of liquid are known in advance and are determined by changes in the lithosphere relief, the model built allows to estimate qualitatively the density and velocity of liquid along the flow line using the relief profile of the day surface above the flow line. The model makes possible to obtain pressure dependence on density from the data on lithosphere relief above the vortex area and, thus, to obtain information on the moving substance state in the intra-mantle vortex.

Keywords Intra-mantle vortex · Diagnostic model · Motion of ideal compressible fluid · Earth mantle · Lithosphere relief · Evaluation of vortex substance state in mantle by lithosphere relief

1 Introduction

According to the hypothesis stated in Kasyanov (2012), in the 3rd millennium BC the fall of the previously captured temporary Earth satellite occurred. Orbital motion of the satellite led to onset of the modern layer of asthenosphere, and its fall caused the formation of a vortex ring (vortex torus) in the mantle, the collapse of which under the Earth's rotation influence resulted in the formation of a modern system of intensive vortices in Earth's mantle. According to the development of mentioned hypothesis (Kasyanov and Samsonov 2019), in the modern system of intensive vortices in the mantle three energy-bearing vortices are distinguished: vortex (1) of Himalayas+Tien Shan—Kolyma (HTS-K), vortex (2) of Kolyma—Yellowstone (K-Y) and vortex (3)

S. Yu. Kasyanov (✉)
N.N. Zubov State Oceanographic Institute, Roshydromet, 6, Kropotkinsky lane, 119034 Moscow, Russia

T. Chaplina (ed.), *Processes in GeoMedia—Volume IV*, Springer Geology,
https://doi.org/10.1007/978-3-030-76328-2_11

of Yellowstone—Hudson Bay—Eastern Mediterranean Sea (Y-HB-EMS) (Fig. 1). The substance moving in the vortices is partially liquid, partially gaseous state. This substance, which includes a substance of former temporary Earth satellite, mainly consists of a mantle substance and contains significant amounts of SiO_2 and Al_2O_3 (Kasyanov and Samsonov 2019).

Vortex (1) (HTS-K) (Fig. 2) is in contact with the liquid Earth core, which manifests itself on the day surface in the form of the Baikal rift and lithosphere relief around Lake Baikal, which could be formed by isostatic compensation for stationary hydrodynamic disturbances arising from the interaction of the vortex with the liquid core and mantle.

The movement of the medium in the vortex is two-phase, clockwise, if viewed from the South. The center of the Southern vortex outcome of the HTS-K lies near a point (39.8°N, 88.7°E) at a depth of about 1200 km, the vortex outcome radius is about 1300 km in the mantle or about 1600 km on the surface. The vortex substance moves in an inclined plane at large depth and hydrostatic pressure variations. The wavy line is a manifestation on the day surface of a stationary condensation-evaporation front of the substance when it moves in a vortex.

Data on changes in geophysical parameters during 1997–1998 surge at the gravitational resonance in the solar system allowed to estimate the maximum vortex (1) velocity as of 4131 m/s (Kasyanov and Samsonov 2019).

There is the hypothesis that the relief inside the Himalayas + Tien Shan ring is associated with the hydrodynamic pressure in the intensive vortex located under it. The relief of lithosphere inside the Himalayas+Tien Shan ring contains abrupt height variations from 5000 m in Tibet to 1300 m in Takla-Makan desert (Fig. 2).

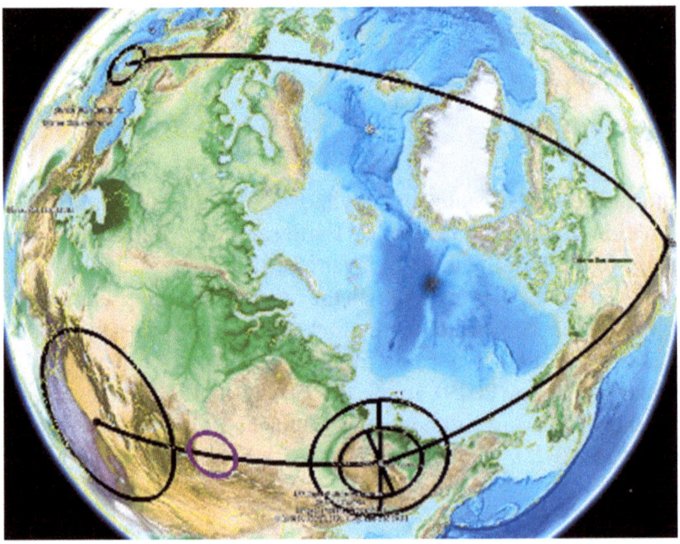

Fig. 1 The structure of the system of three related intensive vortices in the Earth's mantle

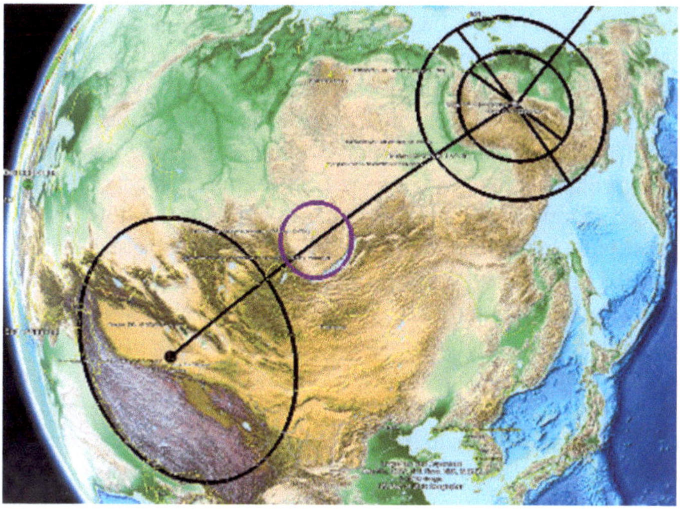

Fig. 2 Schematic image of the vortex 1 (Himalayas + Tien Shan Kolyma)

Based on the presence of symmetry elements relative to the meridian in the relief structure inside the Himalayas + Tien Shan ring, the vortex axis is assumed to lie in the meridional plane. In the relief of the Eastern part of the Himalayas + Tien Shan mountain ring stand out three mountain ranges, close to arcs of three concentric circles, whose radius increases to the East. It seems this ridges can be traces of three consecutive positions of the boundary of the vortex (1) outcome when turning the vortex from its original position in towards the meridional plane. Turning of vortex (1) outcome could occur under the joint influence of gyroscopic and buoyancy forces. The angle of vortex axis inclination to the vertical is approximately equal to the initial angle of vortex axis of Himalayas + Tien Shan—Kolyma, namely 62°.

2 One-dimensional Diagnostic Model of the Motion of Matter in a Vortex

The stationary motion of a single-velocity compressible ponderous medium is considered along the flat circular flow lines lying in the inclined plane inside mantle. The center of circular flow lines lies at a depth of 1200 km. The plane normal to the flow line makes angle $\psi = 62°$ with vertical and lies in the meridional plane. The diagnostic one-dimensional stationary problem of determining the moving medium density by a given pressure distribution along the flow line is to be solved.

One-dimensional movement along the flow line of radius $s = 557$ km in the mantle at the given density ρ_0 in the lower point of the trajectory and constant along the trajectory value of $\rho v_\varphi = Q$ (where ρ is the density, and v_φ is the speed) this is under study. Projection of flow line in the mantle with radius $s = 557$ km on the day surface is almost equivalent to white circle line with radius of 686 km in Fig. 3. The considered trajectory on an inclined plane in the mantle, the which projection to the Earth's sphere is shown in Fig. 3, actually deviates from circular, but this difference is not crucial for one-dimensional motion dynamics along the trajectory.

Assumptions of absence of viscosity and of the direct contact of moving fluid with the undisturbed mantle are adopted. As the pressure on the flow line approximate hydrostatic pressure in the mantle at the point of trajectory passage, which is created by a layer of the undisturbed spherical mantle (values of pressure according to the PREM model (Dziewonski and Anderson 1981) and the elevations of the day surface relief along white circle with radius of 686 km in Fig. 3 are used.

On the plane of the vortex outcome unit vectors $\mathbf{e}_1$ and $\mathbf{e}_2$ are introduced, such that $\mathbf{e}_1$ is directed strictly to the East, and $\mathbf{e}_2$ lies in the meridional plane and is directed to the North. Cylindrical coordinate frame with the origin located at depth of 1200 km on the vortex outcome plane and axis overlapping the vortex axis is introduced. The azimuth angle φ is counted from the South direction clockwise, if looking from the South.

Then the radius-vector of a point on a single circle looks like

$$\mathbf{a} = -\mathbf{e}_1 \sin\varphi - \mathbf{e}_2 \cos\varphi$$

where φ is azimuth angle of a point on the trajectory in cylindrical coordinate frame, and unit-vector of the curved trajectory velocity is

$$\mathbf{e}_\varphi = d\mathbf{a}/d\varphi = -\mathbf{e}_1 \cos\varphi + \mathbf{e}_2 \sin\varphi$$

At the points of the trajectory representing circle with radius s, the projection of gravity acceleration $\mathbf{g}$ on the curved trajectory velocity is found.

Here $\mathbf{r}$—radius-vector of trajectory point in Greenwich Cartesian geocentric coordinate frame, $\mathbf{r}_c$—radius-vector of vortex outcome center, then $\mathbf{s}$—radius-vector of the trajectory point relatively to the vortex outcome center:

$$\mathbf{s} = s\mathbf{a} = -s\mathbf{e}_1 \sin\varphi - s\mathbf{e}_2 \cos\varphi$$

Then $\mathbf{r} = \mathbf{r}_c + \mathbf{s}$, and also

$$\mathbf{s}\frac{d\mathbf{s}}{d\varphi} = 0, \mathbf{e}_1\mathbf{r}_c = 0, \mathbf{e}_2\frac{\mathbf{r}}{|\mathbf{r}|} = \sin\psi,$$

where ψ—vortex axis angle with vertical in the center of the vortex outcome. Since.

$$\mathbf{g} = -\frac{g}{r}(\mathbf{r}_c + \mathbf{s}), \mathbf{s}\frac{d\mathbf{a}}{d\varphi} = 0, \Rightarrow$$

$$\mathbf{g} \cdot \mathbf{e}_\varphi = \mathbf{g}\frac{d\mathbf{a}}{d\varphi} = -\frac{g}{r}(\mathbf{r}_c + \mathbf{s})\frac{d\mathbf{a}}{d\varphi} = -\frac{g}{r}\mathbf{r}_c\frac{d\mathbf{a}}{d\varphi} = -\frac{gr_c}{r}\sin\psi\sin\varphi$$

Now the equations of stationary motion along a given circular trajectory with radius s can be written:

$$v_\varphi\frac{\partial v_\varphi}{\partial\varphi} = -\frac{1}{\rho}\frac{\partial p}{\partial\varphi} + s\mathbf{g}\cdot\mathbf{e}_\varphi$$

And the continuity equation is reduced to $\partial(\rho v_\varphi)/\partial\varphi = 0$. From the relation: $\rho v_\varphi = Q = \text{const}$.

$$p = \int_r^R \rho_m(z)gdz + \rho_0 gh,$$

where $\rho_m(z)$—mantle density at geocentric radius z on the PREM model, R—Earth radius, $\rho_0 = 2600$ kg/m^3—density of rocks composing the terrain, g—9.81 m/s^2 acceleration of gravity assumed constant, h—elevation of the relief above sea level. Then:

$$\frac{\partial p}{\partial\varphi} = \rho_0 g\frac{\partial h}{\partial\varphi} - g\rho_m(r)\frac{\partial r}{\partial\varphi}$$

$$r^2 = r_c^2 + 2r_c\cdot s + s^2 = r_c^2 - 2sr_c\cos\varphi\sin\psi + s^2$$

$$\frac{\partial r}{\partial\varphi} = \frac{sr_c\sin\varphi\sin\psi}{r} \text{ and } \frac{\partial p}{\partial\varphi} = \rho_0 g\frac{\partial h}{\partial\varphi} - g\rho_m(r)\frac{sr_c}{r}\sin\varphi\sin\psi$$

Thus, the equation of motion takes form:

$$v_\varphi\frac{\partial v_\varphi}{\partial\varphi} = -\frac{1}{\rho}\rho_0 g\frac{\partial h}{\partial\varphi} + \frac{1}{\rho}sg\rho_m(r)\frac{r_c}{r}\sin\varphi\sin\psi - sg\frac{r_c}{r}\sin\varphi\sin\psi$$

$$\text{or } v_\varphi\frac{\partial v_\varphi}{\partial\varphi} = \frac{1}{\rho}\left(-\rho_0 g\frac{\partial h}{\partial\varphi} + sg\frac{r_c}{r}(\rho_m(r) - \rho)\sin\varphi\sin\psi\right)$$

In the obtained equation of one-dimensional motion along the given trajectory, using the relation $\rho v_\varphi = Q$ being constant along the trajectory, at replacement of the velocity through density $v_\varphi = Q/\rho$ a nonlinear first order differential equation for density is obtained:

$$\frac{Q}{\rho}\frac{\partial(Q/\rho)}{\partial\varphi} = \frac{1}{\rho}\left(-\rho_0 g\frac{\partial h}{\partial\varphi} + sg\frac{r_c}{r}(\rho_m(r) - \rho)\sin\varphi\sin\psi\right),$$

$$-\frac{Q^2}{\rho^2}\frac{\partial \rho}{\partial \varphi} = -\rho_0 g \frac{\partial h}{\partial \varphi} + sg\frac{r_c}{r}(\rho_m(r) - \rho)\sin\varphi \sin\psi \Rightarrow$$

$$\frac{\partial \rho}{\partial \varphi} = \frac{\rho^2}{Q^2}\left(\rho_0 g \frac{\partial h}{\partial \varphi} - sg\frac{r_c}{r}(\rho_m(r) - \rho)\sin\varphi \sin\psi\right)$$

3 Results

For integration along the flow line it is necessary to set density and velocity values at some point of the trajectory. It is most convenient to estimate the values of density and velocity at the bottom point of the trajectory because it is natural to suppose that the pressure and density here should be maximum ones.

First, the value of velocity is estimated at the lower point of the trajectory. As follows from the data for geophysical parameters surge of 1997–1998, velocity on periphery of HTS-K vortex is earlier estimated as 4131 m/s (Kasyanov and Samsonov 2019). It is assumed that the rotation in the vortex core is solid-state with maximum velocity of 4131 m/s at the vortex periphery. Vortex periphery is assumed to approximately correspond the circle with radius about 1300 km on the day surface. Taking into account the solid-state motion in the vortex core, the velocity $v_\otimes$ of the vortex (1) is obtained on the flow line located inside the vortex core, which is presented by white circle with radius of 686 km in Fig. 3. This velocity appears to be $v_\otimes =$ 2180 m/s, this value is accepted as the velocity at the bottom point of the flow line. To integrate the density change equation, it is also necessary to set a constant along the flow line under consideration $\rho v_\varphi = Q = v_\otimes \rho_\otimes$. For this purpose, the density value at the bottom point of the flow line should be set.

At the above specified velocity value $v_\otimes$ density value $\rho_\otimes$ is chosen for density distribution obtained upon the solution from the following considerations: the mass of the substance in a single flow tube along the trajectory is equal to the mass of the substance substituted by it, the density of which is taken from PREM model.

The density distribution obtained as a result of integration of the equation and corresponding velocity distribution are presented in Fig. 4. The obtained density values along the trajectory similarity to the previously obtained density estimation in vortex (2) of the system of intensive modern energy-bearing vortices, equal to 4600 kg/m^3 is noteworthy (Kasyanov and Samsonov 2019).

The curve of pressure dependence on density obtained during the integration of the density equation is of particular interest. It is constructed by pointing out on the graph pairs of density and pressure values at each step of the density equation integration (see Fig. 5). This curve represents the analogue of the equation of state for compressible medium in vortex.

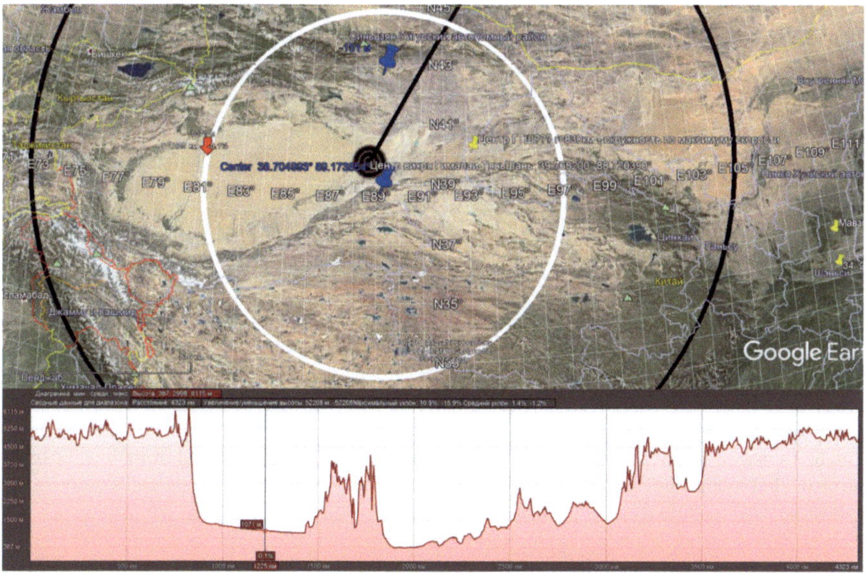

Fig. 3 Relief inside Himalayas+Tien Shan mountain ring. Bottom part shows the change of the day surface relief along the marked white circle (radius 686 km). The distance increases from left to right, from the South, in a clockwise direction. The image obtained within Gooogle Earth (http://www.earth.google.com) environment

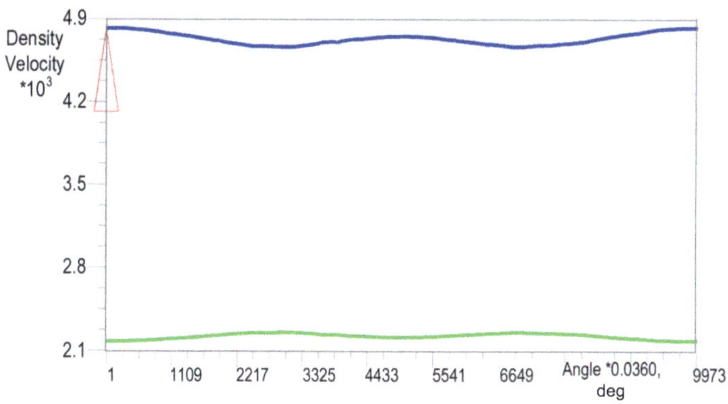

Fig. 4 Density (upper) and velocity (lower) curves depending on azimuth angle along flow line. The angle increases in clockwise direction, zero value in the South, i.e. in the lower point of the trajectory, where the density is the highest and the velocity is lowest (abscissa: Angle in units of 0.036 degree)

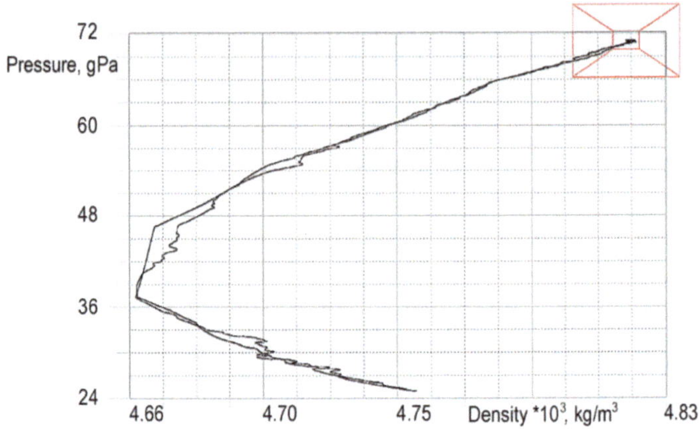

Fig. 5 Density dependence on hydrostatic pressure on the trajectory obtained from the lithosphere surface relief along white circle line in Fig. 3

The curve of pressure dependence on density in the vortex contains, in addition to the expected ascending branch, a descending branch and areas where $dp/d\rho < 0$, which is usually associated with the absolute instability of the medium movement.

The curve of pressure dependence on the density in the vortex is obtained by the lithosphere relief reflecting the movement in the vortex of a real multicomponent substance consisting of the mantle and the immersed body substances. The faults that bound linear segments on the upper branch of the curve, connected in the model with pressure (relief) faults, may be a reflection of phase transitions in the real multicomponent vortex medium. For real substance moving in the vortex, the lower branch of the curve, where $dp/d\rho < 0$, can be interpreted as deceleration of the substance dense phase and increase of its concentration when lifting to the upper point of the trajectory where the pressure of the spherical mantle is minimal. The liquid and gas phase velocities in the vortex may differ in reality.

4 Conclusions

A one-dimensional diagnostic model is built, allowing to estimate (qualitatively) density and velocity distribution and dependence of pressure on density along the flow line. Based on data on the topography of the lithosphere above the vortex area, it is possible to obtain information about the state of the moving matter of the intra-mantle vortex, which provides source for further research.

It is shown that abrupt relief shifts inside the Himalayas+Tien Shan mountain ring can be determined by the movement of compressible fluid at the outcome of the intra-mantle intense vortex boundary under the mountain ring.

References

Dziewonski AM, Anderson DL (1981) Preliminary reference Earth model. Phys Earth Planet Inter 25:297–356

https://www.google.com/earth/

Kasyanov SY (2012) Simulation of orbital motion of a large asteroid at decrease with deepening into lithosphere and mantle of the Earth. Physical Problems of Ecology (environmental Physics). 18:151–164 [in Russian]

Kasyanov SY, Samsonov VA (2019) On the modern system of three energy-bearing vortices in the Earth's mantle. In: Physical and mathematical modeling of processes in geo media 5th international science conference school of young scientists. Ishlinsky institute for problems in mechanics RAS, Moscow, pp 77–79 [in Russian]

Underground Haline Convection Caused by Water Evaporation from the Surface of the Earth

E. B. Soboleva

Abstract Natural haline convection of underground water containing dissolved admixture in soil is considered. Liquid water transits to gas and evaporates from the surface of the Earth whereas a dissolved admixture remains in brine. A denser layer is formed leading to gravity-driven convection to develop. Convective flows and mass transfer are investigated by means of numerical simulation. Regimes with salt precipitation near the front of phase transition are analysed.

Keywords Soil salinization · Porous medium · Haline convection · Darcy's filtration · Dissolved admixture · Evaporation · Numerical simulation

1 Introduction

Groundwater flows are important for natural processes in the Earth as well as for human activity. Transportation of dissolved admixture by groundwater can lead to soil salinization that is accumulation of water-soluble salts in the soil. A high level of salinization impacts negatively on environmental health, agricultural production, and economics.

Salinization can occur naturally during water evaporation from the surface of the Earth. Due to water evaporation, some amount of groundwater should be delivered to the surface of the Earth by underground upflow. Moving upward, water transits from liquid to gas phase because of pressure stratification (a decrease in the pressure with the height). While vapor is leaving a domain filled with aqueous salt solution, dissolved salts are remaining into it and accumulating under the front of phase transition. So, a profile of salt concentration which increases with approaching to the front of phase transition is formed. The solution density grows with the salt concentration, so that more dense fluid lies over less dense one. However, salt outflow due to diffusion counteracts salt inflow with ascending water. If diffusion is able to balance a salt supply with solution upflow, the concentration profile remains stable

E. B. Soboleva (✉)
Ishlinsky Institute for Problems in Mechanics RAS, Moscow, Russia
e-mail: soboleva@ipmnet.ru

T. Chaplina (ed.), *Processes in GeoMedia—Volume IV*, Springer Geology,
https://doi.org/10.1007/978-3-030-76328-2_12

in the gravity field. On the contrary, small density disturbances grow leading to a development of natural haline convection. Instability of salinity profile in a porous medium induced by evaporating upflow is investigated by different analytical methods; relevant studies can be found in (Duijn et al. 2001; Il'ichev et al. 2008; Tsypkin 2020) and referred literature. Some regimes and features of haline convection have been obtained by numerical simulation in (Duijn et al. 2001; Soboleva and Tsypkin 2014; Soboleva 2016, 2017a).

In the present work, we investigate unsteady convective flows and mass transfer in a porous medium developed due to instability of salinity profile. Numerical simulation of dynamic processes under the front of phase transition on the basis of the hydrodynamic model including the continuity, Darcy's and admixture transport equations is conducted. If brine becomes supersaturated, salt dissolved starts to precipitate. The most attention is paid to regimes with salt precipitation.

2 Methodology

2.1 Problem Under Study

The problem under study can be reduced to the problem in a two-dimensional horizontal porous layer with the upper boundary modelling the front of phase transition. The porous layer is filled with water solution of salt moving initially upward at the constant velocity $\mathbf{U}_0 = (0, V_0)$. Based on the fluid velocity $\mathbf{U} = (U, V)$, the filtration velocity $\mathbf{u} = (u, v)$ is defined as $\mathbf{u} = \phi\mathbf{U}$. Here, ϕ is the porosity of solid phase. So, the filtration velocity at the initial time is $\mathbf{u}_0 = (0, v_0)$. The spatial distribution of dissolved admixture is initially uniform. We will operate the admixture density ρ_c (mass of dissolved admixture per volume) denoted at the initial time as ρ_{c0}. At the lower boundary, the velocity $\mathbf{u}_0$ and admixture density ρ_{c0} are held during all time. At the upper boundary, the velocity $\mathbf{u}_0$ is held as well because water passes through this boundary and evaporate above it. The upper boundary condition for the density is the following: the mass flux of admixture incoming to this boundary with water flow is equal to the mass flux outgoing due to diffusion. The part of horizontal layer is considered. The vertical boundaries do not pass both water and salt. The problem setup is detailed in Fig. 1.

2.2 Mathematical Model

Dynamics and mass transfer of saline water in a porous medium is described by the hydrodynamic model. The continuity, motion and admixture transport equations are as follows (Bear and Cheng 2010).

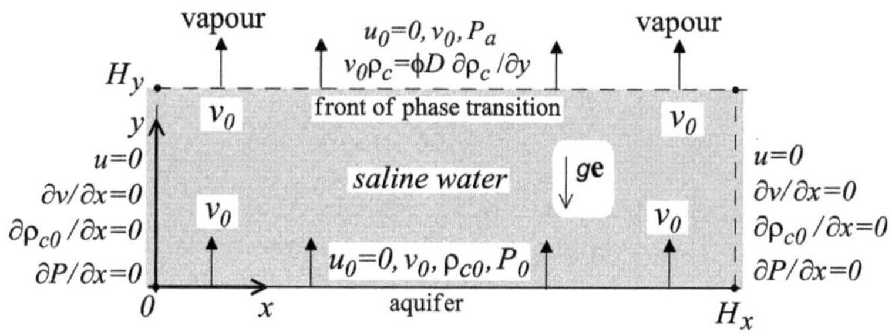

Fig. 1 Sketch of the problem

$$\nabla \cdot \mathbf{u} = 0 \tag{1}$$

$$\mathbf{u} = -\frac{k}{\mu}(\nabla P - \rho g \mathbf{e}) \tag{2}$$

$$\phi\frac{\partial \rho_c}{\partial t} + \mathbf{u} \cdot \nabla \rho_c = \nabla \cdot (\phi D \nabla \rho_c) \tag{3}$$

The values ρ, P, k, μ, D, g, $\mathbf{e} = (0, -1)$ are the density of saline water, pressure, permeability, dynamic viscosity, admixture diffusion coefficient, modulus of gravity force acceleration and unit vector co-directed with the gravity force. The equation of state

$$\rho = \rho^0 + \alpha\rho_c \tag{4}$$

closes the set of governing equations. Here, ρ^0 is density of pure water; the superscript "0" denotes pure water. The constant $\alpha = 0.815$ is taken for water. The quantities ρ^0, k, ϕ, μ, D are supposed to be constant. We will also consider the concentration (mass fraction) of admixture c defined as $c = \rho_c/\rho$. In saturated solution, $c^s = 0.265$; the superscript "s" denotes that solution is saturated.

If the fluid phase is pure water, the Darcy's Equation (2) at $\mathbf{u} = 0$ gives rise to the hydrostatic equation:

$$\nabla P^0 = \rho^0 g \mathbf{e} \tag{5}$$

Integrating the last equation, one can obtain the pressure distribution in motionless water: $P^0 = P_a + \rho^0 g(H_y - y)$, with P_a to be the atmosphere pressure. The magnitude of P^0 increases with distancing from the surface of the Earth due to stratification.

Equation (5) is subtracted from Eq. (2) so that the variables $P - P^0$ and $\rho - \rho^0$ become under consideration. Next, the set of governing equations are converted to a dimensionless form. The scales are the fluid velocity V_0, height of domain H_y, time V_0/H_y, pressure $\alpha\rho_c^s g H_y$, density $\alpha\rho_c^s$. We find the following dimensionless equations

$$\nabla \cdot \mathbf{w} = 0 \tag{6}$$

$$\mathbf{w} = -Ra^s\phi(\nabla\Pi - S\mathbf{e}) \tag{7}$$

$$\frac{\partial S}{\partial t} + \mathbf{w}\cdot\nabla S = \frac{1}{Pe}\Delta S \tag{8}$$

including the filtration velocity $\mathbf{w} = \mathbf{u}/V_0$, pressure $\Pi = (P - P^0)/\alpha\rho_c^s g H_y$, and density $S = \rho_c/\rho_c^s$. The governing equations contain the Rayleigh-Darcy Ra^s and Peclet Pe numbers:

$$Ra^s = \frac{kg\alpha\rho_c^s}{\phi\mu V_0},\quad Pe = \frac{H_y V_0}{D} \tag{9}$$

The Ra^s number is based on the admixture density in saturated solution ρ_c^s. According to the equation of state Eq. (4) written for saturated brine: $\rho^s = \rho^0 + \alpha\rho_c^s$, the complex $\alpha\rho_c^s$ is the most difference between densities of brine and water. The other parameter responsible for the fluid phase behavior is the initial admixture concentration c_0. Two quantities of admixture mass dissolved in water c and S are related to each other by the formula: $c = Sc^s/(1 - \alpha(1 - S))$.

Equations (6)–(8) together with the boundary and initial conditions are solved numerically with the use of the author's numerical code designed for the problems about haline convective flows and mass transfer in porous media. The code based on the finite-difference method is employed effectively diring several years (Soboleva and Tsypkin 2014; Soboleva 2016, 2017a, b, 2018, 2019a, b).

2.3 Physical and Numerical Parameters

Physical properties of water and dissolved salt are described by the following parameters: $\rho^0 = 10^3$ kg m^3, $\mu = 10^{-3}$ Pa s, $D = 1.6 \times 10^{-9}$ m^2 s^{-1}. The surface of the Earth can evaporate from zero to 800 mm of water during a year that results in the mean groundwater upflow velocity $V_0 = 0 \div 2.5 \times 10^{-8}$ m s^{-1}. If a deep aquifer is overpressured and the upflow is driven also by the additional pressure gradient, the groundwater velocity can reach the magnitude of 80 mm per day or $V_0 = 9 \times 10^{-7}$ m s^{-1} in high-permeable soils (Il'ichev et al. 2008). The admixture concentration in groundwater varies usually between 10^{-4} (so called ultra fresh water) and 3×10^{-2} (concentrated brine) so one should take $c_0 = 10^{-4} \div 3 \times 10^{-2}$. The soil porosity ϕ is about $0.1 \div 0.3$. The most variable physical parameters along with the upflow velocity are the permeability k and geometrical scale H_y changing in several orders of magnitude. We have roughly $k = 10^{-17} \div 10^{-12}$ m^2 and $H_y = 10^{-1} \div 10^3$ m. For estimations we take the range of upflow velocity $V_0 = 10^{-10} \div 9 \times 10^{-7}$ m s^{-1}. One can find that the magnitudes of Ra^s and Pe vary in the very wide ranges: $Ra^s = 10^{-3} \div 3 \times 10^2$ and $Pe = 6 \times 10^{-3} \div 6 \times 10^6$ indicating different regimes of flows and mass transfer.

3 Numerical Results

3.1 Regimes Without Salt Precipitation

The stability of salinity profile formed diring water evaporation from aqueous salt solution and regimes of natural haline convection induced in the case of developing of instability have been investigated in (Soboleva and Tsypkin 2014; Soboleva 2016, 2017a). We give shortly the main results which are necessary to understand the occurring processes.

The Pe number is the ratio of upflow velocity to diffusion velocity. If Pe is small, diffusion plays a significant role. It is the case of very slow ascending flows when a stable profile of admixture concentration can form. In the stable state, an amount of salt delivered the domain by solution upflow through the lower boundary is equal to an amount of salt leaving the domain due to diffusion through the same boundary. Convection in this case does not develop. The regime of stable state can occur at $Pe \leq 2.2$. The parameter separating stable and unstable regimes is $Ra^s PeK$ with K to be the coefficient depending on c_0, c^s, and c^*; here, c^* is the calculated salt concentration at the upper boundary of domain. The coefficient K varies from 0 at $c^* = c_0$ to 1 at $c^* = c^s$.

At higher Pe, the state of salinity profile becomes unstable and convective motions develop. The motions can transform to steady convection with periodic rolls occupying the computation domain totally. In the other case, small saline drops are generated under the upper boundary and a steady regime does not reach. At $Pe > 44$ only stochastic regime of saline drops which elongate, deformate and merge together occurs.

An increase in Pe corresponds with that the contribution from diffusion falls or the geometric scale H_y grows. In a horizontal layer with the large height H_y the convective process does not reach the lower boundary in realistic time and develops in the same fashion that in a semi-infinite domain. The approach of semi-infinite domain works well at $Pe > 44$. In this case, a convective process does not change with changing in Pe but only is observed in a different geometric scale. At $Pe > 44$, the parameters controlling the dynamic process are $Ra^s K$ and c_0. Note that these parameters do not include the heigh H_y.

3.2 Regimes with Salt Precipitation

When upflow delivering dissolved salt to the upper boundary is intensive and natural convection does not manage to lead it away, solution becomes supersaturated and salt precipitates. However, the early version of numerical code employed in (Soboleva and Tsypkin 2014; Soboleva 2016, 2017a) did not allow us to simulate a salt precipitation process. Calculations have been stoped if the current concentration c exceeded the concentration of saturated brine c^*. Using the modern version of code,

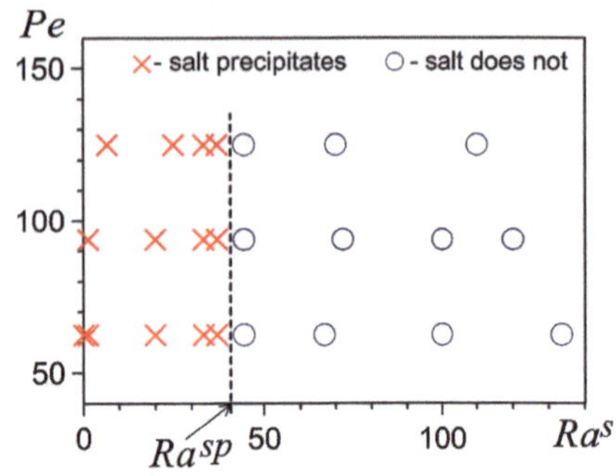

Fig. 2 Map of dynamic regimes. Cross marks indicate regimes with salt precipitation, circle marks indicate regimes without precipitation

we simulate salt precipitation. The code is applicable to the early stage of salt precipitation process because the mathematical model does not take into consideration a decrease in porosity when solid salt is accumulated in a pore space. We are planning to include the effect of changing in porosity into consideration in the future.

We conducted numerical simulations at $c_0 = 10^{-4}$, $Pe > 44$ and the variable Rayleigh-Darcy number: $Ra^s = 0.13 \div 134$. Simulations have been performed in 10×1 or 5×1 computation domains. The non-uniform 1000×190 space grid with a smaller spatial step near the upper boundary was typically used. The step of integration in time was usually $\tau = 3 \times 10^{-5}$.

The Ra^s number is the ratio of convective velocity to upflow velocity showing a role of natural haline convection relative to a forced ascending motion. One can anticipate that salt precipitates at rather small Ra^s when convective flows are not sufficiently intensive. Results of simulation are shown in Fig. 2. We mark differently our calculations demonstrating salt precipitation and the other in which brine does not reach a supersaturated state and all salt remains in solution. As obvious, salt precipitates at $Ra^s < Ra^{sp}$ and does not at $Ra^s > Ra^{sp}$. Here, $Ra^{sp} \approx 37 \div 45$ is the threshold Rayleigh-Darcy number. We examined three magnitudes of $Pe > 44$ and obtained that the threshold Ra^{sp} number is not influent by the Pe number according to our prediction. One can assume that the Ra^{sp} number increases with a growth of the initial concentration c_0.

Figure 3 exhibits evolution of concentration field and zones of salt precipitation at $Pe = 63$ and $Ra^s = 13.4$. These dimensionless parameters correspond, for example, with the set: $V_0 = 10^{-7}$ m s^{-1}, $\phi = 0.3$, $k = 2 \times 10^{-13}$ m^2, $H_y = 1$ m. The time scale is $H_y/V_0 \approx 116$ days. As shown, the salinity profile becomes unstable that leads to a development of convection. We see that periodic drops of brine are formed under the upper boundary (a). Then, drops become larger, they move along the boundary and merge together (b). The precipitation process starts not from the beginning of convection but when a sufficient amount of dissolved admixture accumulates under the upper boundary. The start is in the time $t \approx 25$. As demonstrated, solid salt accumulates under brine drops therefore zones of salt precipitated locate quasi-periodically and a length of these zones correlate with a size of brine drops

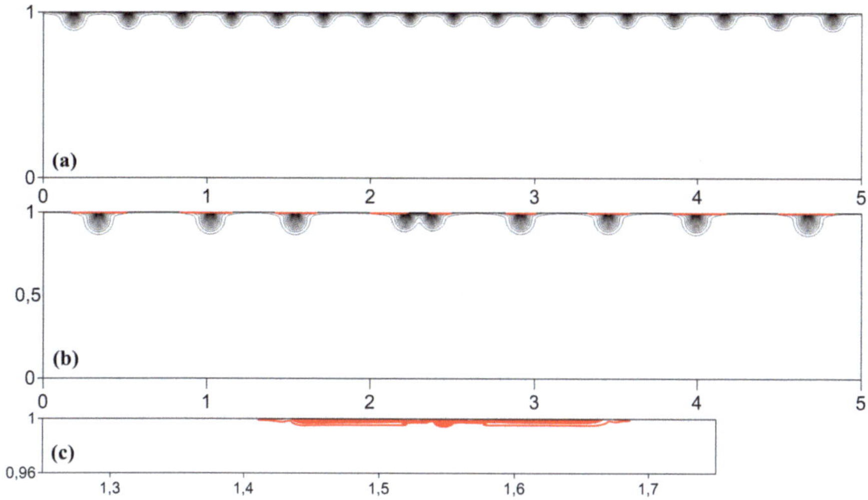

Fig. 3 Field of dissolved salt concentration c at the times $t = 24$ (**a**), 40 (**b**). Pattern (**b**): red color indicates the salt deposition zones. Pattern (**c**): isolines of salt precipitated in an enlarged scale at the time $t = 40$

Fig. 4 Profile of admixture concentration c under the boundary at different times t

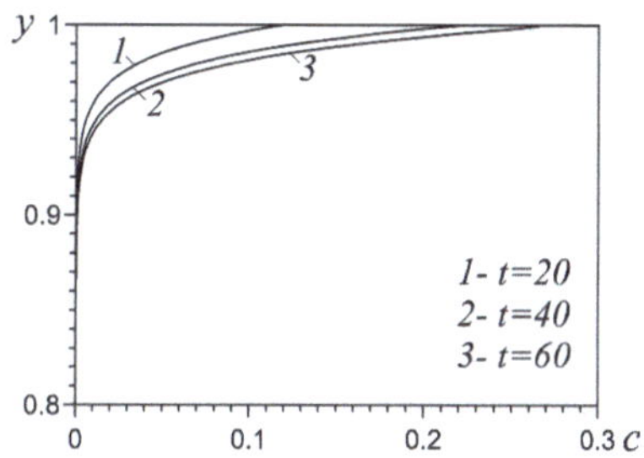

(b). A zone of salt deposition elongates along the boundary with the most amount of salt in the center (c).

Our calculations show that salt can precipitates without a development of convective motions. At $Ra^s = 1.3$ and smaller magnitudes of Ra^s, a salinity profile formed due to ascending flow remains stable. A salt concentration at the upper boundary c^* grows monotonously up to the maximal quantity c^s corresponding with saturated brine. When a condition $c^* = c^s$ reaches, salt starts to precipitate uniformly at the upper boundary. The profile of admixture concentration c in the upper part of computation domain at $Pe = 63$ and $Ra^s = 1.3$ is shown in Fig. 4. The start of precipitation is in the time $t \approx 53$. In this case, we do not observe a quasi-periodic salt deposition as in Fig. 3. Solid salt lays under the upper boundary continuously.

4 Conclusion

We investigated natural haline convection in a porous medium developed due to an unstable salinity profile which is formed by upflow of aqueous salt solution with water transition to the gas phase. Our study is focused on regimes with salt precipitation under the front of phase transition. The problem is solved numerically at the rather high Peclet Pe number corresponding with that the approach of semi-infinite domain is valid. In this case, only the Rayleigh-Darcy Ra^s number and initial admixture concentration c_0 are responsible for the regime induced. It was obtained that at $c_0 = 10^{-4}$ the threshold Rayleigh-Darcy Ra^{sp} number separating regimes with salt precipitation is $Ra^{sp} \approx 37 \div 45$. Salt precipitates under the boundary modelling the front of phase transition at $Ra^s < Ra^{sp}$ because convection does not manage to lead concentrated brine away. At small $Ra^{sp} \leq 1.3$, the salinity profile remains stable and salt precipitates uniformly along the boundary.

Acknowledgements The author would like to thank G. G. Tsypkin for many fruitful discussions. This work was supported by the Government program (contract AAAA-A20-120011690131-7).

References

Bear J, Cheng A (2010) Modeling groundwater flow and contaminant transport. Springer, New York

Duijn CJ, Wooding RA., Pieters GJM, Ploeg A (2001) Stability criteria for the boundary layer formed by throughflow at a horizontal surface of a porous medium. Reports on applied and numerical analysis, Eindhoven University of Technology, the Netherlands, 0112

Il'ichev AT, Tsypkin GG, Pritchard D, Richardson ChN (2008) Instability of the salinity profile during evaporation of saline groundwater. J Fluid Mech 614:87–104. https://doi.org/10.1017/S0022112008003182

Soboleva E (2017a) Numerical investigations of haline-convective flows of saline groundwater. J Phys Conf Ser 891. https://doi.org/10.1088/1742-6596/891/1/012104

Soboleva E (2017b) Numerical simulation of haline convection in geothermal reservoirs. J Phys Conf Ser 891. https://doi.org/10.1088/1742-6596/891/1/012105

Soboleva EB (2018) Density-driven convection in an inhomogeneous geothermal reservoir. Int J Heat Mass Transf 127(part C), 784–798. https://doi.org/10.1016/j.ijheatmasstransfer.2018.08.019

Soboleva EB (2019a) Numerical study of haline convection in a porous domain with application for geothermal systems. In: Karev V, Klimov D, Pokazeev K (eds) Physical and mathematical modeling of earth and environment processes (2018), pp 63–74. Springer Proceedings in Earth and Environmental Sciences. Springer, Cham. https://doi.org/10.1007/978-3-030-11533-3_7

Soboleva EB (2019b) A method for numerical simulation of haline convective flows in porous media applied to geology. Comput Math Math Phys 59(11):1893–1903. https://doi.org/10.1134/S0965542519110101

Soboleva EB, Tsypkin GG (2014) Numerical simulation of convective flows in a soil during evaporation of water containing a dissolved admixture. Fluid Dyn 49(5):634–644. https://doi.org/10.1134/S001546281405010X

Soboleva EB, Tsypkin GG (2016) Regimes of haline convection during the evaporation of groundwater containing a dissolved admixture. Fluid Dyn 51(3):364–371. https://doi.org/10.1134/S001546281603008X

Tsypkin GG (2020) Instability of a light fluid over a heavy one under the motion of their interface in a porous medium. Fluid Dyn 55(2):213–219. https://doi.org/10.1134/S0015462820020135

The Problem of Hydrodynamic Potentials Introduction for the Space of Different Dimensions

A. V. Kistovich

Abstract In the article presented the problem of introduction of potential functions for arbitrary space dimension on the base of unified symmetric approach is considered. The physical sense of the received potentials is shown in explicit from.

Keywords Stream function · Toroidal and poloidal potentials

The problem of the introduction of the physical potentials in the description of the solenoidal but not potential fields, suitable for hydrodynamic problems solving is still actual for the case of three dimensional space. To get closer to the solution of this problem it is convenient to start from the cases of the less dimensions.

The solenoidal velocity field is defined by the condition

$$\nabla \cdot \mathbf{v} = 0 \tag{1}$$

where ∇ is standard Hamiltonian operator.

In the 1-D case when $\nabla = \mathbf{e}_x \partial_x$ and $\mathbf{v} = \mathbf{e}_x \mathrm{v}(x)$ the solution of (1) is $\mathrm{v}(x) = \mathrm{v}_0 = \mathrm{const}$ and v_0 is required value of the velocity «potential» where its dimension is $[\mathrm{v}_0] = \mathrm{sm/s}$.

In the 2-D case $\nabla = \mathbf{e}_x \partial_x + \mathbf{e}_y \partial_y$, $\mathbf{v} = A(x, y)\,\mathbf{e}_x + B(x, y)\,\mathbf{e}_y$ and Eq. (1) reforms into the equation $A'_x + B'_y = 0$, on the base of it the standard stream function $\Psi(x, y)$ is introduced so

$$A = \Psi'_y, \quad B = -\Psi'_x \tag{2}$$

where $[\Psi] = \mathrm{sm}^2/\mathrm{s}$.

At the same time the result (2) can be presented in the form

$$A = F'_y, \quad B = G'_x, \quad F + G = 0 \tag{3}$$

A. V. Kistovich (✉)
Ishlinsky Institute for Problems in Mechanics of the Russian Academy of Science, Moscow, Russia

T. Chaplina (ed.), *Processes in GeoMedia—Volume IV*, Springer Geology,
https://doi.org/10.1007/978-3-030-76328-2_13

from which the result (2) follows and also $[F] = \text{sm}^2/\text{s}$, $[G] = \text{sm}^2/\text{s}$.

In the 3-D case $\nabla = \mathbf{e}_x \partial_x + \mathbf{e}_y \partial_y + \mathbf{e}_z \partial_z$, $\mathbf{v} = A(x, y, z)\,\mathbf{e}_x + B(x, y, z)\,\mathbf{e}_y + C(x, y, z)\,\mathbf{e}_z$ and Eq. (1) reforms into $A'_x + B'_y + C'_z = 0$. The sequence of arguments analogues to the sequence in the 2-D case leads to the presentation of velocity field components

$$A = F''_{yz}, \quad B = G''_{xz}, \quad C = H''_{xy}, \quad F + G + H = 0, \tag{4}$$

where $[F] = \text{sm}^3/\text{s}$, $[G] = \text{sm}^3/\text{s}$, $[H] = \text{sm}^3/\text{s}$.

From the obtained result (4) it follows that in the 3-D case only two scalar functions are sufficient for the solenoidal velocity field defining. The choice of these functions in the concrete problem is due to the convenience of the physical fields description. For example, let the z-axis is the preferred axis of the flow in the sense that the z-component of the velocity field is defined by the one scalar function and the other x- and y- components are defined by two functions. So the any solenoidal velocity field permits the representation of the form

$$\begin{aligned} \mathbf{v} = \nabla \times \mathbf{S} = \left(R'_y - Q'_z\right)\mathbf{e}_x + \left(P'_z - R'_x\right)\mathbf{e}_y \\ + \left(Q'_x - P'_x\right)\mathbf{e}_z, \quad \mathbf{S} = P\,\mathbf{e}_x + Q\,\mathbf{e}_y + R\,\mathbf{e}_z \end{aligned} \tag{5}$$

then the comparison of (4) and (5) leads to the relations

$$F''_{yz} = R'_y - Q'_z, \quad G''_{xz} = P'_z - R'_x, \quad H''_{xy} = Q'_x - P'_y \tag{6}$$

The left part of the third relation of (6) namely the value H''_{xy} is invariant with respect to exchange $x \rightleftarrows y$. The same property should have the right side of the relation mentioned. In addition, in accordance to the condition the z-axis preference the right side should be described by means of one scalar function. This result can be achieved by the two manners:

$$1.\ \ Q = -\Psi'_x, \quad P = \Psi'_y, \quad 2.\ \ Q = \Psi'_y, \quad P = -\Psi'_x \tag{7}$$

In the first case from (6, 7) it follows

$$\begin{aligned} A = F''_{yz} = \Phi'_y + \Psi''_{xz}, \quad B = G''_{xz} \\ = -\Phi'_x + \Psi''_{yz}, \quad C = H''_{xy} = -\Psi''_{xx} - \Psi''_{xy} \end{aligned} \tag{8}$$

where the symbol Φ is used instead of R.

The result (8) may be presented in a reduced vector form

$$\mathbf{v} = \nabla \times (\Phi\,\mathbf{e}_z) + \nabla \times \nabla \times (\Psi\,\mathbf{e}_z) \tag{9}$$

coinciding with the representation of the velocity field by means of a toroidal-poloidal potential (Chandrasekhar 1961).

In the second case, the relations have place

$$A = F''_{yz} = \Phi'_y - \Psi''_{yz}, \quad B = G''_{xz} = -\Phi'_x - \Psi''_{xz}, \quad C = H''_{xy} = 2\Psi''_{xy} \tag{10}$$

The vector presentation of (10) has the from

$$\mathbf{v} = \nabla \times \left(-\Psi'_x \mathbf{e}_x + \Psi'_y \mathbf{e}_y + \Phi\, \mathbf{e}_z\right) \tag{11}$$

or may be presented in the form

$$\begin{aligned}\mathbf{v} &= \nabla \times (\Phi\, \mathbf{e}_z) + \nabla \times \nabla \times (\Psi\, \mathbf{e}_z) + \nabla \times \left(\left(\Psi'_x + \Psi'_y\right)\left(\mathbf{e}_y - \mathbf{e}_x\right)\right)\\ &= \nabla \times (\Phi\, \mathbf{e}_z) + \nabla \times \nabla \times (\Psi\, \mathbf{e}_z) + \nabla \times \left(\left(\Psi'_x + \Psi'_y\right) \mathbf{e}_z \times \left(\mathbf{e}_x + \mathbf{e}_y\right)\right)\end{aligned} \tag{12}$$

which is symmetric to the exchange $x \rightleftarrows y$.

The result (12) is not described in the scientific issues.

To show of the physical sense of the potentials obtained, it is necessary to consider of the mass flux of the homogeneous liquid of the constant density transported across the bounding surface per unit time.

In the 1-D case, the bounding surface is the point and the value v_0 is the flux of the unit mass per unit time across this point. In the 2-D case, the bounding surface is the curve with two ends A and B. The flux through the curve (AB), which described in the (x, y) plane by the equation $f(x, y) = 0$ is defined by the relation $\mathrm{St} = \int\limits_{(AB)} \mathbf{n} \cdot \mathbf{v}\, ds$, where ds is the differential element of the arc. As the relations $df(x, y) = f'_x\, dx + f'_y\, dy = 0, ds = \sqrt{dx^2 + dy^2} = |\nabla f| / f'_y\, dx, \mathbf{n} = \nabla f / |\nabla f|$ and $\mathbf{v} = \Psi'_y \mathbf{e}_x - \Psi'_x \mathbf{e}_y$ are valid then the following relation have place

$$\begin{aligned}\mathrm{St} &= \int\limits_{(AB)} \mathbf{n} \cdot \mathbf{v}\, ds = \int\limits_{(AB)} \frac{\nabla f \cdot \mathbf{v}}{|\nabla f|} \frac{|\nabla f|}{f'_y}\, dx = \int\limits_{(AB)} \left(\Psi'_y \frac{f'_x}{f'_y} - \Psi'_x\right) dx\\ &= \int\limits_{(AB)} \left(-\Psi'_y \frac{dy}{dx} - \Psi'_x\right) dx = -\int\limits_{(AB)} \left(\Psi'_x dx + \Psi'_y dy\right)\\ &= -\int\limits_{(AB)} d\Psi = \Psi(A) - \Psi(B)\end{aligned} \tag{13}$$

The results obtained means that the flux through the curve (AB) equals to the difference of the stream function's values on the ends of the curve.

In the 3-D case, when the solenoidal velocity fields is defined by the relations (4) or by the equivalent representation

$$\mathbf{v} = \frac{1}{3}\nabla \times \left[(G'_x - H'_x)\mathbf{e}_x + (H'_y - F'_y)\mathbf{e}_y + (F'_z - G'_z)\mathbf{e}_z\right] \tag{14}$$

the flux through the surface S bounding by the curve L is defined by the value

$$\int\limits_S \mathbf{v} \cdot d\mathbf{S} = \frac{1}{3}\oint\limits_L \left[(G'_x - H'_x)\mathbf{e}_x + (H'_y - F'_y)\mathbf{e}_y + (F'_z - G'_z)\mathbf{e}_z\right] \cdot d\mathbf{L} \tag{15}$$

Let the curve L is described in the space by two functions

$$L: \quad \varphi(x, y, z) = 0, \psi(x, y, z) = 0,$$

so the relations $\varphi'_x dx + \varphi'_y dy + \varphi'_z dz = 0, \quad \psi'_x dx + \psi'_y dy + \psi'_z dz = 0$ are valid.

By the means of the announced relations the flux (15) can be presented in the form

$$\int\limits_S \mathbf{v} \cdot d\mathbf{S} = \frac{1}{3}\oint\limits_L \left[sign(J_{yz})(G'_x - H'_x)\, dx + sign(J_{zx})(H'_y - F'_y)\, dy + sign(J_{xy})(F'_z - G'_z)\, dz\right] \tag{16}$$

where $J_{ik} = \varphi'_i \psi'_k - \varphi'_k \psi'_i, \quad J_{ik} = -J_{ki}$.

As it follows from (16), in the 3-D case the liquid mass flux through 2-D bounding surface S is defined by the values of two potential functions (with regard of (4) these pairs are F and G or F and H, or G and H) on the contracting curve L of the mentioned surface.

The presented approach to the solenoidal field description may be used also in electrodynamic of homogeneous space, in particularly for the waveguide electromagnetic fields description, and permits to allow the well-known solutions by the very simple actions.

Reference

Chandrasekhar S (1961) Hydrodynamic and hydromagnetic stability. International series of monographs on physics. Clarendon, Oxford, p 622

Interannual Variability of El Nino

I. A. Martyn, Y. A. Petrov, S. Y. Stepanov, and A. Y. Sidorenko

Abstract The article examines the nature and the possibility of predicting the onset and strength of the El Niño climatic phenomenon. The analysis of the internal structure of the surface temperature of the ocean and air, pressure, amount of precipitation in the region of the tropical Pacific Ocean during the onset and absence of the El Niño (La Niña) phenomenon is carried out, according to the interactive system Giovanni, the spatial and temporal variability of these characteristics is considered. The results obtained do not contradict theoretical hypotheses.

Keywords El Nino · Southern oscillation · La Nina · Climatic anomaly · Statistical analysis

1 Introduction

At present, more and more often, various anomalies associated with an increase or decrease in temperature, an increase or decrease in precipitation, etc., prescribe the El Niño phenomenon. Thus, in the face of climate change, the problem of the possibility of assessing El Niño is increasingly gaining popularity among scientists. By definition, El Niño is extremely catastrophic in the region of direct action, as well as has long-range responses in other parts of the Earth, so it is very important to be able to predict the El Niño offensive, as well as to assess its possible strength and duration. That is why the study of El Niño is an important and urgent task not only for the Pacific Ocean region, but also for the whole world (Bondarenko 2006; Gustoev 1991).

I. A. Martyn (✉) · Y. A. Petrov · S. Y. Stepanov · A. Y. Sidorenko
Russian State Hydrometeorological University, Saint-Petersburg, Russia

T. Chaplina (ed.), *Processes in GeoMedia—Volume IV*, Springer Geology,
https://doi.org/10.1007/978-3-030-76328-2_14

2 Description Phenomenon

El Niño (Spanish El Nin˜o—baby, boy) is a climatic anomaly that occurs in the Pacific Ocean. The name South Oscillation is also used. The opposite phase of the oscillation is La Niña (Spanish: La Nin˜a—baby, girl) (Bondarenko 2006).

The first signs of the emergence of El Niño are:

1. Increasing pressure over Indonesia and Australia;
2. Pressure drop over Tahiti;
3. Weakening of the trade winds in the South Pacific Ocean or their termination and change of direction from east to west;
4. Warm air mass in Peru. Rains in the Peruvian deserts.

Off the coast of Peru during El Niño, upwelling weakens (Martyn et al. 2019a). In the normal season, the depth of the thermocline is 50–60 m; during El Niño, the depth increases to 100–120 m (Fig. 1).

This affects the increase in the heat content of the waters. The mechanism of El Niño formation consists in a change in pressure over the city of Darwin and the island of Tahiti, which causes air mass movements and is caused by manifestations of Walker's circulation, different in strength (Fig. 2). Walker's circulation is ocean and atmosphere dependent.

In this circulation, the decisive factor is the different water temperature. Cold and dry air is located above cold water, this air is carried to the west by the southeastern trade winds. During this transport, the air heats up and absorbs moisture and rises over the western Pacific (Fig. 2a). Part of the heated air moves to the pole, forming a Hadley cell (Fig. 3). Another part of the heated air goes down and its circulation ends (Fig. 4).

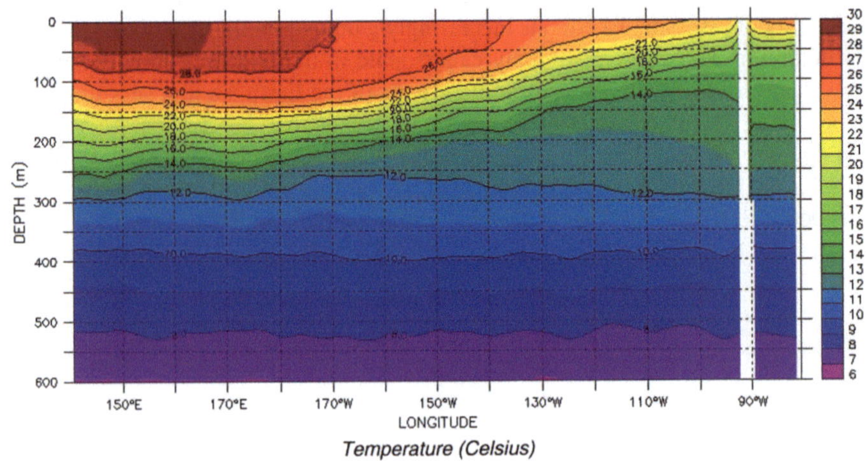

Fig. 1 Distribution of water temperature by depth in the section

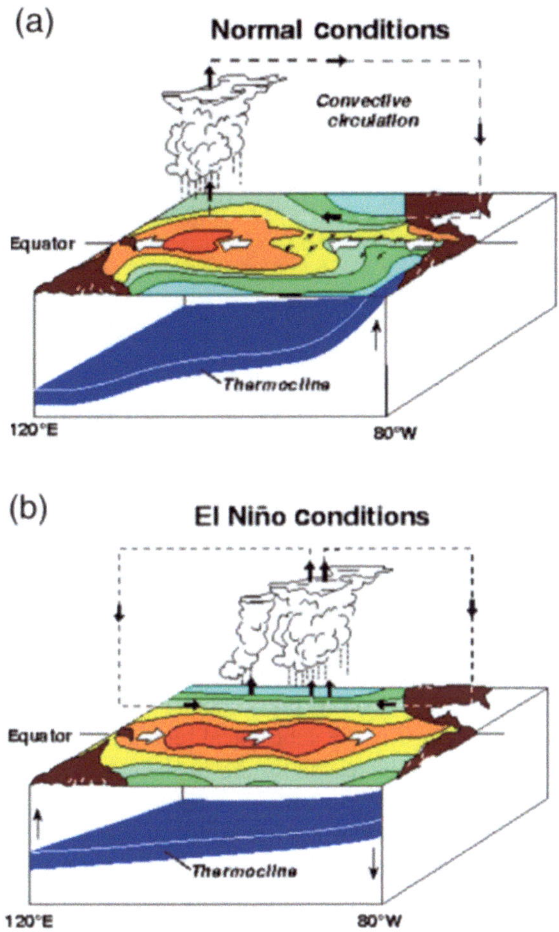

Fig. 2 Walker's cell under normal ocean conditions (**a**) and El Niño (**b**)

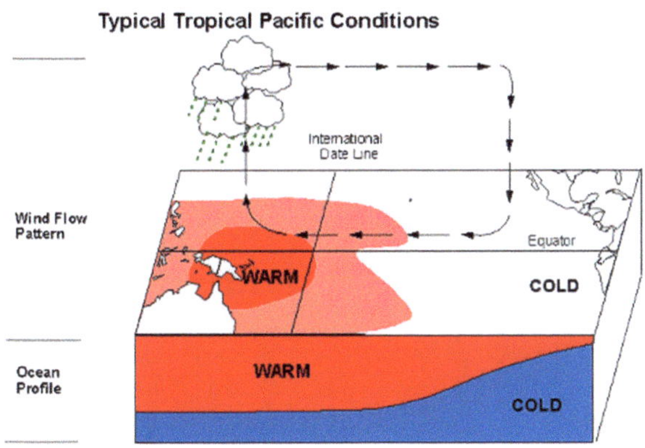

Fig. 3 Atmospheric circulation under normal ocean conditions

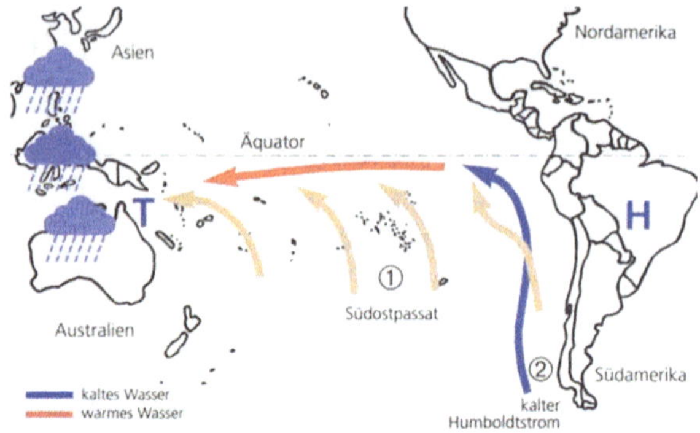

Fig. 4 Scheme of currents and winds in the normal state of the ocean

A feature of Walker's circulation is that it is not deflected by the Coriolis force, but passes through the equator, where the Coriolis force is absent. The mechanism of this circulation depends on the subtropical pressure zone. If it is strongly pronounced, then strong southeast trade winds will be observed. The trade wind increases the activity of the area of the lift of the coast of South America and decreases the temperature of surface waters near the equator. This state of the ocean is called La Niña, which is the reverse phase of the Southern Oscillation. Walker's circulation is also driven by the cold temperature of surface waters (Fig. 2b). In this case, there is a low pressure in the city of Jakarta (Indonesia) and is associated with a small amount of precipitation over Polynesia. Due to the weakening of the Hadley cell (Fig. 5), a decrease in pressure occurs in the subtropical regions, where the high pressure area is usually located, this is the reason for the weakening of the trade winds. The lifting force off the coast of South America decreases, as a result of which the water temperature rises in the equatorial zone, and it is at this moment that El Niño is most likely to appear (Bondarenko 2006).

The tongue of warm water off the coast of Peru, during El Niño, is responsible for the weakening of Walker's circulation. In Polynesia, a large amount of precipitation falls at this time, and in Indonesia there is a drop in atmospheric pressure. The latter

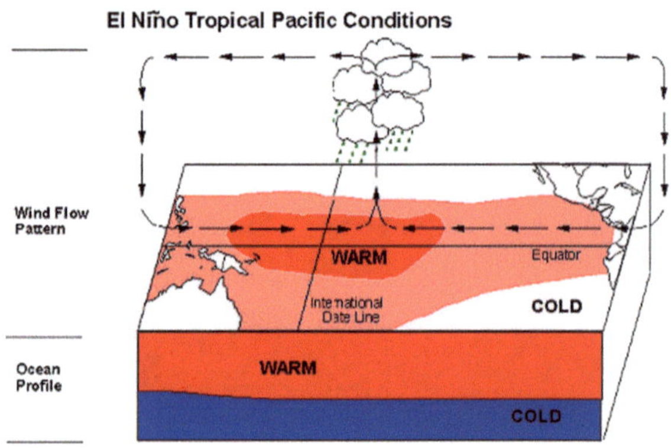

Fig. 5 Atmospheric circulation during El Niño

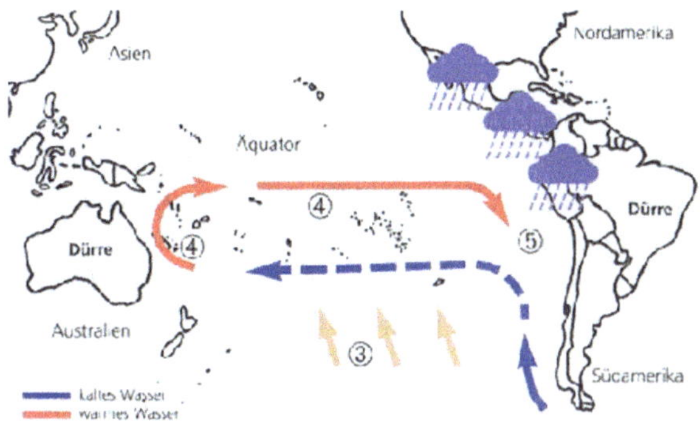

Fig. 6 Diagram of currents and winds at El Niño

increases the circulation of Hadley, because of this, an increase in pressure occurs in the subtropical zone. This effect explains the alternation of El Niño and La Niña (Fig. 6).

Currently, there are many definitions of the El Niño phenomenon and different points of view of its origin, so it was decided to determine how El Niño manifests itself and the reasons for its occurrence.

Objective. The aim of this work is to determine patterns in the interannual variability of the El Niño phenomenon, as well as to assess the possibility of predicting its onset, strength and duration of action.

The definition of the phenomenon and the possible causes of its occurrence were determined using the values of water and air temperature, precipitation, pressure, wind speed and direction, and the Southern Oscillation index.

Methodology. The initial data for the work was taken from the NOAA website. For the study, statistical and probabilistic methods were used, which include: primary analysis, autocorrelation analysis, spectral analysis in the Statistica program, cross-correlation analysis (cross-correlation) (Martyn et al. 2020a).

The arithmetic mean of the statistical series ($\overline{X}$). It characterizes the center of gravity of a characteristic or its point of equilibrium under various fluctuations.

$$\overline{X} = \frac{1}{N}\sum X_i \tag{1}$$

where N is the length of the statistical series.

Median (Me)—the value of a feature in the middle of a ranked (ascending) series. The main property is that the sum of the absolute deviations of the members of the series from the median is the smallest value:

$$\sum |X_i - Me| = \min \tag{2}$$

For short series (N < 30), the median is used instead of the arithmetic mean.

Variance (D) and the associated standard (root-mean-square) deviation (σ). The mean dispersion of the values of the series from the arithmetic mean is characterized.

$$D = \frac{1}{N}\sum (X_i - \overline{X})^2, \quad \sigma = \sqrt{D} \tag{3}$$

The distribution function shows the relationship between the possible values of a random variable and the probabilities of their occurrence. The empirical function is called the distribution function calculated from the sample (EGF). In hydrometeorology, EFR is called "recurrence".

EFR calculation consists of the following steps:

Determining the number of intervals

$$K_{\max} = 5\lg(n) \tag{4}$$

where k is the number of intervals;

Round to whole number of intervals.

Calculate the range of variation

$$\Delta X_k = \frac{X_{\max} - X_{\min}}{K_{\max}} \tag{5}$$

Define interval boundaries (C_1, C_2), (C_2, C_3),..., $(C_{K\max}, C_{K\max+1})$, где $C_1 =$ min, $C_2 = C_1 + \Delta X_k$, $C_3 = C_2 + \Delta X_k$ etc. so that $C_{K\max+1} =$ max.

Calculate the middle of each interval X_k -.

Estimate the frequency (repeatability) m_k as the number of members of the sample that fall into each interval. Plot a frequency distribution histogram that represents the empirical distribution function.

The internal structure of the series is investigated using autocorrelation.

Autocorrelation is the correlation of a statistical series with itself at different time shifts. Autocorrelation coefficients r (τ) at each shift is calculated by the formula:

$$r(\tau) = \frac{1}{\sigma^2 x(N - 1 - \tau)}\sum (X_i - \overline{X})(X_{i+\tau} - \overline{X} \tag{6}$$

where N is the length of the series, τ is the shift.

The maximum number of shifts depends on the row length. If the row is short (N ~ 30), then $\tau_{\max} = \frac{1}{3}N$. If the row is long ($N$ ~ 1000) then $\tau_{\max} = \frac{1}{10}N$.

If all the correlation coefficients R (τ) are plotted against the shift τ, then we get the autocorrelation function.

Each correlation coefficient should be tested for significance using the Student's t test. A null hypothesis is put H_0: r = 0 by solving a quadratic equation for r, the critical value r_crit is calculated, corresponding to tcrit at a significance level α = 0.05 and the number of degrees of freedom ν = N-τ-1.

$$r_{\kappa pum} = -\sqrt{N - \tau - 1} + \sqrt{N - \tau - 1 + 4t_{\kappa pum}} \tag{7}$$

where N is the length of the series, the τ-shift of the ACF, t_crit (α, ν = N-τ-1) is the Student's test.

These calculated r_crit values are also plotted on the ACF graph and are the lines of significance levels. Autocorrelation coefficients in absolute value exceeding the significance level are significant, and the values between the significance levels are considered equal to zero.

Cross-correlation is used to determine whether two characteristics are related in asynchronous communication. Cross-correlation analysis (cross-correlation) is the correlation of two statistical series with each other at different shifts.

The cross-correlation coefficients are calculated by the formula

$$r(\pm\tau) = \frac{1}{\sigma_x \sigma_y}(N - 1 - \tau)\sum\left[\left(X_i - \overline{X}\right)(y_{i\pm\tau} - \overline{y})\right] \quad (8)$$

where N is the length of the row, τ is the shift, varies from -τmax to τmax.

The maximum number of shifts is determined depending on the row length. If the row is short (N ~ 30), then τmax = 1/3 N. If the row is long (N ~ 1000) then τmax = 1/10 N.

We check the correlation coefficients for significance or calculate the significance level of the CCF at each shift, as for the ACF. All CCF values in absolute value exceeding the significance level are significant, and between the significance levels are equal to zero.

The construction of maps of wind fields was carried out using the Surfer program, according to the data of the interactive system Giovanni, the spatio-temporal variability of water temperature is considered (Martyn et al. 2019b, 2020b).

3 Results

The SOI South Oscillation Index is most commonly used to determine the El Niño onset. The index has negative values during the onset of El Niño (Fig. 7), which proves the reliability of using this index to assess the onset of the phenomenon.

One of the main manifestations of El Niño is an increase in water and air temperature in the water area of the tropical Pacific Ocean. Figures 8 and 9 clearly shows the maximum temperature relative to the monthly average in the time intervals 1997–1998. and 1982–1983. This indicates a positive anomaly in water and air temperatures in these years, it was in these years that the strongest El Niño phases were noted. From which we can conclude that El Niño directly affects the increase in temperature, the water temperature anomalies are greater than the air temperature anomalies.

With increased temperature anomalies, negative pressure anomalies are observed (Fig. 10), but these anomalies are not directly related to the El Niño force. For example, the most powerful El Niño was observed in 1997–1998, while the largest pressure anomalies were observed in 1969 and 1972.

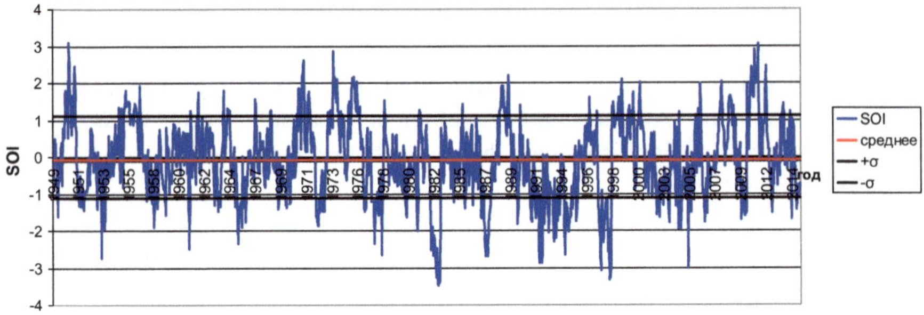

Fig. 7 Southern Oscillation Index for the period January 1949—December 2014 in the tropical Pacific

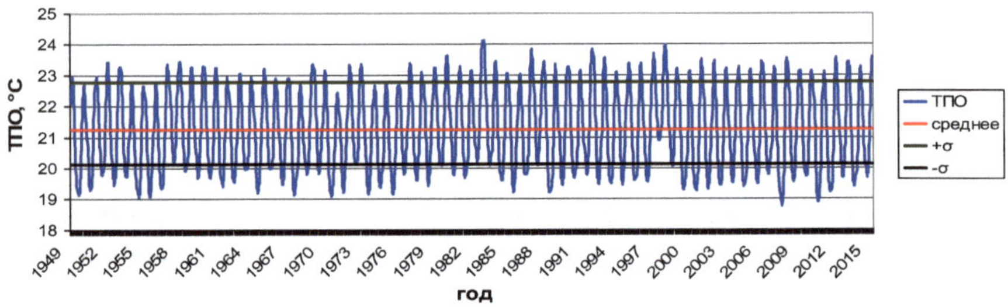

Fig. 8 SST for the period January 1949—March 2015 in the tropical Pacific

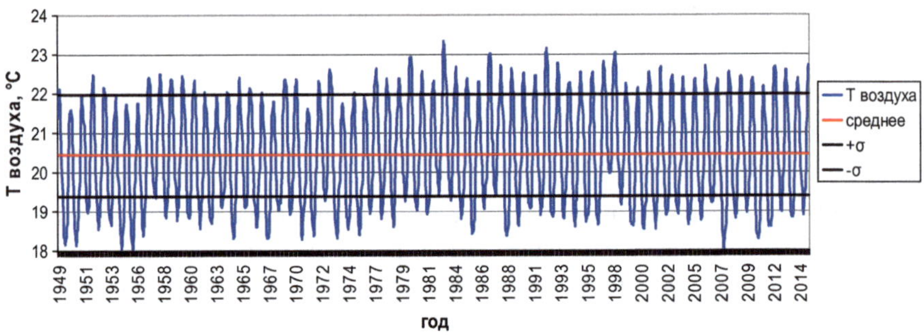

Fig. 9 Air temperature for the period January 1949—March 2015 in the tropical Pacific

Since it was found above that positive SOI values can be taken as El Niño, we will continue to analyze this particular characteristic. However, over the studied region there are zones of both low and high background pressure, the negative Southern Oscillation index is observed more often than the positive one (Fig. 11), which suggests that the El Niño state should be observed more often than the normal ocean state, but in fact it is well known that during the studied time interval (1949–2015) the ocean was more often in a normal state.

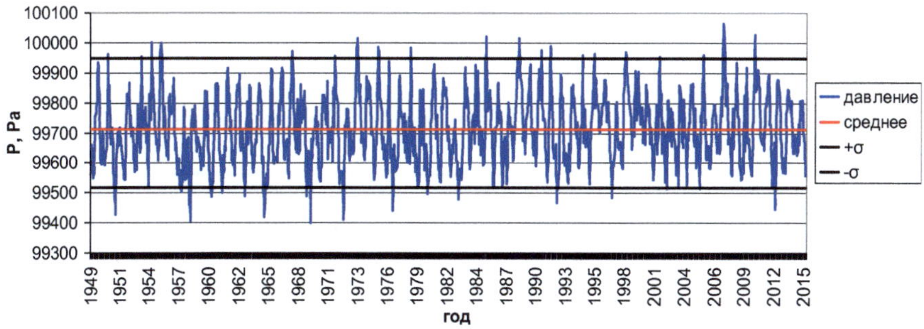

Fig. 10 Pressure for the period January 1949—March 2015. in the tropical Pacific

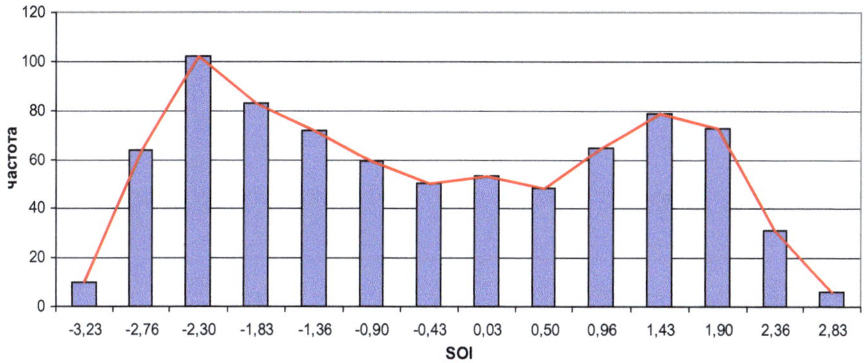

Fig. 11 The empirical distribution function for the Southern Oscillation index for the period January 1949—December 2014

A negative Southern Oscillation index well describes the presence and strength of El Niño, but the presence of a negative index does not always mean the presence of El Niño. The frequency of the index is approximately 5 years (Fig. 12), which corresponds on average to the frequency of the phenomenon of the most popular versions, where the frequency of the phenomenon is considered to be from 3 to 8 years.

After analyzing Figs. 13, 14, 15 and 16, we can conclude that the cause of El Niño is a change in pressure and an increase in water and air temperature, and immediately with the arrival of El Niño, an increase in precipitation occurs. The theory about the baric cause of the formation of El Niño may well be valid.

One of the important signs of an El Niño offensive is a change in wind direction, a weakening of the trade winds or their complete cessation, which contributes to the fact that the warm waters that were previously driven from the coast do not disappear and thereby increase the temperature of the coastal waters of Peru and Ecuador. Plotting wind fields will help you track how wind changes in coastal Peru and Ecuador with and without El Niño. Comparison of January 2011 (Fig. 17) and

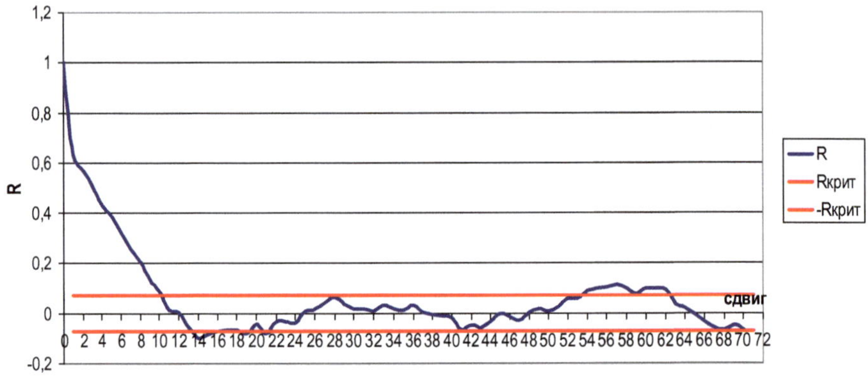

Fig. 12 Autocorrelation function for the Southern Oscillation index for the period January 1949—December 2014

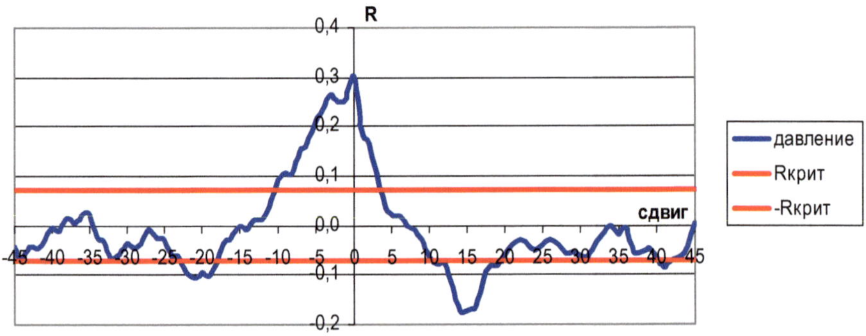

Fig. 13 Correlation of pressure with the South Oscillation index for the period 1949–2014

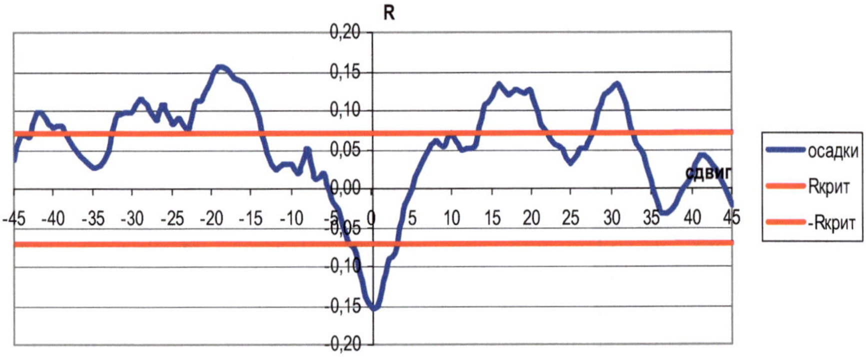

Fig. 14 Correlation of precipitation with the Southern Oscillation index for the period 1949–2014

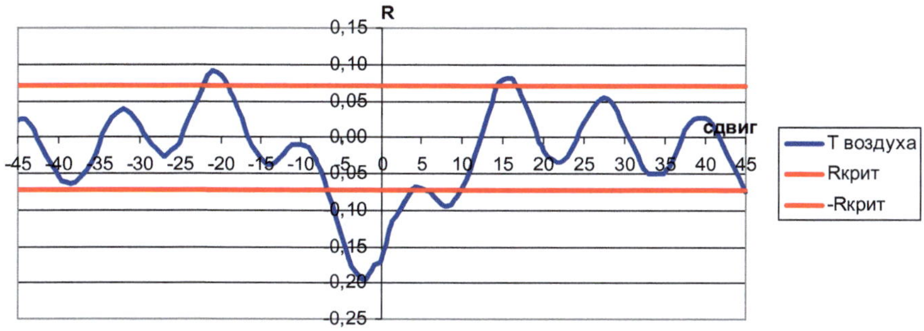

Fig. 15 Correlation of T air with the Southern Oscillation index for the period 1949–2014

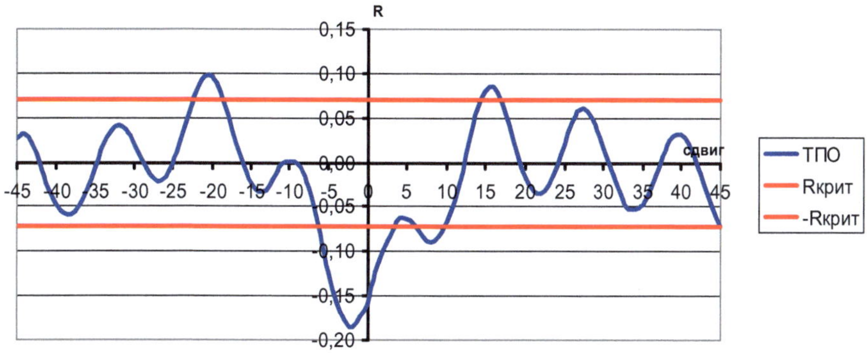

Fig. 16 Correlation of SST with the Southern Oscillation Index for the period 1949–2014

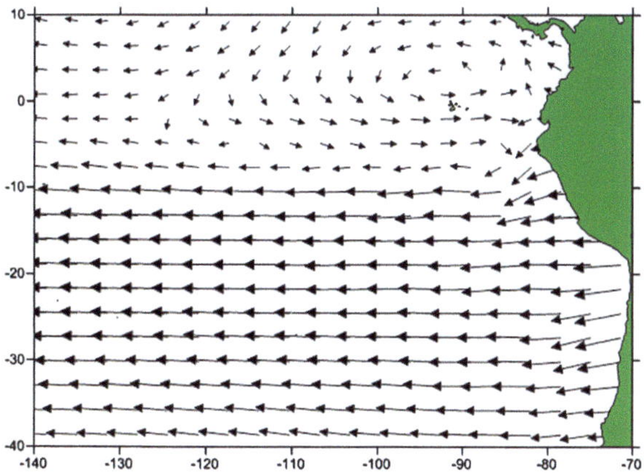

Fig. 17 Wind direction in the tropical Pacific region in January 2011

January 1998 (Fig. 18) showed that during El Niño (Fig. 18) the wind is strictly eastward throughout the region.

In this case, we do not observe a change in wind direction, but the wind speed during El Niño is less than a normal year, which confirms the weakening of the trade winds during El Niño and, therefore, water draining is less efficient and an increase in the temperature of coastal waters is expected.

The most modern method for determining El Niño and its manifestations, or rather changes in water temperature in Peru and Ecuador, is the satellite method. Using satellite images for February–March 2005 (Fig. 19) and 2010 (Fig. 20), as well as May–June 2010 (Fig. 21), the distribution of surface water temperature is shown.

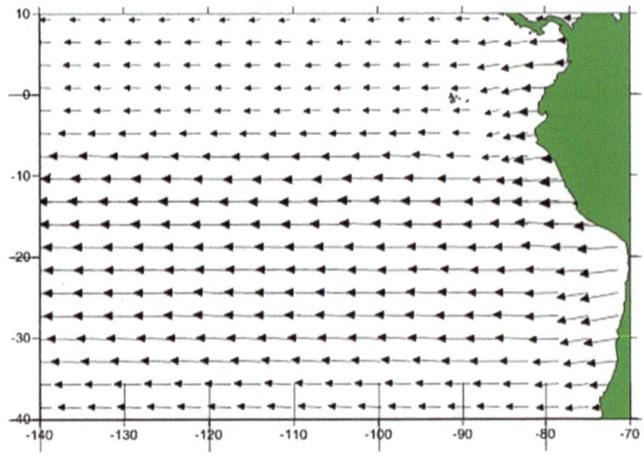

Fig. 18 Wind direction in the tropical Pacific region in January 1998

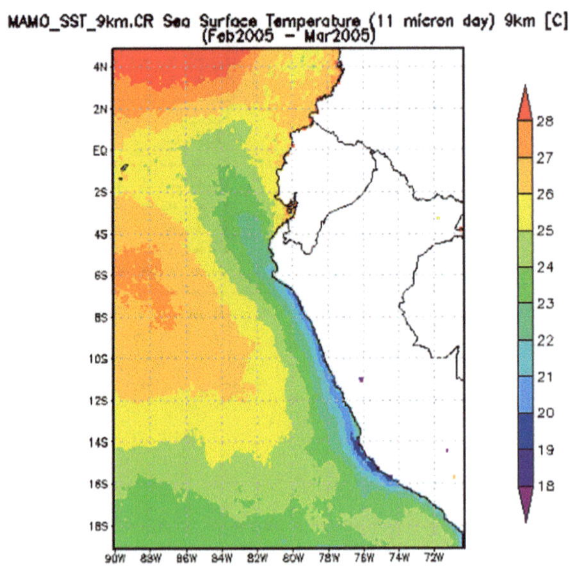

Fig. 19 SST under normal ocean conditions for the period February 2005—March 2005

Fig. 20 SST during El Niño, February 2010—March 2010

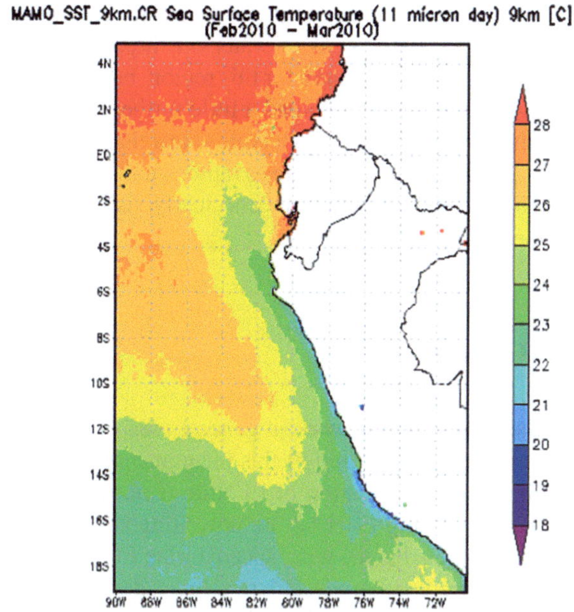

Fig. 21 SST during La Niña, May 2010—June 2010

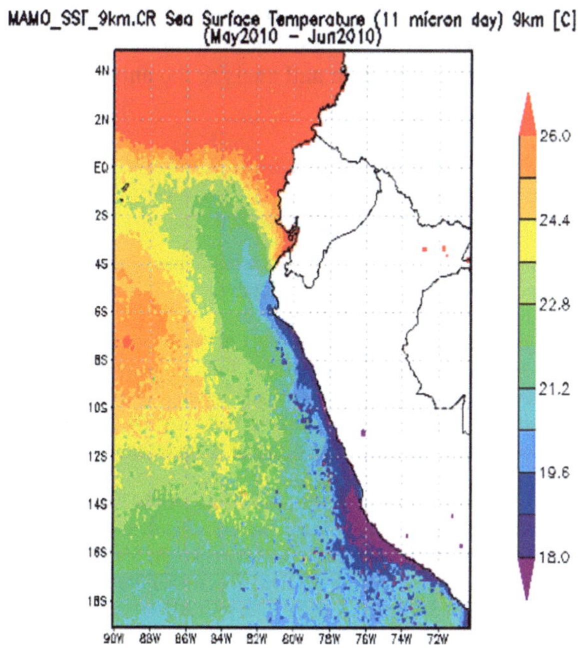

We can say that many of the images that can be found in various sources show how rather large coastal areas of water off the coast of South America are occupied by anomalously warm waters for this region, but these images show that during El Niño (Fig. 20) warm waters spread a little further than the equator and do not even reach the shores everywhere, and it is also clear that upwelling does not completely cease to exist, it weakens and occupies a rather small area, but still does not disappear. During La Niña, a cold water anomaly is clearly visible (Fig. 21) relative to the normal state of the ocean.

4 Conclusion

In the course of the work, a general idea of the phenomenon was compiled. The reason for the formation of this phenomenon was identified, which is of a pressure nature, but the formation of El Niño is not excluded due to several factors. It was found that during El Niño the upwelling stops, there was no noticeable change in wind direction, but the wind speed during El Niño is lower, therefore, the trade winds weaken. The Southern Oscillation Index describes the El Niño offensive well. It was revealed that the strength of the phenomenon directly affects the temperature and precipitation anomaly. The frequency of the phenomenon is approximately 5 years. When studying the structure of the process, it was noted that it was required to get rid of the annual course and analyze the anomalies directly.

References

Bondarenko AL (2006) El Niño—La Niña: mechanisms of formation. Nature 5:39–47

Gustoev DV (1991) El Niño predictability. In: Proceedings of the leningrad hydrometeorological institute, issue 112, L., LGMI

Martyn IA, Kosyuk MA, Petrov YA (2019a) Reduction of rainforest area as a consequence of the El Niño climate phenomenon. In: Topical issues in forestry. Materials of the III international scientific and practical conference of young scientists, pp 250–252

Martyn IA, Kraeva EV, Sidorenko AY, Petrov YA, Stepanov SY (2019b) Application of satellite remote sensing data to assess the El Niño offensive. Inf Technol Syst Manag Econ Transp law 4(36):99–101

Martyn I, Petrov Y, Stepanov S, Sidorenko A (2020a) Spatial-temporal variability of ice cover of the Bering sea. In: IOP conference series: earth and environmental science, vol. 539(1), p 012198

Martyn I, Petrov Y, Stepanov S, Sidorenko A, Vagizov M (2020b) Monitoring forest fires and their consequences using MODIS spectroradiometer data. In: IOP conference series: earth and environmental science, vol. 507(1), p 012019

Propagation of a Single Long Wave in the Bays with U-Shaped Cross-Section Form

A. Yu. Belokon and V. V. Fomin

Abstract In the framework of the nonlinear long wave theory, the evolution of a single wave propagating in the bays with U-shaped cross-section is studied. A good conformity was found between the numerical and analytical estimates of the wave height variation along the bay axis. It is shown that the influence of the shape of the bay cross-section on the wave field is manifested in the sea level rise with the approach to the bay periphery. Estimates of vertical run-up and drainage depth of the shore at the bay top with different cross-sectional shape were obtained. It is found that in bays with a triangular cross-sectional shape there are the greatest run-up height and the greatest drainage depths from the shore. The distance traveled by the wave from the bay entrance to the point of wave breaking is the largest for the bay with cross-sectional shape is approximate to rectangular.

Keywords Nonlinear long waves · Narrow bays with U-shaped cross-section · Numerical solution · Wave amplification · Run-up of long waves on the shore

1 Introduction

The study of the long wave propagation in narrow bays (canals) followed by run-up on a coast is of practical importance. Wave propagation features would be taken into account during the developing measures to protect the coast, which is relevant because of the dense building of the coast, leading economic activities and the existence of recreational areas near the sea. The change in the basin width, depth and the cross-sectional shape significantly influences the amplification, weakening and spatial structure of the wave field. This problem was studied, for example, in Didenkulova et al. (2012), Bazykina and Dotsenko (2015, 2016). During wave propagation in narrow bays (canals) the rise of the sea level and wave run-up differ to a large extent from the case of a flat slope. The real wave heights in such bays significantly exceed the values calculated for waves propagating on a flat slope (Didenkulova 2013; Levin and Nosov 2016; Friedrichs and Aubrey 1994). The availability of sidewall

A. Yu. Belokon (✉) · V. V. Fomin
Marine Hydrophysical Institute RAS, Sevastopol, Russia

T. Chaplina (ed.), *Processes in GeoMedia—Volume IV*, Springer Geology,
https://doi.org/10.1007/978-3-030-76328-2_15

boundaries can causes the focusing of the wave energy within a bay and a significant increase of the waves during their propagation and run-up. In some cases the depth of run-down of the bottom can even exceed the run-up height. Therefore, there is a need for a detailed study of the dynamics of long waves in narrow bays.

Studies of wave propagation in bays with an U-shaped cross-section, carried out in Didenkulov et al. (2015), Didenkulova and Pelinovsky (2011a), Rybkin et al. (2014), Garayshin et al. (2016), Didenkulova and Pelinovsky (2011b), are scientific and applied interest. In these articles analytical functions of the amplitude characteristics are obtained within the framework of the one-dimensional model. To describe the wave propagation in narrow bays and canals, the water flow can be considered uniform at the cross-section, therefore the application of one-dimensional models is justified. The use of cross-section-averaged sea level distributions and horizontal flow velocity allows reducing two-dimensional long-wave equations to a one-dimensional model. Such models are easy to implement, they can be used to find a number of accurate analytical solutions (Didenkulova et al. 2012; Friedrichs and Aubrey 1994; Didenkulov et al. 2015; Didenkulova and Pelinovsky 2011a; Rybkin et al. 2014), but they don't take into account the influence of the basin cross-section shape on the wave characteristics, which can be important during studying the height of the wave run-up on the shore.

In the present work, within the framework of the numerical long-wave model, the changes of the amplitude characteristics of single waves propagating in bays with an U-shaped cross-section are investigated. A comparison of the results of numerical simulation with analytical solutions is carried out and the influence of the basin cross-section shape on changes in the height of a single wave is studied; estimates of the run-up height and estimates of the depth of run-down at the bay top are obtained depending on the ratio of the bay length to the wavelength.

2 Mathematical Formulation of the Problem and Method of Solution

A Cartesian coordinate system (*x*, *y*) is introduced. Consider a model basin with an U-shaped cross-sectional shape of length *L* and width *W*. The relief of the bottom of the basin is described by the following equations:

$$h(x, y) = -h_{\max} + \alpha x^m + \beta(W/2 - y)^m, \ (0 < x \le L, 0 < y \le W/2), \quad (1)$$

$$h(x, y) = -h_{\max} + \alpha x^m + \beta(y - W/2)^m, \ (0 < x \le L, W/2 < y \le W), \quad (2)$$

where $h_{max} = h(0, W/2)$, $\alpha = h_{max}/(L)^m$, $\beta = h_{max}/(W/2)^m$, m is an arbitrary positive number.

Figure 1 shows the basin geometry with $m = 1, 2, 5$, and $h_{max} = 100$ м. The value $m = 1$ corresponds to the triangular cross-sectional shape of the basin; $m = 2$—parabolic cross-sectional shape of the basin; $m = 5$ is the cross-sectional shape of the basin, which is approximates to rectangular.

The problem is to study a single long wave propagation in an U-shaped bay. To describe the wave propagation in the bay, the following equations are used:

$$\frac{\partial u}{\partial t} + u\frac{\partial u}{\partial x} + v\frac{\partial u}{\partial y} + g\frac{\partial \zeta}{\partial x} = 0 \tag{3}$$

$$\frac{\partial v}{\partial t} + u\frac{\partial v}{\partial x} + v\frac{\partial v}{\partial y} + g\frac{\partial \zeta}{\partial y} = 0 \tag{4}$$

$$\frac{\partial \zeta}{\partial t} + \frac{\partial (Hu)}{\partial x} + \frac{\partial (Hv)}{\partial y} = 0 \tag{5}$$

where $u = u(x, y, t)$ and $v = v(x, y, t)$ are depth-averaged projections of horizontal velocities on the x and y axes respectively; t is time; g—gravitational acceleration;

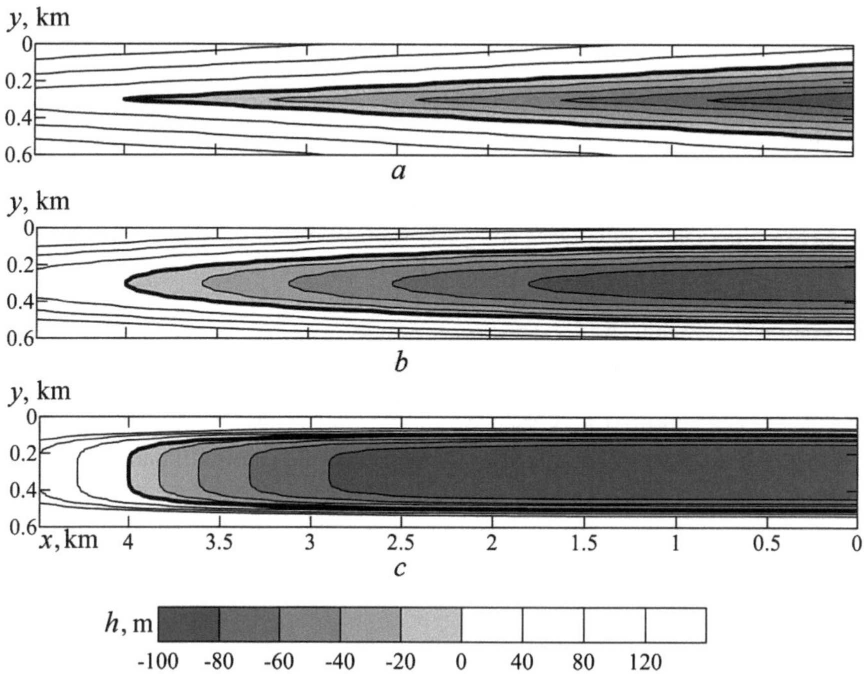

Fig. 1 The relief of U-shaped bays with different values of m: $a - m = 1$; $b - m = 2$; $c - m = 5$

$\zeta = \zeta(x, y, t)$ is the displacement of the free surface of the liquid; $H(x, y, t) = h(x, y) + \zeta(x, y, t)$ dynamic depth of the basin.

At the initial moment of time, the liquid in the bay is in an unperturbed state, i.e.:

$$u = v = 0, \zeta = 0 (t = 0). \tag{6}$$

Through the open end ($x = 0$; $W_1 < y < W_2$, where W_1, W_2 are points of water edge along the y axis in the unperturbed state) a single wave with a height a_0 in the shape of a half-sine with a horizontal dimension λ_0 enters the bay (hereinafter—wavelength). The incoming wave is simulated by the boundary conditions:

$$\zeta = a_0 \sin\left(\frac{\pi t}{T_0}\right), u = \frac{g}{C_0}\zeta, v = 0$$
$$(x = 0, \quad W_1 < y < W_2, \quad 0 < t \leq T_0), \tag{7}$$

where $T_0 = \lambda_0 / C_0$—time of wave propagation across section x = 0; $C_0 = C_0(0, y) = \sqrt{gh(0, y)}$—phase velocity of the wave entering the bay.

After full entering of the wave into the bay conditions (7) are replaced by conditions:

$$\frac{\partial u}{\partial t} - C_0 \frac{\partial u}{\partial x} = 0, v = 0,$$
$$(x = 0, \ 0 \leq y \leq W, \ t > T_0), \tag{8}$$

which provide free exit of reflected waves from the bay.

The initial-boundary value problem (3)–(8) was solved by the finite-difference method using an explicit-implicit first-order approximation scheme with respect to time (Kowalik 2001; Voltcinger and Pyaskovsky 1968).

The calculations were performed at 900 × 120 grid with spatial resolution $\Delta x = \Delta y = 5$ м. Taking into account the Courant stability condition ($\Delta t < \Delta x/2\ C_0$), the integration step in time $\Delta t = 0.05$ c.

Two solutions to problem (3)–(8) were considered. In the first variant (problem I) the processes of run-up and run-down were modeled using the algorithm proposed in Kowalik and Murty (1993), Sielecki and Wurtele (1970). In the second variant (problem II), a solid wall 5 m deep was placed on the sidewall boundaries of the basin, on which the normal component of the flow velocity was assumed to be zero.

Also, the problem of the propagation of a single wave in an U-shaped bay was solved, in which the equations averaged over the bay cross-section were used (Didenkulova and Pelinovsky 2011a):

$$\frac{\partial u}{\partial t} + u\frac{\partial u}{\partial x} + g\frac{\partial \zeta}{\partial x} = 0, \quad \frac{\partial \zeta}{\partial t} + u\frac{\partial H}{\partial x} + \frac{m}{m+1}H\frac{\partial u}{\partial x} = 0 \tag{9}$$

Details of the numerical algorithm for solving Eqs. (9) are given in Bazykina and Dotsenko (2015); Bazykina and Dotsenko (2016).

3 The Results of Numerical Experiments

Proceed to the analysis of the numerical experiment results. Figure 2 shows the process of transformation of a waveform with a length of $\lambda_0 = 1$ km and a height of $a_0 = 0.1$ m along the axis of the bay with a parabolic cross-sectional shape ($m = 2$). When a wave propagates in the bay, its height increases significantly, and the propagation velocity and wavelength decrease. Due to the narrowing of the bay and the lessening of the cross-sectional area, a partial reflection of the wave occurs as it propagates, however, a significant part of the wave energy reaches the top of the bay. Curve *7* in Fig. 2 illustrates the maximum level rise in a wave run-up stage without breaking.

In the linear approximation, the distribution of the amplitudes of the wave fields obeys the Green law, according to which the wave height is $A \sim h^{-1/4}$ for a given x. For waves in bays with an U-shaped cross section in Didenkulova and Pelinovsky (2011a), the following equation is obtained:

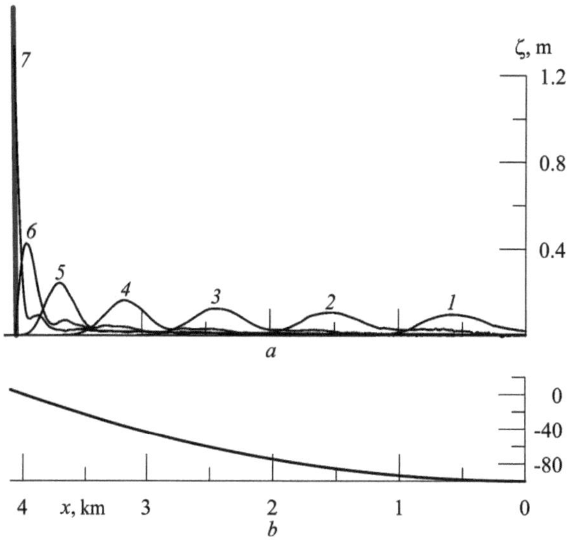

Fig. 2 Evolution of the single wave profile along the axis of the bay (*a*) during wave propagation in the bay with a parabolic cross-sectional shape (for $m = 2$): *1*–*7* correspond to time moments 40, 80, 120, 160, 200, 230, 270 s; depth distribution along the bay axis (*b*). The initial wave height is $a_0 = 0.1$ m, the initial length is $\lambda_0 = 1000$ m

Fig. 3 The dependence of the dimensionless maximum wave height during wave propagation in the bay with a triangular cross-sectional shape ($m = 1$) on the variation in depth along the axis of the bay

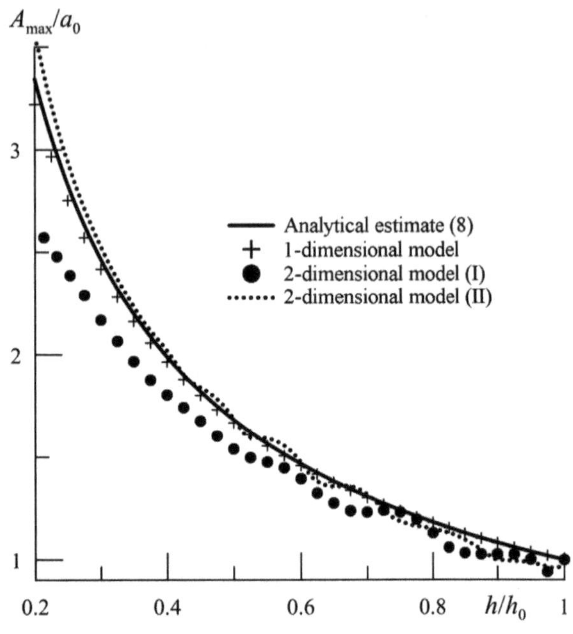

$$A_{max}(x) = a_0 \left(\frac{h(x)}{h_0} \right)^{-\left(\frac{1}{4}+\frac{1}{2m}\right)}, \tag{10}$$

where $A_{max}(x)$—maximum sea level elevation along the x-axis; $h(x)$ is maximum depth along the x-axis; h_0 is maximum depth at the entrance to the bay (at $x = 0$).

Figure 3 presented the dimensionless functions of the maximum height on the depth of the wave propagating in the bay with a triangular cross-section (at $m = 1$). Here an analytical estimate is given using formula (10); the curve obtained in the framework of the one-dimensional model (9), and the results of solving two-dimensional problems I and II.

As can be seen, in the one-dimensional case there is a good agreement between the numerical solution of system (9) and analytical dependence (10). In the two-dimensional case, the wave shape is significantly affected by the cross-sectional shape of the basin. The wave front gets a curved shape caused the fact that its phase velocity is maximal along the bay axis and decreases with approach to the sidewall boundaries. As the basin narrows and its depth decreases, the wave height increases and its propagation rate decreases. At the same time, as it spreads, the wave partially reflects from the lateral boundaries of the basin, which leads to a certain decrease in sea level in the section. This can explain the "wave-like" nature of the curves. The dependencies found in solving problem II are quite well described by an analytical estimate. The curve obtained by solving problem I lies lower than the others, and with a decrease in h/h_0, this difference increases. This behavior of the curve is due to the fact that when $m = 1$, the bay has flatter shores.

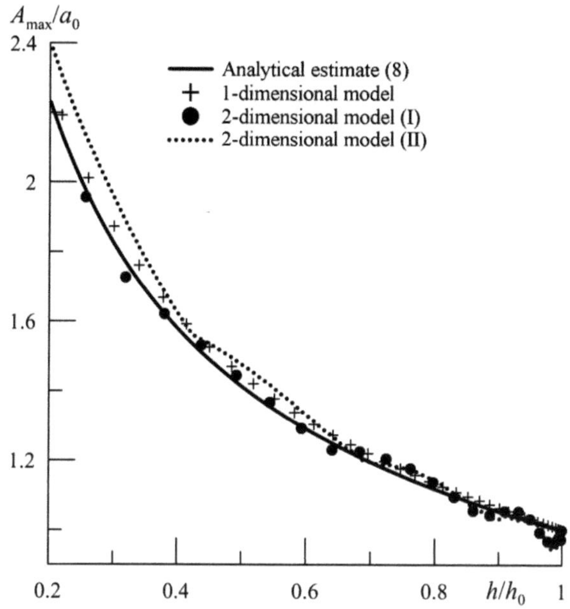

Fig. 4 The dependence of the dimensionless maximum wave height during wave propagation in the bay with a parabolic cross-sectional shape ($m = 2$) on the variation in depth along the axis of the bay

Changes of the maximum wave height during its propagation in the bay with a parabolic cross-sectional shape ($m = 2$) are shown in Fig. 4. The dependences obtained for the sea level distribution along the bay axis within the framework of problem I agree well with the accurate solution. The curve that describes the maximum values of the free surface along the bay axis, obtained by solving problem II, lies above the rest. This difference increases with decreasing function h/h_0.

Figure 5 shows the dependences of the dimensionless maximum sea level heights on the depth of the bay at $m = 5$. In this case, the shape of the cross-section of the bay is approximate to rectangular and the amplitude characteristics of the wave increase more slowly compared to the cases discussed above. As can be seen, the analytical and numerical estimates describing the distribution of the sea level along the bay axis coincide well.

It is of interest to consider the influence of the cross-sectional shape of the bay on the run-up height on the coastal slopes. Figure 6 shows the normed values of the run-up maximum R_{max} at different values of m.

The most run-up is observed in the bay with a triangular cross-sectional shape ($m = 1$). When $m = 5$, there is no significant run-up on the coastal slopes, because of this bay has the most steep shores. Such problem can be simplified by replacing the procedure of run-up on the coast by the condition of non-permeability.

Further, for different values of m, the problem of the wave propagation in an U-shaped bay was solved, using Eq. (9) averaged over the bay cross-section. Figure 7 shows the maximum vertical wave run-up on the coast (a) and the depth of the run-down (b) from the ratio of the bay length to initial wavelength L/λ_0.

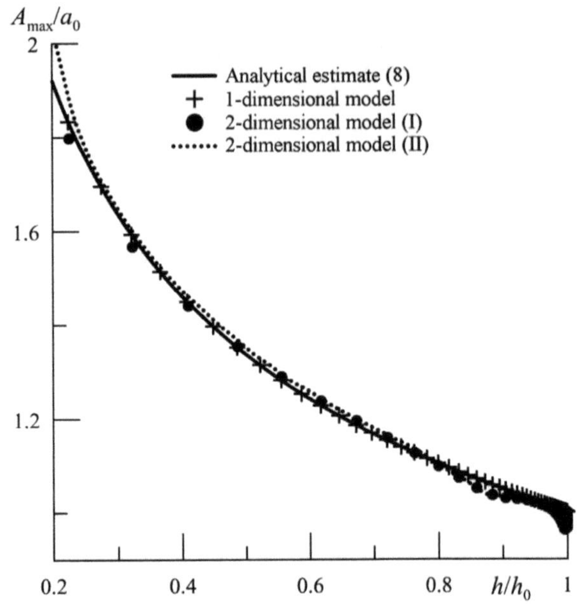

Fig. 5 The dependence of the dimensionless maximum wave height during wave propagation in the bay with a rectangular cross-sectional shape ($m = 5$) on the variation in depth along the axis of the bay

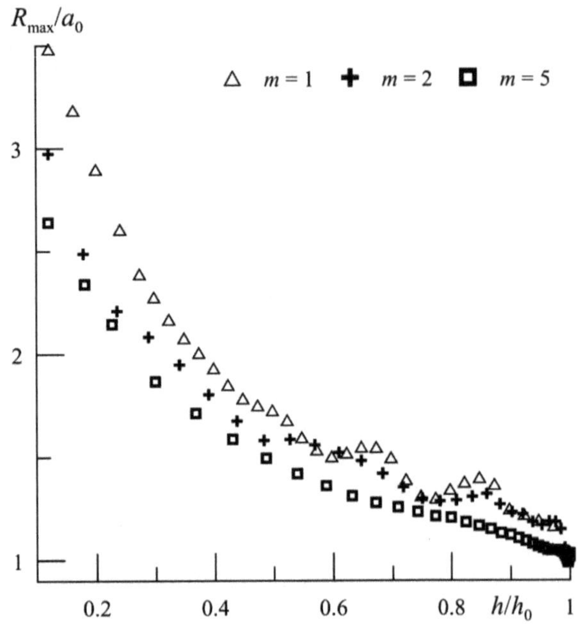

Fig. 6 Dependences of the maximum run-up height during wave propagation in the bay with different cross-sectional shape, on the change in depth along the bay axis

Fig. 7 Dependences of the maximum run-up height (**a**) and the run-down depth (**b**) on the L/λ_0 in bays with different cross-sectional shapes

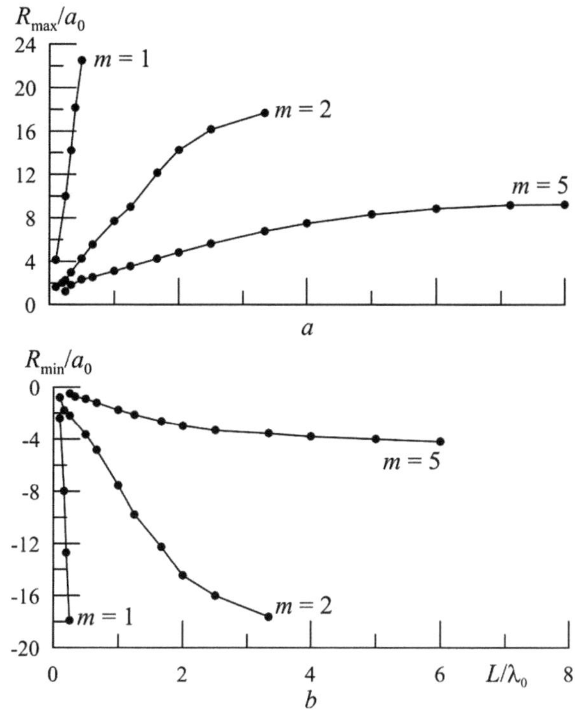

In the bay with a triangular cross-sectional shape ($m = 1$) near the shore the wave heights increase significantly due to the highest rate of change of the basin's geometric parameters. The shore slope at the bay top is quite flat, so the height of the wave run-up can reach extreme values. For relatively small wavelengths (for large L/λ_0), the steepness of the forward wave slope increases rapidly, causes the wave breaking. Increasing L/λ_0 causes exceeding of the run-down depth over the run-up height in terms of absolute value.

In a bay with a parabolic cross-sectional shape ($m = 2$) the run-up height also increases with rising L/λ_0 function. Moreover, the values of the run-up height and the run-down depth are approximately equal in terms of absolute value.

At $m = 5$ the basin has the most flat bottom and the steep coastal slope, therefore, the wave near the shore doesn't increase as much as in the cases discussed above. The run-down depth in such bay is more than two times less than the run-up height in terms of absolute value.

The presented dependences are obtained for non-breaking waves. With a further increasing basin length or a wavelength decreasing the nonlinearity effect affects rising of the wave front slope and its subsequent breaking. The distance of the wave propagation before the breaking increases with growth of number m.

4 Conclusion

The propagation of single waves in model bays with an U-shaped cross-section has been studied. Estimates of amplitude characteristic variation during wave propagation in basins with a triangular, parabolic, and near-rectangular cross-sectional shape are found. A comparative analysis of the dependences obtained numerically in the framework of one-dimensional and two-dimensional nonlinear models with available analytical estimates is carried out. A good agreement was identified between numerical solutions and analytical estimates of the variation of wave height along the bay axis. It is shown that the influence of the cross-sectional shape of the basin on the amplitudes of the wave fields is manifested in the level rise approaching the periphery of the basin.

Estimates of run-up height and run-down depth at the top of the bay with different cross-sectional shapes are presented. It is shown that in the basins with a triangular cross-sectional shape, the most run-up heights are observed, and the values of the run-down depths can exceed the run-up heights in absolute value. In bays with a parabolic cross section, the most run-up heights and run-down depths are equal in absolute value. In basins with a rectangular cross-sectional shape, the run-up heights are the minimal. The run-down depth in such bays is two times less in terms of absolute value than the run-up height, and the distance of the wave propagation before the breaking is the most.

The investigation is carried out within the framework of the state task on the theme No. 0827–2019-0004 “Complex interdisciplinary investigations of the oceanologic processes conditioning functioning and evolution of the Black and Azov seas' ecosystems of the coastal zones” (code “Coastal investigations”).

References

Bazykina AYu, Dotsenko SF (2015) Nonlinear effects at propagation of long surface waves in the channels with a variable cross-section. Phys Oceanogr 4:3–13

Bazykina AY, Dotsenko SF (2016) Vliyanie nelineynosti I donnogo treniya na dlinnye volny v kanalah peremennogo poperechnogo secheniya [Influence of non-linearity and bottom friction on long waves in channels of variable cross-section]. Processy v Geosredah 2:97–103 (in Russian)

Didenkulov OI, Didenkulova II, Pelinovsky EN, Kurkin AA (2015) Influence of the shape of the bay cross section on wave run-up 2015. Izv AN Fiz Atmosfery i Okeana 51(6):741–747

Didenkulova I (2013) Tsunami runup in narrow bays: the case of Samoa 2009 tsunami. Nat Hazards 65(3):1629–1636

Didenkulova II, Pelinovsky DE, Tyugin DY et al (2012) Begushchie dlinnye volny v vodnyh pryamougolnyh kanalah peremennogo secheniya [The running long waves in the water rectangular channels of variable section]. Vestn MGOU 5:89–93 (in Russian)

Didenkulova I, Pelinovsky E (2011a) Runup of tsunami waves in U-shaped bays. Pure Appl Geophys 168:1239–1249

Didenkulova I, Pelinovsky E (2011b) Nonlinear wave evolution and runup in an in-clined channel of a parabolic cross-section. Phys Fluids 23(8):086602

Friedrichs CT, Aubrey DG (1994) Tidal propagation in strongly convergent channels. J Geoph Res 99(C2):3321–3336
Garayshin VV, Harris MW, Nicolsky DJ, Pelinovsky EN, Rybkin AV (2016) An analytical and numerical study of long waves run-up in U-shaped and V-shaped bays. Appl Math Comput 279:187–197
Kowalik Z (2001) Basic relations between tsunamis calculations and their physics. Sci Tsun Hazar 19(2):99–115
Kowalik Z, Murty TS (1993) Numerical simulation of two-dimensional tsunami runup. Mar Geodesy 16:87–100
Levin BV, Nosov MA (2016) Physics of tsunamis. Springer, p 388
Rybkin A, Pelinovsky E, Didenkulova I (2014) Nonlinear wave run-up in bays of arbi-trary cross-section: generalization of the Carrier-Greenspan approach. J Fluid Mech 748:416–432
Sielecki A, Wurtele M (1970) The numerical integration of the non-linear shallow-water equations with sloping boundaries. J Comp Phys 6:219–236
Voltcinger NE, Pyaskovsky RV (1968) The main oceanological problems of the shal-low water theory, L. Gidrometeoizdat 300

Temperature Dependencies of Compressional Wave Velocity and Attenuation in Hydrate-Bearing Coal Samples

G. A. Dugarov, M. I. Fokin, and A. A. Duchkov

Abstract Gas hydrates are widespread in nature. Hydrate-bearing sands are well studied due to their high permeability and potential as a new resource of natural gas. Other hydrate-bearing deposits are studied less, in spite of the fact that earlier it was hypothesized the possible presence of gas hydrates in coal. But the possibility of the formation of gas hydrate from bound water in coal was confirmed in experiments recently. This work considers the results of a series of experiments to study the acoustic properties of coal samples containing methane hydrate. As a result, the temperature dependence of compressional wave velocity and attenuation was revealed.

Keywords Gas hydrates · Acoustic properties · Coal samples · Hydrate-bearing media

1 Introduction

Natural gas hydrates are widespread in offshore bottom sediments and permafrost. They are stable under high pressure and low temperature and formed from free water and a hydrate-forming gas (for example, natural gas). In the earth's crust, suitable pressures and temperatures exist within the so-called hydrate stability zone. For permafrost rocks, the hydrate stability zone have a thickness of 100–600 m starting from the depth of 200 m (Collett 2002). In addition, natural coals contain water and methane formed during the metamorphism of coal matter. Makogon first proposed the hypothesis about the possible presence of gas hydrates in coal deposits (in the case of suitable thermobaric conditions) (Makogon 1974). This hypothesis was also discussed later by different authors (Oparin and Skritskiy 2012; Dyrdin et al. 2013). But recently the possibility of the hydrate formation from bound water in natural

G. A. Dugarov (✉) · M. I. Fokin · A. A. Duchkov
Trofimuk Institute of Petroleum Geology and Geophysics SB RAS, Novosibirsk, Russia
e-mail: dugarovga@ipgg.sbras.ru

A. A. Duchkov
Novosibirsk State University, Novosibirsk, Russia

T. Chaplina (ed.), *Processes in GeoMedia—Volume IV*, Springer Geology,
https://doi.org/10.1007/978-3-030-76328-2_16

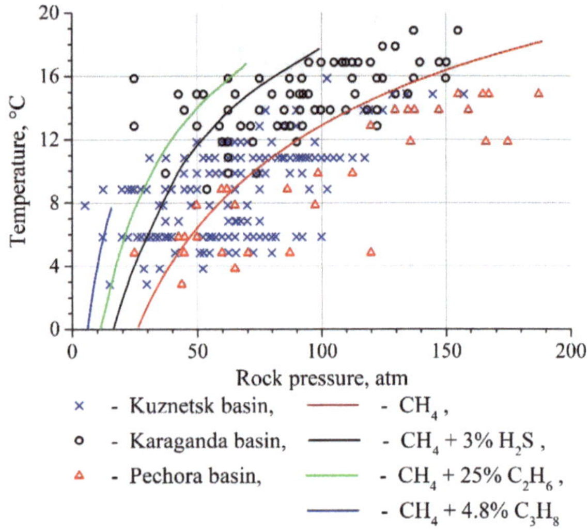

Fig. 1 Thermobaric conditions for some Russian coal basins in comparison with phase boundaries of hydrates of methane and its mixture with other gases (Smirnov 2014)

coals was confirmed by experiments (Bustin et al. 2015; Smirnov et al. 2016, 2018, 2020; Turakhanov et al. 2020).

Besides, several coal basins in Russia have PT-conditions suitable for gas-hydrate formation (Smirnov 2014), see Fig. 1. This is especially true for coal basins in the Northern regions. In Fig. 1, one can see thermobaric conditions for different locations within three coal basins: Kuznetsk (crosses), Karaganda (circles) and Pechora (triangles). Lines show the stability fields of gas-hydrate formation from pure methane and its mixtures with other gases. A marker location below a line indicates that corresponding gas-hydrate is stable in these thermobaric conditions. One can see that majority of markers lay below the lines for a mixture of methane with other gases. For pure methane, there are still many in the gas-hydrate stability zone, especially for the Pechora Basin. The Pechora Basin is located in the permafrost region and its gas component has a mixture of methane and heavy hydrocarbons.

The physical properties of hydrate-bearing coal samples are not well studied yet. This paper discusses the results of the series of experiments on the formation and acoustic measurements on the hydrate-bearing samples formed from bituminous coal. For this, we used a specialized setup (Duchkov et al. 2016, 2017), on which a significant amount of the experiments with sand samples were previously carried out (Duchkov et al. 2018, 2019; Dugarov et al. 2019).

2 Methods and Procedures

The procedure of hydrate-bearing media formation consists in preparing samples with water and a free hydrate-forming gas and maintaining thermobaric conditions within the hydrate stability zone (Duchkov et al. 2017; Dugarov et al. 2019). In these

Table 1 Parameters of coal samples. $S_{w,ads}$—saturation by adsorbed water, $S_{w,free}$—saturation by free water, S_h—hydrate saturation, $S_{h,max}$—maximum hydrate saturation if all water transformed to hydrate

ID	Porosity	$S_{w,ads}$	$S_{w,free}$	S_h	$S_{h,max}$
C05	0.22	0.15	0.51	0.36	0.83
C06	0.21	0.16	0.23	0.28	0.50
C07	0.21	0.16	0	0.08	0.20

experiments we used methane. Low volatile bituminous coal from bed XXVII of "Berezovskaya" coalmine, Kemerovo region of the Kuznetsk coal basin, was used as a rock matrix. Technical analysis of the coal revealed a large number of functional groups (Smirnov et al. 2020), which should lead to increased amount of bound water. The density of coal was estimated by pycnometric method as 1.30 g/cm^3.

For homogeneous formation of the samples we first crashed coal into a powder with a particles size of 0.2–1 mm. Then the coal powder was dried at 110 °C under vacuum. After that dry coal were placed in a desiccator with clean water at the bottom. Coal after this stage of sample preparation contains only adsorbed water (adsorbed from the air), see $S_{w,ads}$ in Table 1. We also added different amount of water before placing the sample into a cell. This added water we will call as "free" water, see $S_{w,free}$ in Table 1. There were no signs of free water leaking from the samples during manipulations with them.

The cell with the sample was placed into the chamber, which allows to create axial and lateral pressure on the sample (Duchkov et al. 2016, 2017). In our experiments, the values of axial and lateral pressure were the same and equal to 300 atm, in order to avoid specimen deformation. Brass mesh was used at the cell ends to provide methane supply to the sample after loading into the chamber. The initial gas pressure was 150 atm. Gas pressure decreased during adsorption by coal and hydrate formation, no additional gas was added during the experiment.

We started the hydrate formation process by decreasing the temperature to the constant value within the methane hydrate stability zone. The compressional (P) wave velocity increased with increasing of hydrate amount in the sample. Stopping of the velocities increase was an indicator of the end of hydrate formation.

Hydrate saturation of the sample was estimated at the end of the experiment. After the end of methane hydrate formation the temperature was increased above phase boundary of methane hydrate stability. During the hydrate dissociation a sharp increase of gas pressure and a drop of P-wave velocity was being observed. Hydrate saturation of the sample was estimated from the measured gas volume corresponding to the recorded gas pressure increase, see S_h in Table 1.

From initial water saturation we can also estimate maximum possible hydrate saturation if all water had transformed to hydrate, see $S_{h,max}$ in Table 1. One can see that only about 45% of the water transformed to hydrate in coal samples. While in sand samples almost all of the water transformed to hydrate (Duchkov et al. 2018, 2019; Dugarov et al. 2019).

3 Results

For each sample, after the hydrate formation was finished we studied acoustic properties of hydrate-bearing samples at different temperatures. As a result, the temperature dependence of the P-wave velocity was revealed.

For sample C05, it was frozen to −11 °C with a further stepwise increase in temperature with holding at each value about 5–6 h, see Fig. 2, green line. In this case, with a change in temperature, a new value for the P-wave velocity was established (black line in Fig. 2). In the time interval from 45 to 55 h, a sharp velocity drop is observed associated with the dissociation of methane hydrate. For each experiment we measured this released gas volume to estimate the final hydrate saturation.

For further experiments, we used slow linear growth of temperature at the rate of 2 °C per hour. The obtained temperature dependences are shown in Fig. 3. At the higher hydrate saturation (sample C06) the larger velocity difference is observed in the presence and absence of hydrates. The differences are 0.27 and 0.82 km/s for samples C07 and C06 respectively.

The similar dependence is observed for the P-wave attenuation for the same samples, see Fig. 4. The attenuation was estimated as inverse quality factor by spectral ratio method. One can also see that the sample less saturated by hydrate (sample C07) have higher attenuation but its estimation was more unstable due to weak consolidation of the sample.

The consistent decrease of acoustic velocities in the sample with increasing temperature can be explained using the existing model of moisture freezing and hydrate formation in rocks (Chuvilin et al. 2011). Due to the interaction with the surface of the rock particles, the water in the rock has a spectrum of energy states. Coal has a complex surface with polar functional groups on it. Such complex surfaces significantly change the energy state of bound water, which leads to a partial formation of hydrate at different temperatures.

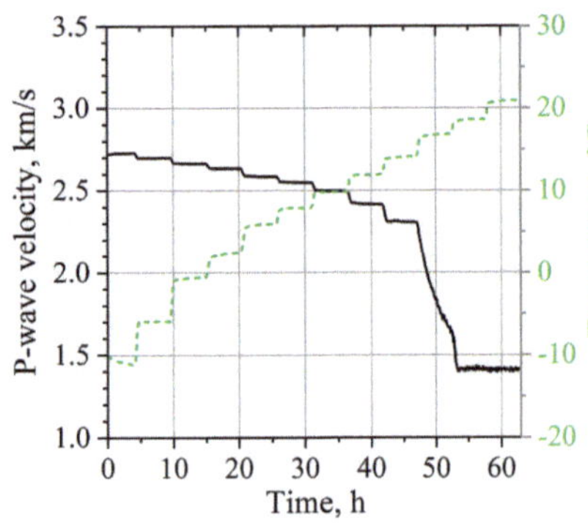

Fig. 2 The changes in P-wave velocity during temperature increase in steps (sample C05)

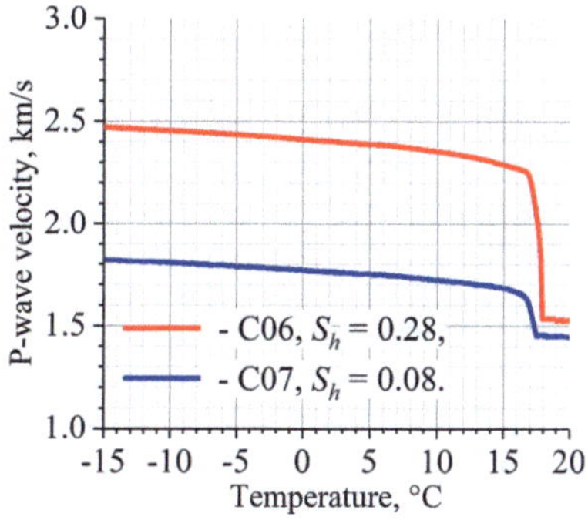

Fig. 3 Temperature dependencies of P-wave velocities in hydrate-bearing coal samples

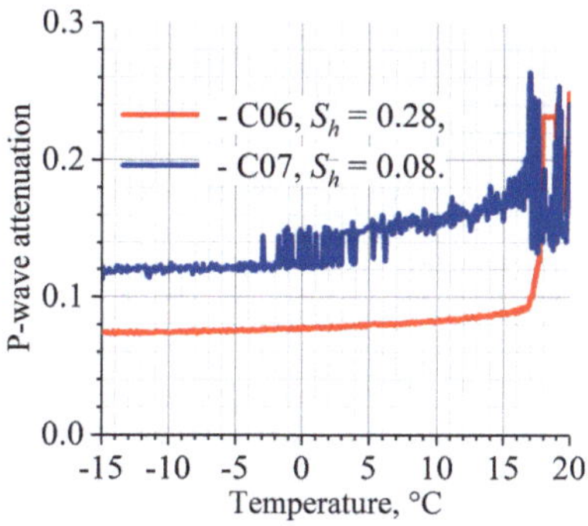

Fig. 4 Temperature dependencies of P-wave attenuation (inverse quality factor) in hydrate-bearing coal samples

4 Conclusion

The results of the experiments on the study of acoustic properties of coal samples formed from bituminous coal containing methane hydrate are discussed in this work. The temperature dependences of P-wave velocity and attenuation within the hydrate stability zone were revealed. We assume that this is due to the presence of a spectrum of water states in the studied samples. Complex coal surface significantly change the energy state of bound water, which leads to a partial formation of hydrate at different temperatures. But we need to conduct more experiments on the measuring acoustic properties of hydrate-bearing coal samples to provide statistically significant data.

Acknowledgements The work was supported by Russian Science Foundation (grant No. 19-77-00068) and Russian Presidential Grant (МК-2647.2019.5).

References

Bustin AMM, Bustin RM, Moudrakovskim IL, Takeya S, Ripmeester JA (2015) Formation of methane clathrate hydrates in coal moisture: Implications for coalbed methane resources and reservoir pressures. Energy Fuels 30:88–97

Chuvilin EM, Istomin VA, Safonov SS (2011) Residual nonclathrated water in sediments in equilibrium with gas hydrate: comparison with unfrozen water. Cold Reg Sci Technol 68:68–73

Collett TS (2002) Energy resource potential of natural gas hydrates. AAPG Bull 86(11):1971–1992

Duchkov AD, Golikov NA, Duchkov AA, Manakov AYu, Permyakov ME, Drobchik AN (2016) Equipment for the studies of the acoustic properties of hydrate-containing samples in laboratory conditions. Seismic Instruments 52(1):70–78

Duchkov AD, Duchkov AA, Permyakov ME, Manakov AY, Golikov NA, Drobchik AN (2017) Acoustic properties of hydrate-bearing sand samples: laboratory measurements (setup, methods, and results). Russ Geol Geophys 58(6):727–737

Duchkov AD, Duchkov AA, Dugarov GA, Drobchik AN (2018) Velocities of ultrasonic waves in sand samples containing water, ice, or methane and tetrahydrofuran hydrates (laboratory measurements). Dokl Earth Sci 478(1):74–78

Duchkov AD, Dugarov GA, Duchkov AA, Drobchik AN (2019) Laboratory investigations into the velocities and attenuation of ultrasonic waves in sand samples containing water/ice and methane and tetrahydrofuran hydrates. Russ Geol Geophys 60(2):193–203

Dugarov GA, Duchkov AA, Duchkov AD, Drobchik AN (2019) Laboratory validation of effective acoustic velocity models for samples bearing hydrates of different type. J Nat Gas Sci Eng 63:38–46

Dyrdin VV, Smirnov VG, Shepeleva SA (2013) Parameters of methane condition during phase transition at the outburst-hazardous coal seam edges. J Min Sci 49(6):908–912

Makogon YF (1974) Hydrates of natural gases [in Russian]. Nedra, Moscow, Russia

Oparin VN, Skritskiy VA (2012) Analytical review of methane explosions in the mines of Kuzbass. Coal 2:29–32

Smirnov VG (2014) Research of phase transitions of methane gas hydrates in the porous structure of coal [in Russian]. Doctoral dissertation, Kuzbass State Technical University, Kemerovo, Russia

Smirnov VG, Manakov AYu, Ukraintseva EA, Villevald GV, Karpova TD, Dyrdin VV, Lyrshchikov SYu, Ismagilov ZR, Terekhova IS, Ogienko AG (2016) Formation and decomposition of methane hydrate in coal. Fuel 166:188–195

Smirnov VG, Manakov AY, Dyrdin VV, Ismagilov ZR, Mikhailova ES, Rodionova TV, Villevald GV, Malysheva VY (2018) The formation of carbon dioxide hydrate from water sorbed by coals. Fuel 228:123–131

Smirnov VG, Dyrdin VV, Manakov AYu, Ismagilov ZR (2020) Decomposition of carbon dioxide hydrate in the samples of natural coal with different degrees of metamorphism. Chin J Chem Eng 28(2):492–501

Turakhanov AH, Shumskayte MY, Ildyakov AV, Manakov AY, Smirnov VG, Glinskikh VN, Duchkov AD (2020) Formation of methane hydrate from water sorbed by anthracite: an investgation by low-field NMR relaxation. Fuel 262, 116656

Numerical Simulation of a Single Wave Interaction with Submerged Breakwater in a Model Basin

S. Yu. Mikhailichenko, E. V. Ivancha, and A. Yu. Belokon

Abstract This work is focused on numerical simulation of the interaction of a solitary surface wave with a submerged rectangular breakwater in the constant depth model basin using the non-hydrostatic hydrodynamic model SWASH (Simulating WAves till SHore). The features of the transformation of a wave passing over an obstacle with the width and height changes of the breakwater were investigated. Based on numerical experiments, the transformation coefficients for the soliton and the zone of its attenuation area behind a breakwater were calculated. The localization of the region of maximum attenuation of the wave passing beyond the breakwater is determined. The analysis of the spatial structure features of free-surface fluctuations caused by the interaction of a soliton with a breakwater is carried out. The depth-averaged orbital fluid velocities are calculated and the dependence of their values and directions on the geometric parameters of the underwater obstacle is determined.

Keywords SWASH model · Soliton · Transformation coefficients · Submerged breakwaters · Orbital fluid velocities

1 Introduction

The problem of potential surface waves description is the classical problem of hydrodynamics in the course of two centuries. The well-known results of Boussinesq The most intense load on the coastal zone is exerted by surface waves, which can lead to the destruction of the coast, flooding of coastal areas and, as a consequence, to significant material damage (Repetin et al. 2003). One of the ways to protect the coast, to preserve the aesthetic appeal of coastal areas, is to construct the submerged breakwaters, which can significantly affect the parameters of waves approaching the coast. Therefore, there is a need for a detailed study of the process of interaction of waves with this kind of hydraulic structures. To represent wind waves in the work, a mathematical model of a soliton was chosen (Pelinovsky 1996), since it is known that at a certain stage of wave deformation by shallow water wave crests

S. Yu. Mikhailichenko (✉) · E. V. Ivancha · A. Yu. Belokon
Marine Hydrophysical Institute, Russian Academy of Sciences, Sevastopol, Russian Federation

T. Chaplina (ed.), *Processes in GeoMedia—Volume IV*, Springer Geology,
https://doi.org/10.1007/978-3-030-76328-2_17

become narrower and steeper, alternating with long and flat troughs (Kononkova and Pokazeev 1985). Many studies have been devoted to the interaction of solitary waves with various underwater obstacles, among them (Lamb 1945; Kotelnikova et al. 2012, 2018; Lin 2004; Wang et al. 2018; Korzinin 2017; Mikhailichenko and Ivancha 2017). Lamb (1945) analytically calculated the transmission and reflection coefficients for a long wave passing over the bottom step. The decisions are based on the assumption that the incident wavelength is large compared to the water depth both in front of and above the step. An experimental study of the interaction of a solitary wave with underwater obstacles in the form of a bottom step and a vertical plate was carried out in Kotelnikova et al. (2012), Kotelnikova et al. (2018). It was identified that depending on the wave amplitude, channel depth and obstacle height, two possible interaction scenarios are realized. In one case, the transformation of the soliton occurs without the formation of a reflected wave (or the reflected wave has a low intensity), in the other the wave is divided into the reflected and transmitted ones above the breakwater. In (Lin 2004; Wang et al. 2018), a complex study of the interaction of solitary waves with a rectangular breakwater was carried out. The analysis of the forces acting on the side of the passing wave on the underwater obstacle is carried out. The coefficients of reflection, transmission and dissipation of wave energy have been calculated for various combinations of the width and height of the breakwater. It is shown that the height of the obstacle makes the main contribution to the transformation of a solitary wave. In (Korzinin 2017), the conditions were determined under which the maximum attenuation of various types of surface waves is achieved during its interaction with an underwater coastal protection structure.

In all of the above works, the interaction of waves with rectangular breakwaters in the two-dimensional case in the *Oxz*-plane was studied. The length of the obstacles was, in fact, infinite. In reality, all breakwaters have a finite length, so when waves interact with them, the effects of diffraction and interference are observed when going around the edges of these structures (Mikhailichenko and Ivancha 2017). These effects can have a certain influence on the amplitude characteristics of both regular and solitary waves, creating local zones of attenuation and amplification of waves.

In this work we studied the interaction of a solitary wave with an underwater rectangular coastal protection structure in a two-dimensional case in the *Oxy*-plane using the SWASH model (SWASH User Manual 2012). The influence of the relative width and height of the breakwater on the transformation of the soliton was analysed. The fields of the depth-averaged orbital velocities of the fluid were also calculated and analysed.

2 The Mathematical Problem Statement

The SWASH model is based on nonlinear shallow water equations, which include a term with non-hydrostatic pressure:

$$\frac{\partial \zeta}{\partial t} + \frac{\partial Hu}{\partial x} + \frac{\partial Hv}{\partial y} = 0 \quad (1)$$

$$\frac{\partial u}{\partial t} + u\frac{\partial u}{\partial x} + v\frac{\partial u}{\partial y} + g\frac{\partial \zeta}{\partial x} + \frac{1}{H}\int_{-d}^{\zeta} \frac{\partial q}{\partial x}dz + c_f\frac{u\sqrt{u^2+v^2}}{H}$$
$$= \frac{1}{H}(\frac{\partial H\tau_{xx}}{\partial x} + \frac{\partial H\tau_{xy}}{\partial y}) \quad (2)$$

$$\frac{\partial v}{\partial t} + u\frac{\partial v}{\partial x} + v\frac{\partial v}{\partial y} + g\frac{\partial \zeta}{\partial y} + \frac{1}{H}\int_{-d}^{\zeta} \frac{\partial q}{\partial y}dz + c_f\frac{u\sqrt{u^2+v^2}}{H}$$
$$= \frac{1}{H}(\frac{\partial H\tau_{yx}}{\partial x} + \frac{\partial H\tau_{yy}}{\partial y}) \quad (3)$$

$$\tau_{xx} = 2\nu_t\frac{\partial u}{\partial x}, \tau_{xy} = \tau_{yx} = \nu_t(\frac{\partial u}{\partial y} + \frac{\partial v}{\partial x}), \tau_{yy} = 2\nu_t\frac{\partial v}{\partial y} \quad (4)$$

where t is time, x, y, z—Cartesian coordinates, $H(x, y, t) = d(x, y) + \zeta(x, y, t)$—dynamic fluid depth, u, v—flow velocity components, $q(x, y, z, t)$—non-hydrostatic pressure additive, g—gravitational acceleration, c_f—bottom friction coefficient, $\tau_{xx}, \tau_{xy}, \tau_{yx}, \tau_{yy}$—components of the horizontal turbulent stress tensor, ν_t—turbulent viscosity coefficient.

Numerical experiments were carried out in a model basin in the form of a square with a side $L = 600$ m (Fig. 1). In the central part of the basin there was a breakwater in the form of a vertical wall 200 m long. The width l and height h of the breakwater varied. In the calculations, we used a rectangular grid with a step along the spatial coordinates $\Delta x = \Delta y = 0.5$ m. The time integration step was $\Delta t = 0.01$ s. The simulation was carried out in a time interval of 90 s. The time profile of a single soliton (Kononkova and Pokazeev 1985) and the initial value of the depth-averaged fluid velocity (SWASH User Manual 2012) were set as the boundary conditions at the input at $x = 0$:

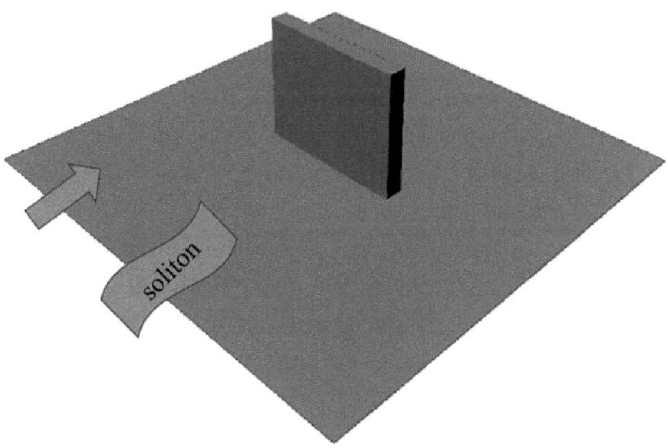

Fig. 1 Scheme of the model basin

$$\zeta_b = a_0 \mathrm{sech}^2(\sqrt{\frac{3a_0}{4H}}\frac{x - x_0 - Vt}{H}),\ u_b = \sqrt{\frac{g}{H}}(2\zeta_b - \zeta)\ x = 0, 0 \le t \le T, \quad (5)$$

where $V = c(1 + \frac{a_0}{2H}) = \sqrt{gH}(1 + \frac{a_0}{2H})$—wave propagation speed, ζ_b—the level of the free surface on the left edge of the model basin, T—the initial period of the wave. The initial wave amplitude a_0 was set equal to 2.0 m. This value was chosen according to the field data, as well as from the calculated concepts of the heights of wind waves modelled in this work by a soliton, which in a strong storm can approach the coastline (Alekseev et al. 2012; Udovik et al. 2016). On the remaining liquid boundaries the conditions of free passage were set:

$$\frac{\partial u}{\partial t} + \sqrt{gH}\frac{\partial u}{\partial x} = 0, x = L \quad (6)$$

$$\frac{\partial v}{\partial t} - \sqrt{gH}\frac{\partial v}{\partial y} = 0, y = 0 \quad (7)$$

$$\frac{\partial v}{\partial t} + \sqrt{gH}\frac{\partial v}{\partial y} = 0, y = L \quad (8)$$

In order to exclude the influence of additional factors on the transformation of a solitary wave as it propagates within the computational domain, the basin depth was set constant ($d(x, y) = 5.0$ m), and bottom friction, vertical and horizontal viscosity were not taken into account in the calculations.

3 Results of Numerical Experiments

Depending on the amplitude of the incident wave and the obstacle size, various variants of the transformation of the soliton above the underwater breakwater are possible. In this regard, in Kotelnikova et al. (2018) a criterion for the interaction of a wave with an obstacle was proposed: $K_{\mathrm{int}} = a/(d - h)$. Here a is the amplitude of the soliton, $(d - h)$ is the depth above the structure. In this work, nine numerical experiments were carried out: for three variants of the interaction criterion ($K_{\mathrm{int}} =$ 0.8, 1.33, 4.0) and three variants of the relative width of the obstacle ($l/\lambda = 0.5$, 1.0, 1.5). The height of the breakwater was $h = 2.5$, 3.5, and 4.5 m, and the width was l = 8, 16 and 24 m.

At the initial time ($t = 0$), the fluid in the basin is in the unperturbed state. At $t >$ 0, a solitary wave enters the computational domain through the left boundary ($x =$ 0) and propagates in the positive direction of the x-axis.

Based on preliminary calculations, the simulated time interval in numerical experiments was chosen equal to 90 s. The entire propagation process of a solitary wave in the model basin can be divided into three stages: the entrance of the soliton into the computational domain and approaching the breakwater, the interaction of the wave with the obstacle, propagation of the transmitted transformed wave behind the breakwater and its exit through the right boundary of the basin (Fig. 2).

At the first stage, when approaching the breakwater, the solitary wave retains its profile. The wave is followed by a dispersion tail (Fig. 2a). From about the 40th second, the soliton begins to interact with the obstacle. In this case, the wave is divided into transmitted and reflected. The reflected wave moves in the opposite direction, to the left boundary, and the passing one is transformed when interacting with the coastal protection structure, its amplitude increases (Fig. 3b—line with rhombuses).

Behind the breakwater, the profile of the passing wave is restored (Fig. 3b—a line with triangles), but its three-dimensional structure changes. At the moment of interaction, the transmitted wave has 2 local peaks near the edges of the obstacle (Fig. 2b), which gradually approach each other and subsequently overlap each other (Fig. 2c). After the expiration of the model time interval, the transmitted wave leaves the computational domain through the right boundary of the basin (Fig. 2d).

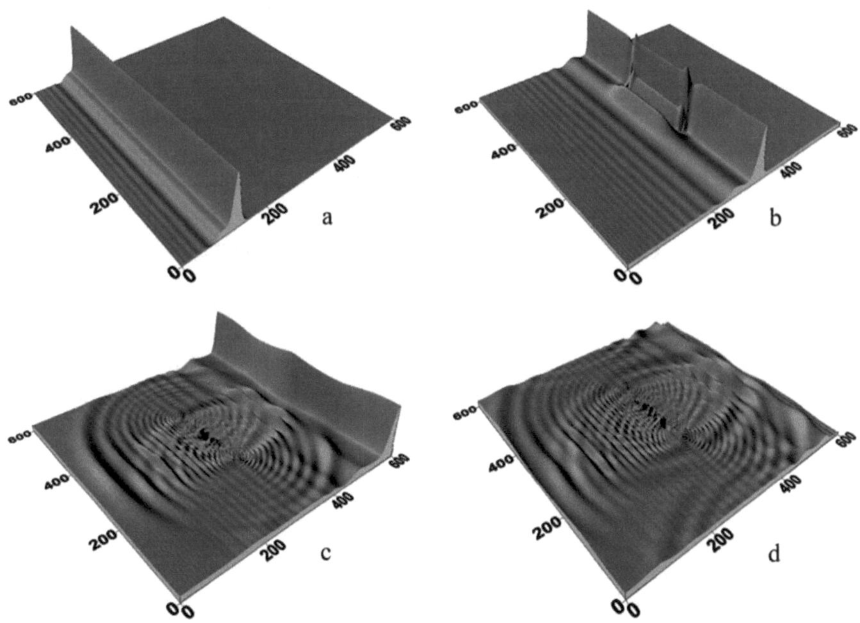

Fig. 2 Position of the soliton at times $t = 20$ s **a**, $t = 44$ s **b**, $t = 75$ s **c** and $t = 85$ s **d** in the case of an underwater breakwater with a relative width $l/\lambda = 1.0$ and height $h/d = 0.7$

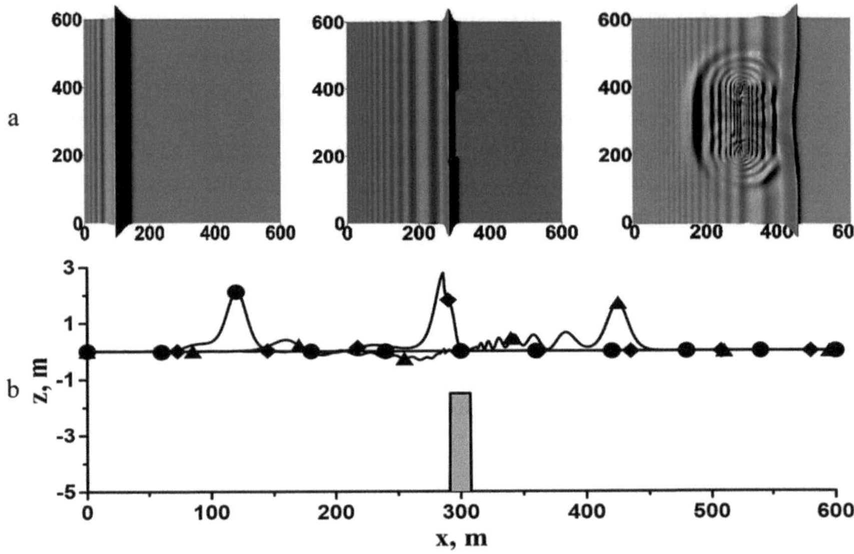

Fig. 3 Three-dimensional fields of the free surface **a** (top view) and profiles **b** of the free surface ($y = 300$ m) at times $t = 20$ s (circles), $t = 40$ s (rhombuses) and $t = 58$ s (triangles) during the propagation of a solution over a submerged breakwater with a relative width $l/\lambda = 1.0$ and a height $h/d = 0.7$

4 Influence of the Width and Height of the Obstacle on a Soliton Transformation

The estimate of the effect of the coastal protection structure width and height on the wave pattern formed by the passing soliton was carried out by calculating the transformation coefficients for each point of the model basin:

$$K_{tr} = a/a_0, \tag{9}$$

where a is the value of the maximum deviation of the free surface level from its unperturbed state, found during the time from the moment the soliton interacts with the obstacle until the end of the simulation time interval, and a_0 is the amplitude of the incoming wave.

The interaction of the soliton with the breakwater leads to the formation of three transformation zones in the model basin—in front of, behind and above the obstacle (Fig. 4). In front of the breakwater there is an area in which the wave reflected from the coastal protection structure propagates. The analysis of the simulation results showed that the amplitude of the reflected wave in the immediate vicinity of the breakwater strongly depends on the height of the hydraulic structure. As the relative height of the obstacle h/d decreases from 0.9 to 0.5, the amplitude of the reflected

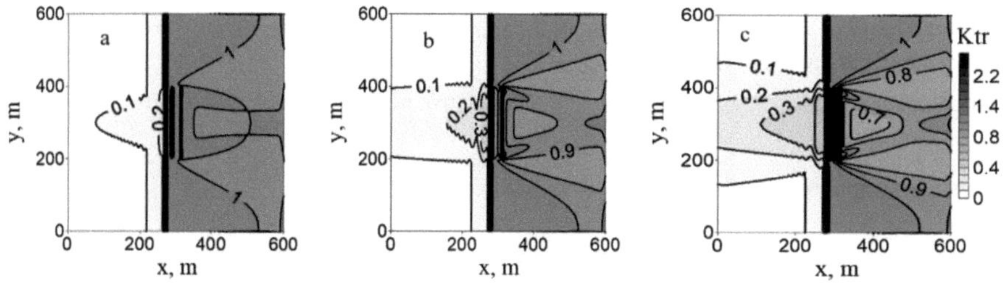

Fig. 4 Fields of transformation coefficients for the case with a breakwater with a relative width l/λ = 1.5 for three variants of the relative height of the obstacle: a—h/d = 0.5, b—h/d = 0.7, c—h/d = 0.9

wave gradually decreases from 0.8–0.9 m (K_{tr} = 0.4–0.45) to 0.4–0.46 m (K_{tr} = 0.2–0.23), which is in good agreement with analytical formulas (Lamb 1945). Variation in the width of the breakwater has little effect on the reflected wave.

Above the breakwater there is an area in which the incident wave is locally amplified with its subsequent separation into reflected and transmitted waves. With an increase in the relative height of the obstacle, the amplitude of the incident wave increases, reaching 3.6 m (K_{tr} = 1.8) at h/d = 0.9 and l/λ = 1.5. The width of the obstacle also affects the amplitude of the incident wave. In particular, in the case of a breakwater with a relative height of h/d = 0.9, an increase in the relative width of the obstacle from l/λ = 0.5 to 1.5 leads to an increase in the amplitude by 14% (from 3.1 to 3.6 m).

The analysis of the fields of transformation coefficients showed that the maximum attenuation of the transmitted wave is achieved in a small-sized area located at a distance of 150–200 m behind the breakwater (Fig. 4b, c). An increase in the height of the coastal protection structure leads to a decrease in the amplitude of the transmitted wave from 1.80–1.90 m (K_{tr} = 0.9–0.95, h/d = 0.5) to 1.30–1.40 m (K_{tr} = 0.65–0.7, h/d = 0.9) (Fig. 4, Table 1). At the same time, there is an increase in the area of the zone of weakening by 31% (Table 2). A change in the width of a hydraulic structure has practically no effect on the amplitude and magnitude of the zone of attenuation of the passing wave (Tables 1 and 2).

Table 1 Minimum values of K_{tr} values for the region of attenuation of the transmitted wave at various geometric parameters of the underwater obstacle

Minimum K_{tr} values		The relative height of the obstacle, h/d		
		0.5	0.7	0.9
Relative width of the breakwater, l/λ	0.5	0.95	0.88	0.70
	1.0	0.94	0.86	0.67
	1.5	0.92	0.84	0.64

Table 2 Relative area of soliton attenuation behind the breakwater in % for various geometric parameters of the underwater obstacle

Attenuation area, %		The relative height of the obstacle, h/d		
		0.5	0.7	0.9
Relative width of the breakwater, l/λ	0.5	52.0	65.0	68.5
	1.0	51.1	64.3	68.4
	1.5	50.3	64.0	68.4

Comparison with the results of works by other authors (Kotelnikova et al. 2012, 2018; Lin 2004) showed good qualitative agreement in the evolutionary dynamics of the soliton. However, for a transmitted wave, the values of the transformation coefficients are noticeably higher than the transmission coefficients. This discrepancy may be due to the difference in setting the initial conditions for a solitary wave.

5 Wave Velocity Field

Before the interaction of the soliton with the obstacle, the depth-averaged orbital velocities of the liquid are observed in the region of the head wave and its dispersive tail (Fig. 5a). The direction of the velocity vectors until the moment of interaction of the wave with the obstacle at all points coincides with the positive direction of the x-axis. When approaching the hydraulic structure, there is a sharp increase in wave velocities near the obstacle, which reach the highest values above the crest of the breakwater and in the areas of the end zones ($V = 5.4$–6.0 m/s at $l/\lambda = 1.5$ and $h/d = 0.9$) (Fig. 5b).

The passing of the soliton over the breakwater top is accompanied by the separation of the incident wave into transmitted and reflected. This leads to changes in

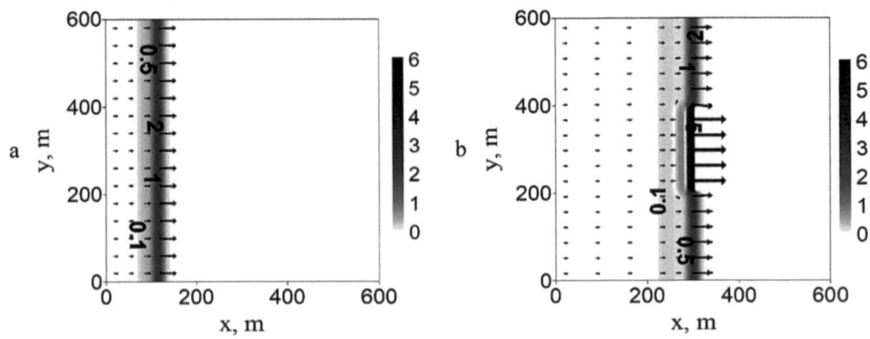

Fig. 5 Values (black-and-white gradient) and directions (arrows) of the wave velocity vectors in the experiment with a breakwater of relative width $l/\lambda = 1.5$ and height $h/d = 0.9$ for times $t = 20$ and 42 s

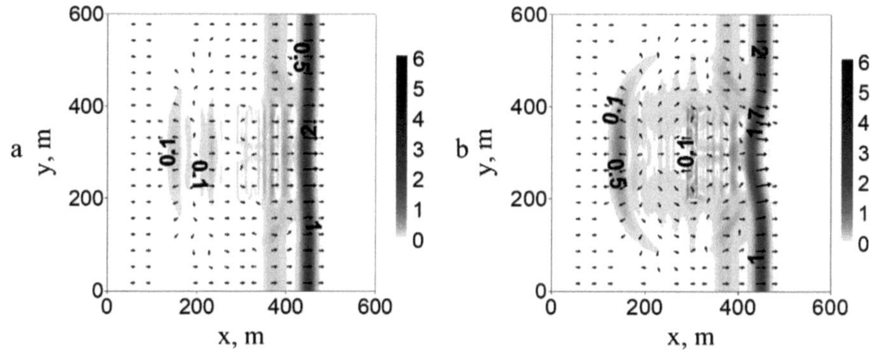

Fig. 6 Values (black-and-white gradient) and directions (arrows) of the vectors of the depth-averaged orbital velocities of the liquid for the time instant $t = 60$ s in experiments with the breakwater $l/\lambda = 1.5$: a—$h/d = 0.5$, b—$h/d = 0.9$

the nature of the distribution of the average fluid velocities. The direction of the vectors in front of the breakwater is reversed and becomes negative. In addition, due to the radial nature of the propagation of reflected waves, a noticeable y-component appears in the velocity vectors (Fig. 6b). Behind the obstacle, the wave velocities retain its sign. The most complex picture is observed in the central part of the basin. The field of depth-averaged fluid velocities in this region is determined by the mutual superposition of reflected waves and is characterized by a significant inhomogeneity of directions.

Analysis of the results of numerical calculations showed that the height of the breakwater has a significant effect on the magnitude of wave velocities. The interaction of a soliton with an obstacle in the central part of the passing wave leads to a decrease in the maximum values of the averaged orbital velocities of the liquid from 2.30–2.35 m/s to 1.70–1.75 m/s with an increase in the height of the breakwater to the maximum value (Fig. 6). In addition, the picture of the velocity distribution becomes more complex. Numerical experiments have shown the absence of a noticeable effect of the width of a hydraulic structure on the value and direction of fluid velocities.

6 Conclusions

A numerical simulation of the interaction of a solitary surface wave with an underwater rectangular breakwater in a model basin of constant depth was carried out using the SWASH non-hydrostatic hydrodynamic model. It is shown that the interaction of a soliton with an obstacle in all cases ($K_{\text{int}} = 0.8, 1.33, 4.0$) leads to the separation of the incident wave into reflected and transmitted ones. The dependences of the amplitude characteristics of both reflected and transmitted waves on the height of the coastal protection structure are obtained. With an increase in the relative height of the breakwater from $h/d = 0.5$ to 0.9, the amplitudes of the reflected and transmitted waves decrease (by 50 and 25%, respectively). The area of the attenuation zone of the transmitted wave increases by 31%. Changing the width of the hydraulic structure does not significantly affect the transformation of the soliton. Analysis of the fields of transformation coefficients has established that the maximum attenuation of a solitary wave is achieved in a small-sized area located behind the breakwater at a distance commensurate with its length. The calculation of the depth-averaged orbital velocities of the fluid showed that when the soliton interacts with the breakwater, the values and directions of the wave velocities change. The noticeable effect of the height of the obstacle on the distribution pattern and values of fluid velocities is revealed. The effect of the width of the coastal protection structure, as in the case of the amplitudes, is minimal.

The work was carried out within the framework of the state assignment on the topic No. 0827–2019-0004 "Comprehensive interdisciplinary studies of oceanological processes which determine the functioning and evolution of ecosystems in the coastal zones of the Black and Azov Seas" (code "Coastal Research").

References

Alekseev DV, Fomin VV, Ivancha EV, Kharitonova LV, Cherkesov LV (2012) Mathematical modeling of wind waves in the Sevastopol Bay. Mar Hydrophysical J 1:75–84

Kononkova GE, Pokazeev KV (1985) Dynamics of sea waves. Moscow State University Publishing House, Moscow, p 298

Korzinin DV (2017) Transformation of waves over a submerged bar according to the data of physical and mathematical modeling. Ecol Saf Coast Shelf Zones Sea. 4:14–21

Kotelnikova AS, Nikishov VI, Srebnyuk SM (2018) Experimental study of the interaction of a solitary surface wave with an underwater step. Hydrodyn Acoust 1(1):42–52

Kotelnikova AS, Nikishov VI, Srebnyuk SM (2012) Interaction of surface solitary waves with underwater obstacles. Reports of the NAS of Ukraine 7:54–59

Lamb H (1945) Hydrodynamics. Dover, New York, p 760

Lin P (2004) A numerical study of solitary wave interaction with rectangular obstacles. Coast Eng 51:35–51

Mikhailichenko SY, Ivancha EV (2017) Numerical modeling of the interaction of surface gravity waves with a single breakwater. Environ Saf Coast Shelf Zones Sea 4:22–29

Pelinovsky EN (1996) Tsunami wave hydrodynamics. N. Novgorod: IAP RAS, p 276

Repetin LN, Belokopytov VN, Lipchenko MM (2003) Winds and waves in the coastal zone of the southwestern part of Crimea. Environ Saf Coast Shelf Zones Integr Use Shelf Resour 9:13–28

SWASH User Manual (2012). The SWASH team. Delft University of Technology, Netherlands 1(10A), p 91

Udovik VF, Mikhailichenko SY, Goryachkin YN (2016) On a possible way to solve the problem of protecting the shores of the reserve "Chersonesus Tauric". Mar Hydrophysical J 2:27–37

Wang J, He G, You R, Liu P (2018) Numerical study on interaction of a solitary wave with the submerged obstacle. Ocean Eng 158:1–14

Wind Waves Modeling in Polar Low Conditions Within the WAVEWATCH III Model

A. M. Kuznetsova, E. I. Poplavsky, N. S. Rusakov, and Yu. I. Troitskaya

Abstract In the work, the preliminary results polar low simulation is presented. For the selected polar low, wind waves on ice-free water using the WAVEWATCH III model are simulated and a conclusion about the optimal set of wind input source term parameterization is made. The effect of ice in the model is taken into account in a number of ways. The use of the simplest models IC0, IS0 and IC1, IS1 demonstrates similar results. They allow one to take into account the presence of ice, but are not adapted to take into account its destruction. Simulation of ice destruction by waves during an intense Arctic storm is implemented using the WAVEWATCH III model with the IS2 module. The simulation results show the destruction and subsequent propagation of waves into the area containing ice.

Keywords Numerical simulation · Wind wave modeling · Polar low · Barents sea · WAVEWATCH III

1 Introduction

The most dangerous marine weather phenomena are considered to be tropical cyclones, which arise and develop over the oceans, mainly in the tropical zone, where wind speeds can exceed 70 m/s. They can have a significant impact on the temperate and subtropical zones due to the extratropical penetration of tropical cyclones . At high latitudes, polar hurricanes, that are intense, rapidly developing atmospheric vortices, similar to tropical cyclones in their formation mechanisms and some morphological features, are observed. The reduction of ice cover in the Arctic over the past decade has resulted in polar hurricanes occurring along the entire Northern Sea Route. Wind speeds in them reach 35–40 m/s, posing a threat to shipping and oil production on the Arctic shelf.

Over the past 20 years, significant progress made in predicting storm trajectories, while the quality of storm intensity forecasting remains poor. This is due to the fact that the intensity (maximum wind speed and minimum pressure in the hurricane)

A. M. Kuznetsova (✉) · E. I. Poplavsky · N. S. Rusakov · Yu. I. Troitskaya
Institute of Applied Physics RAS, Nizhny Novgorod, Russia

T. Chaplina (ed.), *Processes in GeoMedia—Volume IV*, Springer Geology,
https://doi.org/10.1007/978-3-030-76328-2_18

is determined by the interaction of the atmosphere and the ocean, which at high wind speeds has significant uncertainty, especially for the smallest-scale processes: splashes, wave breaking and foam bubbles. Then, hurricane conditions are characterized by very young developing waves, which in its characteristics differ significantly from the developed ones.

Before, the test case with extreme waves in the hurricane conditions on the example of Irma hurricane was considered in WAVEWATCH III (WW3). It was shown that WW3 model reproduces the rotating and motion of the test case of the tropical cyclone Irma. Using the statistics of events associated with the formation of splashes, the contribution of splashes to the exchange of the atmosphere and ocean was estimated. A decrease in the coefficient of surface resistance in the region of strong winds was obtained. The asymmetry of the distribution of surface resistance in the direction of movement of the center of the storm reflected the asymmetry of the distribution of the wind and wave fields. The importance of taking into account the effect of spray and foam in the wind input source term of the wave model for the case of high winds in hurricane conditions was shown.

It can be expected that, for the case of high winds of polar low, taking into account the effect of spray and foam in the function of the wind input source term will also be important. However, first of all, it is necessary to choose the optimal parameterizations of the source function, including taking into account the effects of the presence of ice. Thus, in order to develop the quality of forecasting waves during a polar low, wind waves modeling is carried out in the WW3 model.

2 Methods of Wind Waves Modeling

To simulate wind waves under polar low conditions, the water area of the Barents Sea, where a large number of polar hurricanes are observed, is selected. Among the identified polar hurricanes (Noer and Lien 2010), a hurricane that took place on 02/05/2009 and is observed at the coordinates of 69° N, 40° E, is chosen. The modeling is carried out within the WW3 model (The WAVEWATCH III R Development Group (WW3DG) 2016) (Fig. 1).

A rectangular computational grid of the studied water area is considered, it is assumed that the change in curvature is small at small changes in latitudes (66.5°–71° N, 30°–50° E). When calculating waves in the wave model, the following set of parameterizations is used: schemes ST4 (Ardhuin et al. 2010) and ST6 (Rogers et al. 2012; Zieger et al. 2015) for wind input source term (widely used by the world scientific community); DIA (Discrete Interaction Approximation) scheme for calculating nonlinear interactions; JONSWAP scheme to take into account the influence of bottom friction. The spectral resolution of the model is 24 directions, the frequency range is 32 intervals from 0.0373 to 0.787 Hz. The total time step for integrating the full wave action equation is 150 s, the time step for integrating the functions of sources and sinks of wave energy is 2 s, and the time step for transferring energy across the spectrum is 25 s. A number of parameters are considered to take into account the presence of ice.

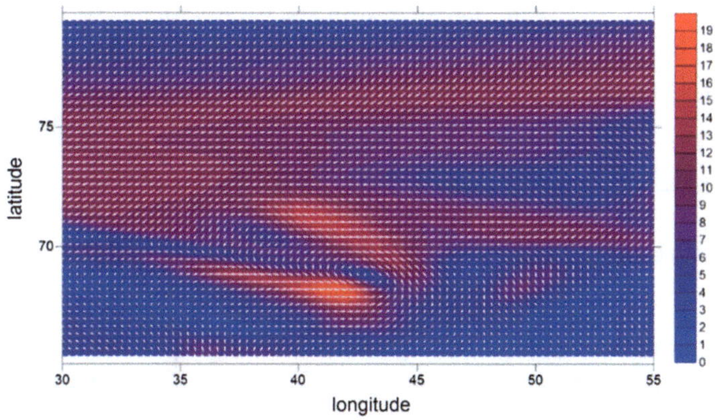

Fig. 1 Wind speed distribution in the studied area from CFSR reanalysis, m/s

3 Results

First of all, wind waves in conditions of polar low are simulated on ice-free water. Analysis of the resulting distribution of significant wave heights using the ST4 and ST6 wind input source terms shows that the ST6 parameterization gives higher significant wave heights than the ST4 parameterization. Parameterization ST6 is Babanin et al. parameterization, BYDRZ (Rogers et al. 2012; Zieger et al. 2015), tested in many cases including in stormy conditions of the North Sea (Vledder et al. 2016). However, as it was shown in Liu et al. (2020), ST6 parameterization was obtained in a specific localization, and has large errors in calculations in the Arctic region. Therefore, ST4 parameterization is selected for the further use (Fig. 2).

The next step is to take into account the influence of ice in the model. In WW3, it can be implemented by using a number of parameterizations: damping processes, activated by switches IC0, IC1, IC2, IC3 and IC4, which can be combined with various variants of the diffusive scattering effects of the sea ice IS0, IS1 and IS2. The IC0 scheme is the simplest scheme for taking into account the effect of ice, where, when the ice concentration is more than 25%, the grid node is considered to be covered with ice, and the wave energy decays exponentially in this node (The WAVEWATCH III R Development Group (WW3DG) 2016). Parameterization IC1 is developed by E. Rogers and S. Zieger, IS1—by S. Zieger. In parameterization IC1, the parameter $C_{ice,1}$ is controlled which is responsible for the exponential decay factor of the waves. In the IS1 parameterization, it is assumed that the size of the ice floe is less than the grid step, and the fraction of the incoming wave energy α_{ice} is scattered isotropically. At each discrete frequency and direction, the wave energy is reduced by α_{ice}, and is redistributed to all directions at the same frequency to conserve energy. The considered ice distribution is shown in Fig. 3.

The simulated significant wave heights in the presence of ice using the parameterizations IC0, IS0 and IC1, IS1 show similar results. They allow one to take into

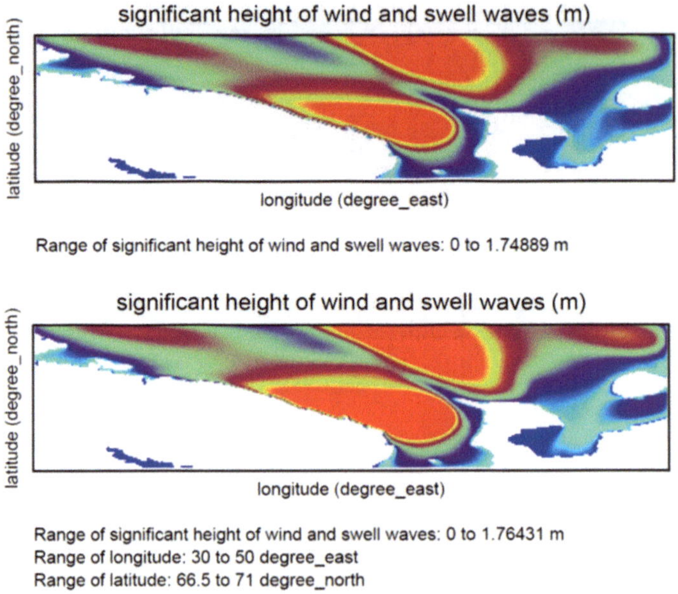

Fig. 2 Distribution of significant wave heights when using the parameterization of wind input ST4 (top), when using the parameterization of wind input ST6 (down)

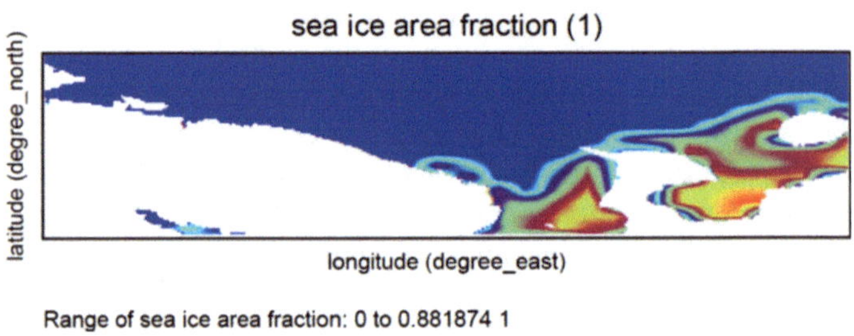

Fig. 3 Distribution of ice concentration

account the presence of ice, but are not adapted to take into account its destruction (Fig. 4).

Simulation of ice destruction by waves during an intense Arctic storm is implemented using the WW3 model with the IS2 module switched on. This module describes the dissipation of waves in the presence of floating ice and is based on the approach of Meylan and Masson (2006), where an estimate of the destruction of ice by waves is added, due to which it became possible to change the maximum size of ice floes. The estimation of the average diameter of ice floes is based on the assumption about the power law of ice size distribution. When estimating the minimum and maximum sizes of ice floes, it is assumed that destruction by waves leads to the formation of ice floes with a size equal to half the wavelength.

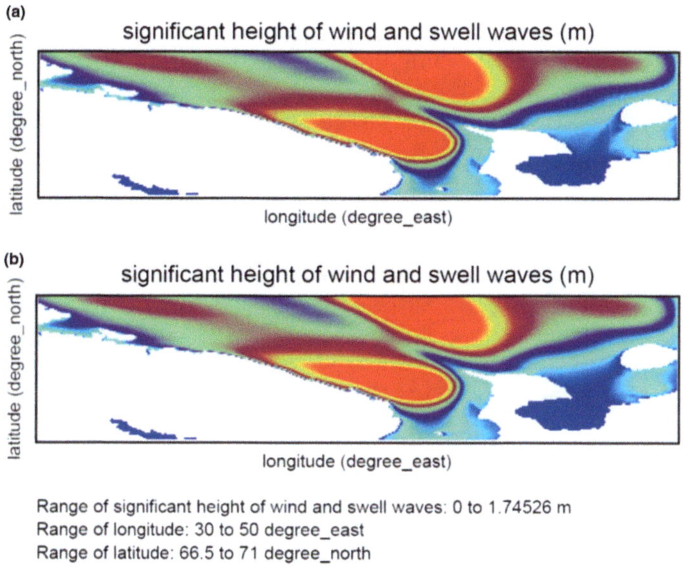

Fig. 4 Distribution of significant wave heights **a** using the parameterizations IC0, IS0, **b** using the parameterizations IC1, IS1

This method is applied together with the IC2 module, where accounting of sea ice requires solving a new dispersion relation equation (The WAVEWATCH III R Development Group (WW3DG) 2016). In Fig. 5 the change of significant wave heights in time is shown. The destruction and subsequent propagation of waves into the area containing ice are demonstrated.

4 Methods of the Results Verification

Satellite altimetry is an effective way to study dynamic processes occurring in the seas of the Arctic shelf including the Barents Sea. Its advantage is the possibility of all-weather monitoring and obtaining information about the surface level throughout the entire water area. At the same time, the change in sea level is closely related to wind-wave processes occurring in the near-surface layer. Analysis of remote satellite sensing data shows that for the considered polar hurricane, there are available measurements from the QuickSCAT scatterometer (October 27, 1999–November 22, 2009). Its second-level processing products provide information on the reconstructed wind speed and wind direction with a spatial resolution of 12.5 km by 12.5 km (Fig. 6). For the further analysis, data from the ERS-2 and Envisat satellites will be used for the polar lows until 2012, and for the later cases, data from the SARAL/AltiKa satellite, as well as, from 2018, the innovative CFOSAT mission.

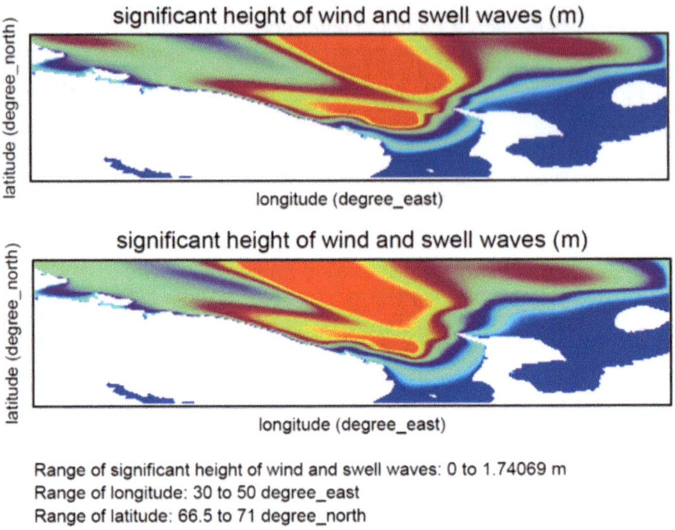

Fig. 5 Distribution of significant wave heights using the parameterization of ice accounting IC2, IS2 at the time t = 1 (top) and t = 2 (down)

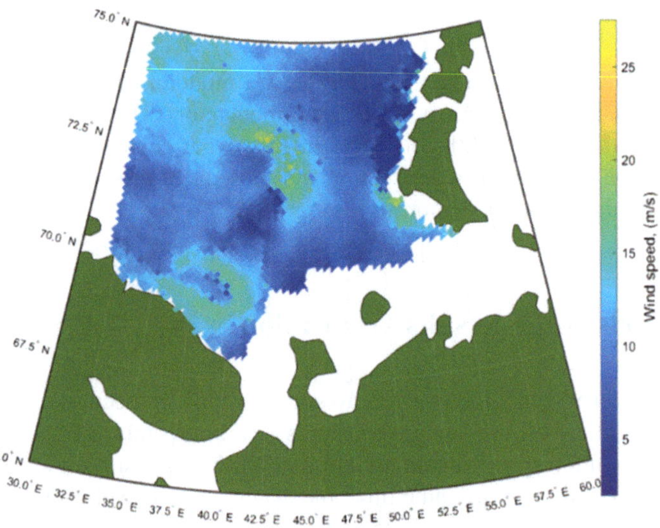

Fig. 6 Data from the QuickSCAT scatterometer covering the Barents Sea region during the polar hurricane (05.02.2009)

5 Conclusions

Moving on to the conclusions, it is worth noting that this work provides preliminary results of the studied of a test case of polar low. For the chosen polar low, wind waves on ice-free water using the WAVEWATCH III (WW3) model are simulated and a conclusion about the optimal set of model parameterizations is made. The effect of ice in the model is taken into account in a number of ways. The use of the simplest models IC0, IS0 and IC1, IS1 demonstrates similar results. They make it possible to take into account the presence of ice, but are not adapted to take into account its destruction. Simulation of the destruction of ice by waves during an intense Arctic storm is implemented using the WW3 model with the IS2 module switched on. The simulation results show the destruction and subsequent propagation of waves into the area containing ice. The following experiments are planned using the proposed methodology for simulating waves in a selected polar low, for which altimetry data will be available. Altimetry measurements will be used to verify the data of model simulations. To analyze the height of the surface of the Barents Sea, data from the ERS-2 and Envisat satellites, and SARAL/AltiKa satellite as well as data from the innovative CFOSAT mission, will be used.

Acknowledgements The work on assessment of the model sensitivity to the ice parameterizations is supported by the RFBR grant 18-05-60299. A.M.Kuznetsova acknowledges RSF grant 21-77-00076 for the support of the modeling of the polar low development.

References

Ardhuin F, Rogers E, Babanin A, Filipot J-F, Magne R, Roland A, Van der Westhuysen A, Queffeulou P, Lefevre J-M, Aouf L, Collard F (2010) Semi-empirical dissipation source functions for ocean waves: part I, definitions, calibration and validations. J Phys Oceanogr 40:1917–1941

Liu Q, Rogers WE, Babanin A, Li J, Guan C (2020) Spectral modeling of ice-induced wave decay. J Phys Oceanogr 50(6):1583–1604

Meylan MH, Masson D (2006) Physics-based parameterization of air-sea momentum flux at high wind speeds and its impact on hurricane intensity predictions. Ocean Mod 11:417–427

Noer G, Lien T (2010) Dates and positions of polar lows over the Nordic Seas between 2000 and 2010. Norwegian Meteorological Institute Rep

Rogers WE, Babanin AV, Wang DW (2012) Observation-consistent input and whitecapping dissipation in a model for wind-generated surface waves: description and simple calculations. J Atmos Oceanic Tech 29(9):1329–1346

The WAVEWATCH III R Development Group (WW3DG) (2016): User manual and system documentation of WAVEWATCH III R version 5.16. Tech. Note 329, NOAA/NWS/NCEP/MMAB, College Park, MD, USA, p 326, + Appendices

van Vledder GP, Hulst ST, McConochie JD (2016) Source term balance in a severe storm in the Southern North Sea. Ocean Dyn 66(12):1681–1697

Zieger S, Babanin AV, Rogers WE, Young IR (2015) Observation-based source terms in the third-generation wave model WAVEWATCH. Ocean Model 1(96):2–5

Granular Geomaterials: Poroperm Properties-Stress Dependence by Unsteady Permeability Tests

Leonid Nazarov, Larisa Nazarova, and Nikita Golikov

Abstract The method for correlation of poroperm properties and effective stresses in weakly coherent rocks has been developed and tested on a lab scale. An appropriate lab testing installation is designed and manufactured; it includes a hydraulic press, a system of force plungers, a polyurethane measurement cell to be filled with sized sand, a compressor and a high-precision autonomous measurement system for fluid flow rate and pressure. The method consists in sequential steady-state and unsteady-state flow tests of a granular geomaterial sample placed in the measurement cell where nonuniform stress state generated by application of the external load to the cell. In the unsteady-state flow tests, the change in pressure was measured during gas release from the cell. In the steady-state flow tests, the flow rates were measured at the varied input pressure. The measured flow rates and pressures were used as the given data in the inverse coefficient problems on determinations of the empirical parameters in the exponential dependence of porosity, permeability on effective stress. Resolvability of the formulated inverse problems within the nonlinear models of mass transfer in a granular media is demonstrated.

Keywords Poroperm properties · Lab test · Granular medium · Non-linear model · Mass transfer · Stress · Pressure · Inverse problem

1 Introduction

Interpretation of geophysical logging data, improvement of reservoir prediction, reduction in geological exploration uncertainty, feasibility studies of technologies for accessing underground mineral resources and reservoir engineering—these and many other challenges require information on poroperm properties of producing

L. Nazarov (✉) · L. Nazarova
Chinakal Institute of Mining, Siberian Branch of the Russian Academy of Sciences, 54 Krasny Prospekt, Novosibirsk, Russia 630091

N. Golikov
Trofimuk Institute of Petroleum Geology and Geophysics, Siberian Branch of the Russian Academy of Sciences, 3 Prospekt Koptyuga, Novosibirsk, Russia 630090

T. Chaplina (ed.), *Processes in GeoMedia—Volume IV*, Springer Geology,
https://doi.org/10.1007/978-3-030-76328-2_19

formations (Corbett 2012; Hsu and Robinson 2019). Producing intervals in crude oil or gas condensate reservoirs are as a rule composed of noncoherent rocks represented by detrital weakly cemented soil or loose sandstone (Selly and Sonnenberg 2015; Borisov and Frolova 2014). The reservoir characteristics of the latter are governed by particle size distribution, by their packing and cementation (Chapuis 2012; Sokolov 2004), as well as by the effective stresses and fluid pressures in a reservoir (Liu et al. 2015; French et al. 2016).

Regarding conventional reservoirs, a large bulk of data on permeability and porosity subject to stresses have been accumulated (Peng and Zhang 2007; Al Ismail and Zoback 2016; Zhang et al. 2007; Roi et al. 2018; Holt 1990; Jia et al. 2017). Concerning granular geomedia which model as a rule weakly consolidated grained reservoir, the experimental evidence is much poorer (Miao et al. 2011; Patino et al. 2019). In the meanwhile, the change in the poroperm properties in the area of vicinity of the well due to the stress redistribution in drilling, or due to the fluid pressure drop during oil-and-gas field operation has a considerable effect on the inversion of log data in quantitative evaluation of poroperm properties of reservoir rocks (Yeltsov et al. 2014; Nazarova et al. 2013; Mikhailov 1998).

This paper describes the method developed using the approach proposed in Nazarov et al. (2020) and proved on a laboratory scale for determining an empirical dependence between porosity, permeability and stresses of granular geomaterials using the data of steady-state and unsteady-state flow tests.

2 Test Installation and Procedure

The test installation (Fig. 1) represents a hydraulic press (force to 3500 N) with ram **4** and a number of force plungers **1**, **2** and **3** to apply loads to different areas of measuring cell **6**. The cell is made of a 4 mm-thick polyurethane rubber and is shaped as a parallelepiped (length $L = 9$ cm, cross-section area $W = 9$ cm^2). The end faces of the cell are covered with aluminum flanges with connecting branches **8** and **9** toward a flow rate meter and a compressor, respectively.

Fig. 1 Lab-scale test installation: **1**, **2** and **3**—plungers; **4** and **5**—ram and legs of the press; **6**—measuring cell; **7**—pan; **8** and **9**—connecting branches

2.1 Steady-State Fluid Flow

Cell was filled with clean and sized sand (fraction 0.18–0.25 mm) and placed in pan to preserve the cell shape under loading. Vertical compression was applied to the cell sites ($i = 1, 2, 3$) with a length $l = L/3$ (Fig. 2a) via the plungers at an incrementally increasing stress σ_m^i controlled by the value of press operating pressure. At each loading step m on the right-hand side of the cell, a constant air pressure p_n was generated, while on the left-hand end, a flow rate Q_{mn}^i was recorded in the steady-state fluid flow mode. The relative accuracy of the flow rate measurement was 1%. The test results are compiled in Tables 1, 2 and 3.

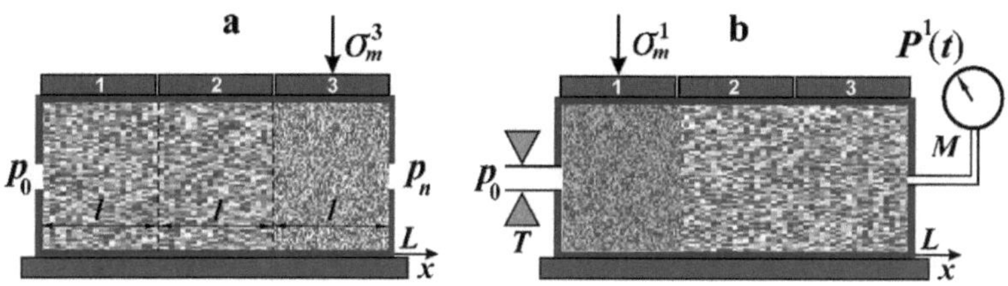

Fig. 2 Experimental design of **a** steady-state flow and **b** unsteady-state flow

Table 1 Flow rate Q_{mn}^1 (ml/min) under loading applied to site 1 ($\sigma_m^2 = \sigma_m^3 = 0$)

σ_m^1, bar	p_n, bar					
	1.05	**1.10**	**1.15**	**1.20**	**1.25**	**1.30**
5	28.1	57.5	88.4	120.7	154.4	189.5
10	26.9	55.1	84.7	115.6	147.9	181.6
20	24.4	50.1	77.0	105.1	134.5	165.1
30	21.9	45.0	69.1	94.4	120.7	148.2

Table 2 Flow rate Q_{mn}^2 (ml/min) under loading applied to site 2 ($\sigma_m^1 = \sigma_m^3 = 0$)

σ_m^2, bar	p_n, bar					
	1.05	**1.10**	**1.15**	**1.20**	**1.25**	**1.30**
5	27.0	55.4	85.1	116.2	148.6	182.4
10	24.9	51.1	78.5	107.1	137.1	168.2
20	21.0	43.1	66.2	90.4	115.7	142.0
30	17.6	36.0	55.3	75.6	96.7	118.6

Table 3 Flow rate Q^3_{mn} (ml/min) under loading applied to site 3 ($\sigma^1_m = \sigma^2_m = 0$)

σ^3_m, bar	p_n, bar					
	1.05	**1.10**	**1.15**	**1.20**	**1.25**	**1.30**
5	26.0	53.4	82.0	112.0	143.2	175.8
10	23.2	47.6	73.1	99.8	127.7	156.7
20	18.4	37.8	58.1	79.3	101.4	124.5
30	14.7	30.0	46.2	63.0	80.6	98.9

2.2 *Unsteady-State Fluid Flow*

After completing measurement of the steady-state fluid flow rate at the maximal value of the vertical stress $\sigma^i_m = 30$ bar, the measurement cell was pressurized up to the constant overpressure $p_* = 1.5$ bar. Then, the tap $\boldsymbol{T}$ on the left-hand side of the cell was turned on (Fig. 2b), and on the right-hand side at times $t_j = j\tau$ the pressure $P^i\left(t_j\right)$ was recorded (by the Keller Leo Record Digital manometer $\boldsymbol{M}$) at an absolute precision of 1 mbar, at a preset sampling frequency $\tau = 0.05$ ms. Such experimental design simulates well performance in case of limited reservoir energy (Cholovskii et al. 2002).

Figure 3 demonstrates the experimental curves $P^i(t)$ at the load $\sigma^i_m = 30$ bar applied to different sites of the measurement cell. Apparently, when loaded sites are located nearer to the outlet, the pressure in the cell is higher; gas outflow time is around 3 s. Table 4 presents the selected experimental data.

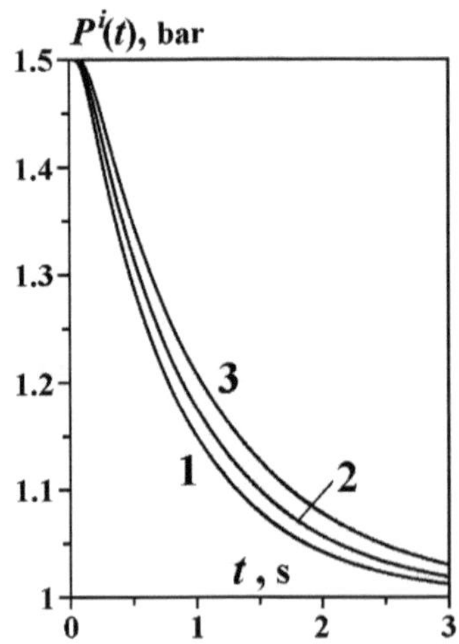

Fig. 3 Pressure variation in time on the right-hand side of the measurement cell; the figures denote the loading site number i

Table 4 Pressure $P^i(t_j)$ (mbar) on the right-hand side of the cell at $\sigma_m^i = 30$ bar

i	t_j, s								
	0.2	0.4	0.6	0.8	1.0	1.5	2.0	2.5	3.0
1	1457	1368	1296	1239	1193	1117	1070	1043	1026
2	1449	1346	1265	1204	1158	1085	1046	1025	1014
3	1435	1313	1223	1160	1115	1053	1023	1010	1005

3 Mathematical Model

The unsteady-state gas flow in the measurement cell is described using a one-dimensional model including:

mass conservation equation

$$(\omega\rho)_{,t} + (\omega V)_{,x} = 0, \tag{1}$$

Darcy's law

$$V = -kp_{,x}/\eta \tag{2}$$

and a constitutive equation

$$p = p_0\rho/\rho_0, \tag{3}$$

where p is the pressure, ρ and η are the density and viscosity of gas, respectively, V is the velocity component along the axis x (Fig. 2), t is the time, ρ_0 is the density of gas under the atmospheric pressure p_0. It is assumed that the permeability k and porosity ω are related with the effective stresses $s = \sigma_m - p$ exponentially (Zoback and Nur 1975; Nazarova et al. 2019; Nazarova and Nazarov 2018).

$$k = k_0 K(s), \quad K(s) = \exp(-\alpha s), \tag{4}$$

$$\omega = \omega_0 \Omega(s), \quad \Omega(s) = \exp(-\beta s), \tag{5}$$

where k_0 and α, ω_0 and β are the empirical constants.

System (1)–(5) may be reduced to a single equation

$$(\Omega p)_{,t} = a\left(pKp_{,x}\right)_{,x} \tag{6}$$

relative to the pressure ($a = k_0/\eta\omega_0$), for which various initial and boundary conditions are formulated in the steady-state and unsteady-state flow modeling (Table 5).

Equation (6) was solved using the unconditionally stable and second-order accurate explicit finite-difference scheme (Samarskii 2001). It should be noticed that in the steady-state flow calculated by a relaxation method, the pressure distribution is independent of porosity, viscosity and value of k_0.

Figure 4 shows the pressure profile at $p_n = 2$ bar in the steady-state fluid flow under loading of various cell sites (curves 0, 1, 2 and 3 fit $\alpha = 0, 0.2, 0.4$ and 0.6 bar^{-1}). As the stress is increased in the vicinity of the left-hand boundary of the computational domain, the gradient pressure grows (Fig. 4a) but the absolute flow rate Q decreases and can be found from the formula

$$Q(t) = \frac{W}{\eta} K(s|_{x=0}) \frac{\partial p(t, 0)}{\partial x}. \tag{7}$$

If loading is applied to site 1 (Fig. 2a), then in the steady-state flow, it follows from (6) and (7) with regard to the boundary conditions in Table 5 that (if $\alpha p \ll 1$)

Table 5 Initial and boundary conditions for Eq. (6)

	Type of experiment	
	Steady-state flow	Unsteady-state flow
Boundary conditions	$p(t, 0) = p_0,\ p(t, L) = p_n$	$p(t, 0) = p_0,\ p_{,x}(t, L) = 0$
Initial conditions	$p(0, x) = p_0$	$p(0, x) = p_*$

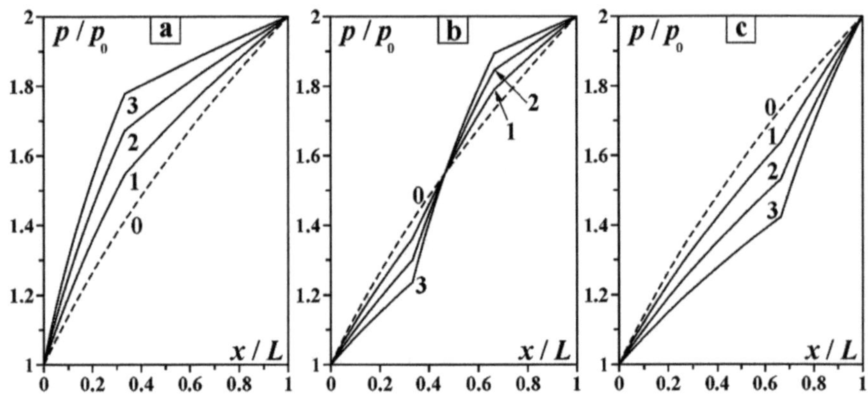

Fig. 4 Pressure distribution in the cell: $\sigma_m^1 = 30$ bar (**a**); $\sigma_m^2 = 30$ bar (**b**); $\sigma_m^3 = 30$ bar (**c**)

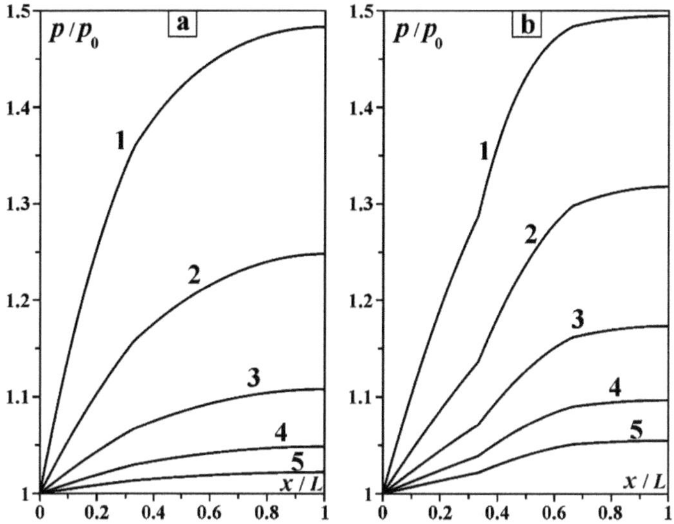

Fig. 5 Pressure profile in the cell at $\sigma_m^1 = 30$ bar (**a**) and $\sigma_m^2 = 30$ bar (**b**) at different times

$$\lim_{t\to\infty} Q(t) = Q_0\left(1 + \frac{2\alpha}{3}\frac{p^2 + pp_0 + p_0^2}{p + p_0}\right)\exp\left(-\alpha\sigma_m^1\right),\ Q_0 = \frac{Wk_0}{\eta L}\frac{p^2 - p_0^2}{2p_0}.$$

Figures 5a, b illustrate evolution of pressure in the measurement cell in the unsteady-state flow, lines 1, 2, 3, 4 and 5 at the curves fit the times t = 0.5, 1.0, 1.5, 2.0 and 2.5 s. The calculations use the values of $\eta = 0.02$ mPa s, $p_* = 1.5$ bar, $\omega_0 = 0.2$, $\beta = 0.02$ bar^{-1}, $k_0 = 150$ mD, $\alpha = 0.02$ bar^{-1}. It can be seen that at the final stage (line 5) of the experiment, the residual pressure in the cell is higher in case of the farther site (Fig. 5b) of load application relative to the outlet.

4 Interpretation of Test Data

4.1 Steady-State Flow Tests

We formulate an inverse coefficient problem as determination of the empirical parameters k_0 and α in relation (4) by the recorded steady-state flow rates Q_{mn}^i (Tables 1, 2 and 3) at the known gas viscosity $\eta = 0.018$ mPa· s (air). The objective function is chosen to be the mean squared error

$$\Phi(k_0, \alpha) = \sqrt{\frac{1}{72}\sum_{i=1}^{3}\sum_{m=1}^{4}\sum_{n=1}^{6}\left[1 - \frac{Q\left(k_0, \alpha, \sigma_m^i, p_n\right)}{Q_{mn}^i}\right]^2}$$

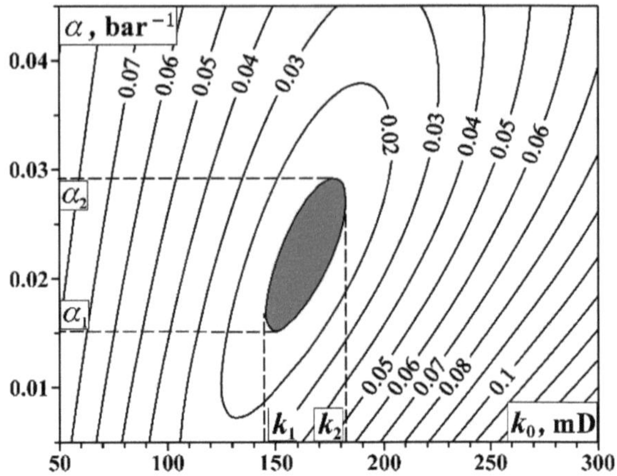

Fig. 6 Contour lines of the objective function Φ

where $Q(k_0, \alpha, \sigma_m^i, p_n)$ is the value of Q calculated at certain k_0 and α from (7) and from the steady-state solution of Eq. (6). Figure 6 shows the contour lines of Φ, which prove resolvability of the inverse problem. The shadowed region is the equivalency domain $\Phi \le 0.01$ to estimate the related ranges of the desired parameters.

$\alpha \in [\alpha_1, \alpha_2]$, $k_0 \in [k_1, k_2]$, $\alpha_1 = 0.015$ bar^{-1}, $\alpha_2 = 0.029$ bar^{-1} and $k_1 = 145$ mD, $k_2 = 181$ mD.

4.2 *Unsteady-State Flow Tests*

Let the parameters k_0 and α in relation (4) be found from the steady-state flow tests. The problem is formulated as determination of the coefficients ω_0 and β in empirical dependence (5) by the measured pressure $P^i(t_j)$ on the right-hand side of the measurement cell (Fig. 2b). We introduce an objective function

$$\Psi(\omega_0, \beta) = \sqrt{\frac{1}{180}\sum_{i=1}^{3}\sum_{j=1}^{60}\left[1 - \frac{p(t_j, \omega_0, \beta)}{P^i(t_j)}\right]^2},$$

where $p(t_j, \omega_0, \beta)$ is the solution of Eq. (6) at certain values of ω_0 and β. Figure 7 demonstrates the contour lines of function Ψ calculated at $\alpha = 0.5(\alpha_1 + \alpha_2) = 0.022$ bar^{-1} and $k_0 = 0.5(k_1 + k_2) = 163$ mD. The shadowed region $\Psi \le 0.01$ is the equivalency domain to find the variation ranges of $\omega_0 \in [\omega_1, \omega_2]$ and $\beta \in [\beta_1, \beta_2]$, $\omega_1 = 0.185$, $\omega_2 = 0.195$ and $\beta_1 = 0.014$ bar^{-1}, $\beta_2 = 0.020$ bar^{-1}. Structurally, this is a ravine function, and it is therefore advisable to use special algorithms of search of the function minimum (Cea 1978; Nazarov et al. 2013).

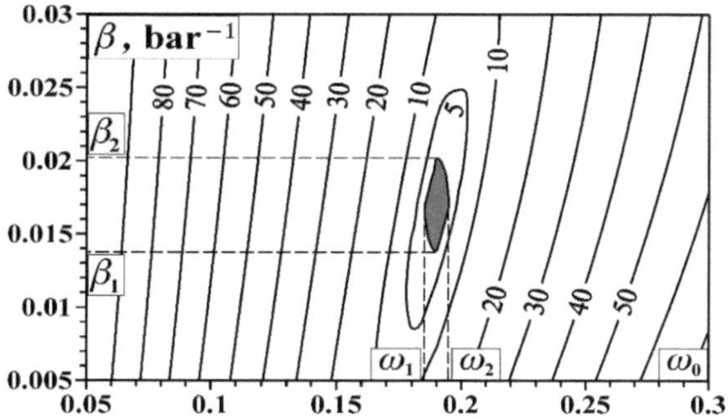

Fig. 7 Contour lines of objective function Ψ (%) at fixed values of k_0 and α

5 Conclusions

The method to find the relationship between the poroperm properties and stresses in granular and weakly coherent rocks and geomaterials has been developed and proved on a laboratory scale. The method consists in successive steady-state and unsteady-state flow tests of a geomaterial sample placed in a special measurement cell in the nonuniform stress state generated by application of external loading. The recorded flow rates and time changes of pressure are the input data for the inverse coefficient problems on determination of constants in the empirical dependences of permeability and porosity on the effective stress. The analysis of the introduced objective functions—error of flow characteristics in the measurements and calculations—proves the resolvability of the formulated inverse problems in the framework of the constructed nonlinear models of mass transfer in granular media.

The further research will be aimed to find the effect of the fractional composition and moisture content on the poroperm properties of geomaterials in the uniform and nonuniform stress state.

Acknowledgements The work was carried out with partial financial support of the Russian Foundation for Basic Research (Project No. 18-05-00830) and Program of Federal Scientific Investigations (Identification Number AAAA-A17-117122090002-5).

References

Al Ismail MI, Zoback MD (2016) Effects of rock mineralogy and pore structure on stress-dependent permeability of shale samples. Philos Trans Ser A Math Phys Eng Sci 374(2078), 20150428

Borisov AG, Frolova EV (2014) Geological and petrophysical classification model of the achime reservoir rocks in Urengoi. GAS Ind Russ 8:12–17

Cea J (1978) Lectures on optimization: theory and algorithms. Tata Institute of Fundamental Research, Bombay

Chapuis RP (2012) Predicting the saturated hydraulic conductivity of soils: a review. Bull Eng Geol Env 71:401–434

Cholovskii IP, Ivanova MM, Gutman IS, Vagin SB, Bragin YuI (2002) Field geology and hydrology of hydrocarbons. Neft gas, Moscow
Corbett PWM (2012) The role of geoengineering in field development. In: Gomes JS (ed) New technologies in the oil and gas industry. InTech, Rijeka, Croatia
French ME, Chester FM, Chester JS, Wilson JE (2016) Stress-dependent transport properties of fractured arkosic sandstone. Geofluids 16(3):533–551
Holt RM (1990) Permeability reduction induced by a nonhydrostatic stress field. SPE Form Eval 5:444–448
Hsu CS, Robinson PR (2019) Petroleum science and technology. Springer International Publishing, Cham, Switzerland
Jia C, Xu W, Wang H, Wang R, Yu J, Yan L (2017) Stress dependent permeability and porosity of low-permeability rock. J Central S Univ 24:2396–2405
Liu JF, Skoczylas F, Talandier J (2015) Gas permeability of a compacted bentonite-sand mixture: coupled effects of water content, dry density, and confining pressure. Can Geotech J 52(8):1159–1167
Miao X, Li S, Chen Z, Liu W (2011) Experimental study of seepage properties of broken sandstone under different porosities. Transp Porous Media 86(3):805–814
Mikhailov NN (1998) Alteration of physical properties of rocks in the well bore zone. Nedra, Moscow
Nazarov LA, Nazarova LA, Karchevskii AL, Panov AV (2013) Estimation of stresses and deformation properties in rock mass based on inverse problem solution using measurement data of free boundary displacement. J Appl Ind Math 7(2):234–240
Nazarov L, Nazarova L, Bruel D, Golikov N (2020) Determination of permeability–porosity–stresses dependence for loose media based on inverse problem solution by lab test data. In: Chaplina T (ed) Processes in GeoMedia II
Nazarova LA, Nazarov LA (2018) Geomechanical and hydrodynamic fields in producing formation in the vicinity of well with regard to rock mass permeability-effective stress relationship. J Min Sci 54(4):541–549
Nazarova LA, Nazarov LA, Epov MI, Eltsov IN (2013) Evolution of geomechanical and electro-hydrodynamic fields in deep well drilling in rocks. J Min Sci 49(5):704–714
Nazarova LA, Nazarov LA, Skulkin AA, Golikov NA (2019) Stress-permeability dependence in geomaterials from laboratory testing of cylindrical specimens with central hole. J Min Sci 55(5):708–714
Patino H, Martinez E, Gonzalez J, Soriano A (2019) Permeability of mine tailings measured in triaxial cell. Can Geotech J 56(4):587–599
Peng S, Zhang J (2007) Stress-dependent permeability. In: Engineering geology for underground rocks. Springer, Berlin, Heidelberg, pp 199–220
Roi R, Paredes X, Holtzman R (2018) Reactive transport under stress: permeability evolution in deformable porous media. Earth Planet Sci Lett 493:198–207
Samarskii AA (2001) The theory of difference schemes. Marcel Dekker Inc., New York
Selly RC, Sonnenberg SA (2015) Elements of petroleum geology. Academic Press, London
Sokolov BA (2004) Geology and geochemistry of oil and gas, 2nd edn. Nauka, Moscow
Yeltsov IN, Nesterova GV, Sobolev AY, Epov MI, Nazarova LA, Nazarov LA (2014) Geomechanics and fluid flow effects on electric well logs: multiphysics modeling. Russ Geol Geophys 55(5–6):775–783
Zhang J, Standifird WB, Roegiers JC, Zhang Y (2007) Stress-dependent fluid flow and permeability in fractured media: from lab experiments of engineering applications. Rock Mech Rock Eng 40(1):3–21
Zoback MD, Nur A (1975) Permeability and effective stress. Bull Am Assoc Pet Geol 59:154–158

T Retracking Skewness of the Sea Surface Elevations from Altimeter Return Waveforms

N. N. Voronina and A. S. Zapevalov

Abstract The present work is devoted to the study of the possibility of remote determination from the satellite of the skewness of the sea surface elevations. The factors limiting the possibility of restoring the skewness of the sea surface distribution by retracking altimeter waveforms are considered. We have shown the existing technique for calculating the distribution parameters could not fully take into account the effect of high wave crests on the shape of the reflected radio impulse. This leads to a distortion of the skewness values obtained from altimetry data. We have also shown the error in determining the skewness depends on the choice of the probability density function model. It also depends on the magnitude of the kurtosis of sea surface elevations.

Keywords Remote sensing satellites · Altimeter return form · Skewness of sea waves

1 Introduction

Currently, the main factor controlling the bias in altimetry measurements of the mean surface level of the sea is the change in the surface state (Gómez-Enri et al. 2006; Pokazeev et al. 2013; Sasamal et al. 2015). The change skewness of sea surface elevations leads to biases in measurements of the level (Rodriguez 1988; Zapevalov 2012; Wang et al. 2015).This is one of the main limitations for oceanographers and geophysicists in using altimeter data.

Dependence of the return waveforms on the state of the sea surface allows one to solve the inverse problem, restore the statistical characteristics of the sea surface from the data of altimetry measurements (Basu and Pandey 1991; Queffeulou 2004; Callahan and Rodriguez 2004). The first altimeter estimates of sea wave skewness demonstrated a large percentage of negative values (Gómez-Enri et al. 2007a) that does not correspond to the field measurements of waves (Jha and Winterstein 2000;

N. N. Voronina (✉) · A. S. Zapevalov
Marine Hydrophysical Institute Russian Academy of Sciences, kapitanskaya Str., Sevastopol, Russian Federation 299011

T. Chaplina (ed.), *Processes in GeoMedia—Volume IV*, Springer Geology,
https://doi.org/10.1007/978-3-030-76328-2_20

Babanin and Polnikov 1995). They have analyzed the reasons of this discrepancy in Gómez-Enri et al. (2007). Two factors were considered: the Hamming filter applied in the altimeter on-board processing and discrepancies between the real mispointing of the altimeter and the value assumed by the retracking algorithm. The analysis showed that the Hamming filter is not the responsible for the negative skewness. It has suggested that second factor plays a key role in the value of the skewness retrieved.

There are other factors that affect the measurement of the sea wave's skewness by retracking altimeter waveforms. These are the known limitations in the use of the Gram–Charlier distribution, if it includes a small number of terms (Kendall and Stuart 1945; Tatarskii 2003). It has been shown that the distribution of the Gram–Charlier describe correctly the probability density function (PDF) of the elevations of the sea surface only in a limited range (Zapevalov 2011).

$$-2.5\,\sigma_\eta < \eta < 2.5\,\sigma_\eta, \tag{1}$$

where η is surface elevation, σ_η is standard deviation of the sea surface elevation. Similar restrictions exist for the modeling of sea surface slopes (Cox and Munk 1954).

The aim of this research is the analysis of the capabilities and limitations for measuring sea wave skewness by retracking altimeter waveforms.

2 The Study Area

The determination of skewness of sea surface elevations from remote sensing data is general physical nature. Obtained here results can be used for the entire World Ocean. An exception is the region along the coast, in which reflections from the land give significant contribution to the signal of the altimeter.

3 Materials and Methods

3.1 Return Waveform

The retracking of geophysical parameters from radar altimeter signals based on the Braun model of return pulse waveform (Brown 1977; Hayne 1980). In Braun model, the return signal is a convolution of the three functions

$$V(t) = F_r(t) * s_r(t) * q(t), \tag{2}$$

where $F_r(t)$ is the average flat surface impulse response, $s_r(t)$ is the radar system point-target response, $q(t)$ is the surface specular point density function written in the altimeter's time domain.

Function $q_s(t)$ is related to the probability density function (PDF) of the sea surface elevation

$$q(t) = \frac{d\eta}{dt} P(\eta(t)), \tag{3}$$

where $P(\eta)$ is PDF of the sea surface elevation. Parameters t and η are related by $t = \eta/(c/2)$, where c is the speed of light.

3.2 Sea Surface Model

It follows from the (2) and (3) that the solution of the inverse problem (determination of the skewness) depends on how well we define the model $P(\eta)$. The sea waves are a weak nonlinear process and the distribution of the surface elevation refers to quasi-Gaussian. Commonly, the PDF of sea surface elevation $P(\eta)$ is described by the truncated Gram–Charlier distribution. When the skewness is determined according to the altimeter data, PDF is used. (Rodriguez 1988; Gómez-Enri et al. 2007a).

$$P_{G-C}(\eta) = \frac{\exp\left(-\frac{\eta^2}{2\sigma_\eta^2}\right)}{\sqrt{2\pi}\,\sigma_\eta}\left[1 + \frac{A_\eta}{6} H_3\left(\frac{\eta}{\sigma_\eta}\right)\right], \tag{4}$$

where σ_η is the standard deviation of sea surface elevation, A_η is skewness, H_3 is the Hermitian polynomial of order 3.

If the Gram–Charlier distribution contains only the first few terms, its scope is limited. The real capabilities of the simulation of the distribution of sea surface elevations using the Gram–Charlier series to analyze in paper (Zapevalov 2011). An obvious example of the limitations is the appearance of the negative values of the PDF. If only the terms of the series not greater than of the third order are used, negative values of the PDF can take place at $|\eta/\sigma_\eta| \geq 2.5$.

Function (4) contains a senior member of the Gram–Charlier series of the order of 3. This function with real values of the sea wave's skewness is shown in Fig. 1. The typical values of the skewness of the sea surface are in the range 0–0.4 (Jha and Winterstein 2000). In storm conditions, the nonlinearity of surface waves is amplified, and the asymmetry can reach a level of 0.5 (Guedes Soares et al. 2003).

The Fig. 1 shows the possibility of the appearance of negative values of the function (4), when $\eta/\sigma_\eta < -2.5$. In this case, the modeling of the return waveform holds a nonphysical effect, which is expressed in the appearance of negative values of return waveform (Pokazeev et al. 2013).

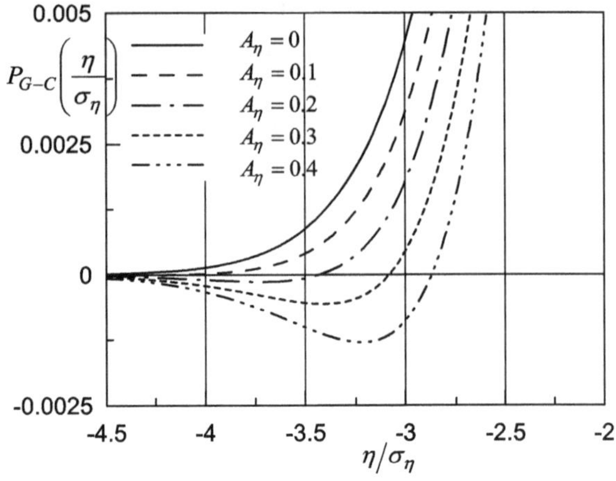

Fig. 1 Negative values in the truncated Gram–Charlier distribution (4)

3.3 The Effect of a Truncated Distribution

The problem of restoring the statistical characteristics of sea surface slopes from radio sounding data is generally similar to the problem of restoring slopes from sounding data in the optical range (Zapevalov 2018). However, there are two fundamental differences. First, since the optical wavelength is much less than the length of any of the sea waves present on the surface, all waves create a specular reflection of light. When sounding in the radio range, there are always waves on the surface whose length is comparable to or less than the length of a radio wave, these waves create diffuse scattering. Second, specular reflection of light occurs at all angles of incidence, while when sounding the sea surface at angles exceeding 20–25°, the resonant scattering mechanism begins to dominate.

The model of the reflected radio impulse corresponds to the truncated distribution of sea surface elevations if the formula (4) uses PDF. This means that regions near the crests of high waves, making a significant contribution to the asymmetry of the full distribution of elevations, are excluded from consideration. The model of the probability density of the sea surface elevations in describing the shape of the reflected radio impulse represented as

$$P^b(\eta) = \begin{cases} 0 & npu\ \eta \le b\,\sigma_s \\ N_b P(\eta) & npu\ b\,\sigma_s < \eta < b\,\sigma_s \\ 0 & npu\ \eta \ge b\,\sigma_s \end{cases} \qquad (5)$$

where the dimensionless parameter b defines the region of values of the surface elevations forming the reflected pulse; N_b is normalizing factor. The shape of the reflected radio impulse, calculated from the truncated distribution of the surface elevations, will be denoted as $V^b(t)$, calculated from the non-truncated distribution—as $V^\infty(t)$.

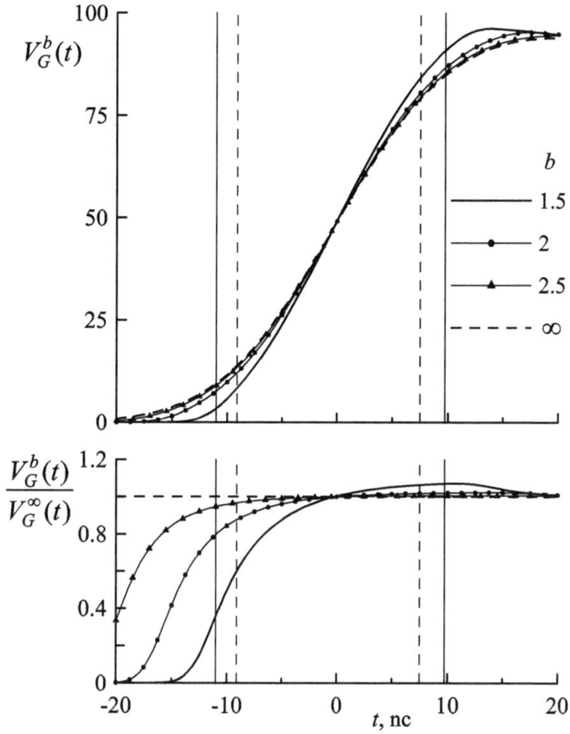

Fig. 2 Altimeter return waveforms, calculated from the models $P_G^b(t)$

Let us consider how the changes in the parameter b affect altimeter return waveform. We start with the case when the field of surface waves is linear. In this case, we can describe the elevation of the sea surface by the Gauss distribution (Longuet-Higgins 1957). The functions $V^b(t)$ obtained for different values of the parameter b are shown in Fig. 2. We assume the temperature noise can be neglected.

The calculations were made with the following values of the parameters determining the shape of the returned waveform: $\theta = 1.6\,°$, $\sqrt{D_r} = 1.327$ ns, $h = 8 \times 10^5$ m. The indicated values correspond to the parameters of the altimeter installed on the spacecraft SEASAT-1 (Mac-Arthur 1976). The vertical lines in Fig. 2 shows the areas inside of which the function $V_G^b(t)$ varies within $0.1 \ \max\left(V_G^\infty\right) \le V_G^\infty(t) \le 0.9 \ \max\left(V_G^\infty\right)$—solid line; within $0.15 \ \max\left(V_G^\infty\right) \le V_G^\infty(t) \le 0.85 \ \max\left(V_G^\infty\right)$—dashed line.

The shape of the pulse, calculated for a nonlinear field of surface waves, is shown in Fig. 3. We can describe the distribution of the elevations of the surface by a truncated Gram–Charlier distribution in the formula (4). The shape and position of the leading edge of the returned waveform vary considerably when the parameter b within the limits from 1.5 to 3. When $b = 5$ the function $V_G^b(t)$ has negative values.

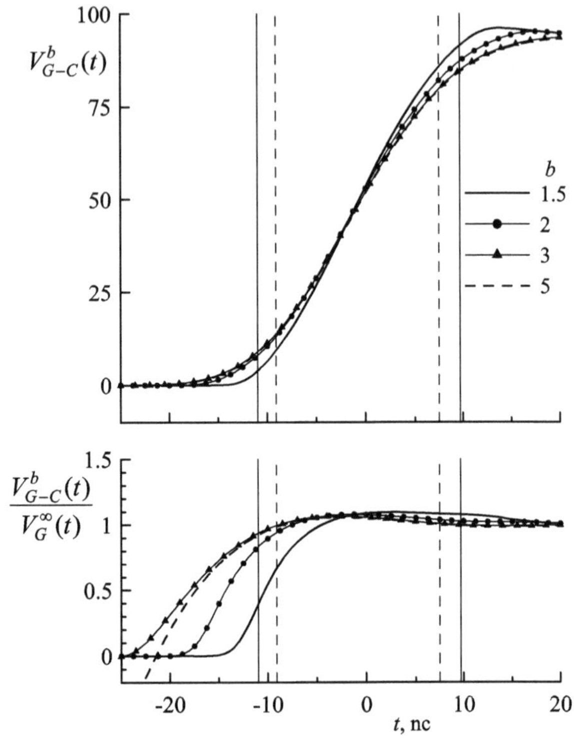

Fig. 3 Altimeter return waveforms, calculated from the models $P^b_{G-C}(t)$

4 Results and Discussion

4.1 The Inverse Problem

Consider the inverse problem—restore of the sea surface elevations skewness.

We can divide the sea surface response for a short radar altimeter pulse into three parts: a thermal noise, a leading edge and a trailing edge (Fig. 4) (Callahan and Rodriguez 2004). The skewness is determined from the change of the slope on the two sections of the leading edge of the returned waveform (Gómez-Enri et al. 2006). The plots are located in areas $[t_{(-)}, t_0]$ and $[t_0, t_{(+)}]$ where $t_0 = 2h/c$ is the time origin corresponding to the midpoint on the waveform leading edge, c is the velocity of light in vacuum.

4.2 Skewness of the Truncated Distribution

We determine the correctness of skewness measurement by how correctly the function $q(t)$ is modeled. If a Gram–Charlier distribution is used for calculations, then

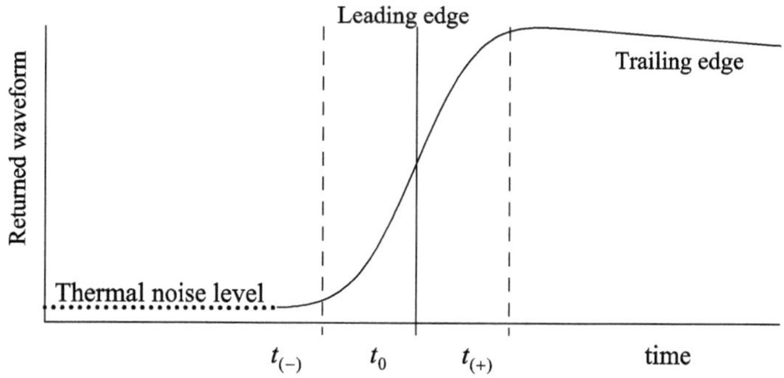

Fig. 4 Schematic of altimeter return waveforms

condition (1) does not allow to fully taking into account the contribution of reflections from the crests of the highest waves (Pokazeev et al. 2013). The temperature noise also prevents the taking into account reflections from the crests.

Let the average value $\eta(t)$ be zero. Then the skewness of the truncated distribution $P^{(b)}_{G-C}(\eta)$ is described as

$$A^{(b)}_{\eta} = \frac{\int\limits_{-\infty}^{\infty} \eta^3\, P^{(b)}_{G-C}(\eta)\, d\eta}{\left[\int\limits_{-\infty}^{\infty} \eta^2\, P^{(b)}_{G-C}(\eta)\, d\eta\right]^{\frac{3}{2}}} \equiv \frac{\int\limits_{-b\,\sigma_\eta}^{b\,\sigma_\eta} \eta^3\, P_{G-C}(\eta)\, d\eta}{\left[\int\limits_{-b\,\sigma_\eta}^{b\,\sigma_\eta} \eta^2\, P_{G-C}(\eta)\, d\eta\right]^{\frac{3}{2}}}. \tag{6}$$

Skewness $P^{(b)}_{G-C}(\eta)$ will differ from the skewness A_η. These discrepancies are shown in Fig. 5. If the parameter b is less ($b < 2.3$), then $P^{(b)}_{G-C}(\eta) < 0$. If $b = 3$ the relation $A^{(b)}_{\eta} \approx 2\, A_\eta$ is satisfied.

Fig. 5 The change in the skewness when integration limits in (6) is change

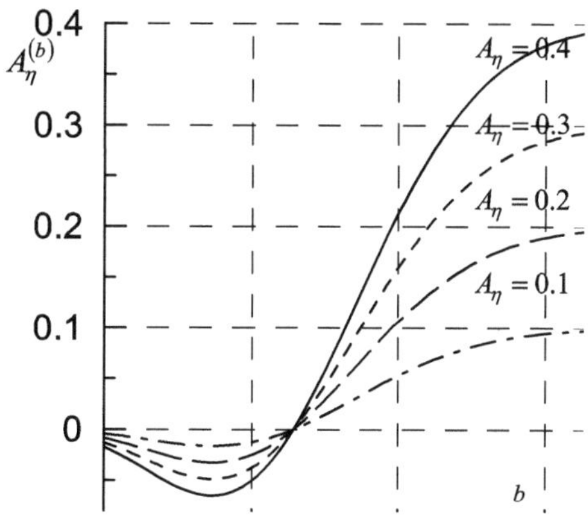

5 Conclusion

The existing method for calculating the skewness of the sea surface elevations is based on measuring the changes in the slope of the individual sections of the leading edge of the altimeter returned waveform. We assumed the distribution of sea surface elevations could be describe by a truncated Gram–Charlier distribution in the Eq. (4). Known limitations in the use of truncated Gram–Charlier series lead to the fact that the skewness obtained based on the data of radio altimetry measurements prove to be incorrect. Such values may be understated or have an opposite sign in comparison with the real skewness of the sea surface wave.

This work was supported by state program of Russia, topic № 0827-2018-0003 "Fundamental studies of oceanological processes that determine the state and evolution of the marine environment under the influence of natural and anthropogenic factors, based on observation and modeling methods".

References

Babanin AV, Polnikov VG (1995) On the non-Gaussian nature of wind waves. Phys Oceanogr 6(3):241–245

Basu S (1991) Pandey PC Numerical experiment with modelled return echo of a satellite altimeter from a rough ocean surface and a simple iterative algorithm for the estimation of significant wave height. In: Proceedings of the Indian academy of sciences-earth and planetary sciences (Earth Planet Science) 100(2):155–163

Brown GS (1977) The average impulse response of a rough surface and its applications. IEEE Trans Antennas Propaga AP-25:67–74. https://doi.org/10.1109/tap.1977.1141536

Callahan PS, Rodriguez E (2004) Retracking of Jason-1 data. Marine Geodesy 27:391–407. https://doi.org/10.1080/01490410490902098

Cox C, Munk W (1954) Measurements of the roughness of the sea surface from photographs of the sun's glitter. J Optical Soc Am 44(11):838–850. https://doi.org/10.1364/josa.44.000838

Gómez-Enri J, Srokosz MA, Gommenginger CP, Challenor PG, Milagro-Pérez MP (2007) On the impact of mispointing error and hamming filtering on altimeter waveform retracking and skewness retrieval. Marine Geodesy 30:217–233. https://doi.org/10.1080/01490410701438166

Gómez-Enri J, Gommenginger CP, Challenor PG, Srokosz, MA, Drinkwater MR (2006) ENVISAT radar altimeter tracker bias. Marine Geodesy 29:19–38. https://doi.org/10.1080/01490410600582296

Gómez-Enri J, Gommenginger CP, Srokosz MA, Challenor PG (2007a) Measuring global ocean wave skewness by retracking RA-2 Envisat waveforms. J Atmos Oceanic Technol 24:1102–1116. https://doi.org/10.1175/JTECH2014.1

Guedes Soares C, Cherneva Z, Antão EM (2003) Characteristics of abnormal waves in North Sea storm sea states. Appl Ocean Res 25:337–344. https://doi.org/10.1016/j.apor.2004.02.005

Hayne GS (1980) Radar altimeter mean return waveforms from near-normal-incidence ocean surface scattering. IEEE Trans Antennas Propag AP-28:687–692. https://doi.org/10.1109/tap.1980.1142398

Jha AK, Winterstein SR (2000) Nonlinear random ocean waves: prediction and comparison with data. In: Proceeding of the 19th International Offshore Mechanics Arctic Engineering Symposium, ASME, Paper No. OMAE 00–п6125

Kendall MG, Stuart A (1945) The advanced theory of statistics. Distribution theory, vol I, Charles Griffin & Company Ltd, London, p 457

Longuet-Higgins MS (1957) The statistical analysis of random moving surface. Phil Trans R Soc Lond A 249:321–387

Mac-Arthur JL (1976) Design of the SEASAT-A radar altimeter applied physics. Lab Laurel MD, SDO-5232. https://doi.org/10.1109/OCEANS.1976.1154217

Pokazeev KV, Zapevalov AS, Pustovoytenko VV (2013) The simulation of a radar altimeter return waveform. Moscow Univ Phys Bull 68(5):420–425. https://doi.org/10.3103/S0027134913050135

Queffeulou P (2004) Long-term validation of wave height measurements from altimeters. Mar Geodesy 27:495–510. https://doi.org/10.1080/01490410490883478

Rodriguez E (1988) Altimetry for non-Gaussian oceans: height biases and estimation of parameters. J Geoph Res 93(C11):14107–14120. https://doi.org/10.1029/jc093ic11p14107

Sasamal SK, Sourabh Bansal, Dutt CBS, Dadhwal VK (2015) Use of SARAL AltiKa Geophysical Products towards the study of cyclone “Phailin”. Int J Eng Sci Innovative Technol (IJESIT) 4(2):125–134

Tatarskii VI (2003) Multi-Gaussian representation of the Cox–Munk distribution for slopes of wind-driven waves. J Atmos Oceanic Technol 20:1697–1705. https://doi.org/10.1175/1520-0426(2003)020<1697:mrotcd>2.0.co;2

Wang X, Miao HL, Wang GZ, Wang YQ, Zhang J (2015) Direct-estimation of sea state bias in Hy-2 based on a merged dataset. In: Proceedings of the international conference on computer information systems and industrial applications CISIA 2015, Bangkok, Thailand. June 28–29, pp. 762–766. https://doi.org/10.2991/cisia-15.2015.207

Zapevalov AS (2011) High-order cumulants of sea surface. Russ Meteorol Hydrol 36(9):624–629. https://doi.org/10.3103/S1068373911090081

Zapevalov AS (2012) Effect of skewness and kurtosis of sea-surface elevations on the accuracy of altimetry surface level measurements. Izv Atmos Ocean Phys 48(2):200–206. https://doi.org/10.1134/S0001433812020120

Zapevalov AS (2018) Determination of the statistical moments of sea-surface slopes by optical scanners. Atmos Oceanic Opt 31(1):91–95. https://doi.org/10.1134/S1024856018010141

Study of the External Influence on Evening Transition in Atmospheric Boundary Layer

E. V. Tkachenko, A. V. Debolskiy, and E. V. Mortikov

Abstract Studying of transitional periods as essential non-stationary processes in the diurnal cycle, has proven to be necessary in order to correctly parameterize them in large-scale climate and weather models. This study focuses on the evening transition (from convective to stably stratified boundary layer) and how it is influenced by external parameters, such as kinematic surface heat flux, geostrophic wind, surface roughness and mesoscale subsidence rate, on its dynamics. The single-column numerical model was used. In particular, the decay of turbulent kinetic energy (TKE), which inevitably takes place during the evening transition in the residual layer, was observed under different model settings and external conditions. The results of some of those experiments were compared with the available LES (Large-Eddy Simulation) data from previous studies. It appears that TKE dynamics in these simulations are very similar to those observed in LES experiments. Different external parameters that were studied influence different characteristics of the TKE decay dynamics (such as rate of decay, time of the decay start etc.). The results of this study show that a better parametrization of turbulent exchange processes is required in order to accurately reproduce evening transition and other non-stationary processes in atmospheric boundary layer.

Keywords Atmospheric boundary layer · Turbulence · Diurnal cycle · Evening transition · Turbulent kinetic energy · Surface heat flux · Geostrophic wind · Numerical experiments

---

E. V. Tkachenko (✉) · A. V. Debolskiy · E. V. Mortikov
Lomonosov Moscow State University, Moscow, Russia

Moscow Center of Fundamental and Applied Mathematics, Moscow, Russia

A.M. Obukhov Institute of Atmospheric Physics of the Russian Academy of Sciences, Moscow, Russia

Marchuk Institute of Numerical Mathematics of the Russian Academy of Sciences, Moscow, Russia

T. Chaplina (ed.), *Processes in GeoMedia—Volume IV*, Springer Geology,
https://doi.org/10.1007/978-3-030-76328-2_21

1 Introduction

The atmospheric boundary layer (ABL) is the lowest part of the atmosphere, with its height ranging from 100 m to 3 km, which directly interacts with the Earth's surface (Stull 1988). It is characterized by the presence of turbulence, which determines transport processes in the ABL, and by the existence of the diurnal cycle due to surface heating during daytime and cooling during nighttime. The latter manifests in the alternation between two states of the ABL: convective (CBL) and stably stratified (SBL) boundary layers. Physical processes that take place on the surface or near it (heat and momentum exchange, evaporation etc.), as well as topography, exert significant influence on the ABL dynamics throughout the diurnal cycle.

Change from CBL to SBL and vice versa happens through transitional periods—the morning transition from SBL to CBL and the evening transition from CBL to SBL. They are accompanied by the change of surface kinematic heat flux, which takes positive values in the daytime and negative—in the nighttime.

In order to correctly reproduce the diurnal cycle in ABL, a clarification of turbulence closures is required. Existing closures do not take into account the effects of non-stationary processes on turbulence (Holtslag et al. 2013), which in some cases leads to incorrect modeling of near surface temperatures (Svensson 2011). This study utilizes numerical modeling of evening transition in order to examine dynamics taking place within it.

During the evening transition, the state of ABL changes, the SBL gets formed near the surface, and the turbulent kinetic energy (TKE) generated during the CBL state, starts to decay. The TKE decay follows the power law $(t) \propto t^{-\alpha}$, where $E(t)$ and t are normalized mean TKE inside BL and time, respectively, and the parameter α can be obtained theoretically or empirically.

The value of $\alpha = 10/7$ was first obtained by Kolmogorov (1941) for homogenous unbounded turbulence, based on the hypothesis of the Loitsanskiy invariant, and then the value of $\alpha = 6/5$ was derived for cases where the Loitsanskiy invariant was allowed to change (Saffman 1967; Oberlack 2002). Numerical experiments show agreement with this value (Nieuwstadt and Brost 1986; Sorbjan 1997), however field measurements and numerical experiments (Nadeau et al. 2011; Rizza et al. 2013) based on those give a substantially different value of $\alpha = 6$.

The ABL dynamics are determined by a large number of processes of various scales. Thus, the majority of numerical studies on the evening transition in ABL use idealized setups. In particular, in the beginning of the experiment, a CBL is formed with prescribed positive heat flux on the surface.

At some moment in time, the surface heat flux gets decreased (abruptly or gradually) to zero, and does not change for the rest of the experiment. Then, the results from the moment when the heat flux starts to decrease to the end of the simulation are examined. This period is alleged to be the evening transition.

Several studies (Nieuwstadt and Brost 1986; Sorbjan 1997; Pino et al. 2006; Guernaoui et al. 2019; Park et al. 2020) delved into the effect of external forces on the evening transition and the TKE decay in particular, however the majority of those studied a very limited range of parameters, using LES (Large-Eddy Simulation) or RANS (Reynolds-Averaged Navier–Stokes) numerical model.

For one, (Sorbjan 1997) explores two LES experiments, where evening transition happens through either abrupt or gradual decrease of surface heat flux. It had been shown there that the TKE decay process is governed by the relation of the external time scale to the convective time scale. Values of $\alpha = 1.2$ for abrupt change and $\alpha = 2$ for gradual change were obtained. A difference in TKE decay dynamics was also observed—in case with gradual heat flux decrease there is a short period when the decay rate is much lower, before it picks up the speed in the main phase.

In (Pino et al. 2006) the wind shear influence is studied for the first time using LES. For that, experiments with and without geostrophic wind were carried out. It is shown that the TKE decay happens to be slower when wind shear is present ($\alpha = 0.7$ when $U_{\text{geo}} = 10$ m s^{-1}, $\alpha = 1.1$ when $U_{\text{geo}} = 0$).

In (Guernaoui et al. 2019) the effect of different time durations of heat flux decrease is studied in depth within LES experiments. Here, a new scaling for TKE was proposed—TKE is normalized by its initial value, rather than using the 'classical' velocity scale proposed by Deardorff (1970) for the CBL turbulence. Time scaling takes heat flux decrease time into account by introducing a new $(t - \tau)/t_*$ time scale (where τ is the time it takes to go from the initial heat flux value to 0), as opposed to the more widely used t/t_*.

One of the more recent studies (Park et al. 2020) revisits the wind shear influence, using a LES model in experiments with $U_{\text{geo}} = \{0.1, 5, 10\}$ m s^{-1}. Again, a decrease in decay rate with the increase of geostrophic wind speed was observed. Moreover, at a certain point in time, the decay ceased and the TKE commenced to grow, which was not observed in the experiments of Sorbjan (1997).

This study examines the influences of changes in heat flux F_s, geostrophic wind speed U_{geo}, surface aerodynamic roughness z_0 and large-scale subsidence rate w_{sub} on TKE decay during evening transition.

A single-column RANS model is used in this study, which allows for estimating the efficacy of existing closures, used in large-scale models, for evening transition simulation.

2 ABL Numerical Model

A Reynolds-Averaged Navier–Stokes system of equations is at the core of the model, which are further reduced due to horizontal homogeneity:

$$\frac{\partial U}{\partial t} + w_{\text{sub}} \frac{\partial U}{\partial z} = f\left(V - V_{\text{geo}}\right) + \frac{\partial}{\partial z}(K_m + \nu)\frac{\partial U}{\partial z},$$

$$\frac{\partial V}{\partial t} + w_{\text{sub}} \frac{\partial U}{\partial z} = -f\left(U - U_{\text{geo}}\right) + \frac{\partial}{\partial z}(K_m + \nu)\frac{\partial V}{\partial z},$$

$$\frac{\partial \theta}{\partial t} + w_{\mathrm{sub}}\frac{\partial \theta}{\partial z} = \frac{\partial}{\partial z}(K_h + \chi)\frac{\partial \theta}{\partial z},$$

where U_{geo} and V_{geo} are geostrophic wind speed components, w_{sub} is large-scale subsidence rate, f is Coriolis parameter, ν and χ are kinematic molecular viscosity and conductivity, z is vertical coordinate, t is time.

K_m and K_h are turbulent viscosity and conductivity coefficients, which depend on turbulent velocity and length scales, determined by TKE and dissipation rate:

$$K_m = S_m\frac{E_k^2}{\varepsilon},$$

$$K_h = S_h\frac{E_k^2}{\varepsilon}$$

Here, for simplicity we assume that the non-dimensional stability functions for momentum $S_m = 0.09$ and heat $S_h = /Pr_t{}^0 = 0.09$ are constant values.

In order to close the system of equations, prognostic equations for TKE and dissi-pation rate are added:

$$\frac{\partial E_k}{\partial t} - \frac{\partial}{\partial z}\left(\left(\nu + \frac{K_m}{\sigma_k}\right)\frac{\partial E_k}{\partial z}\right) = P + B - \varepsilon,$$

$$\frac{\partial \varepsilon}{\partial t} - \frac{\partial}{\partial z}\left(\left(\nu + \frac{K_m}{\sigma_\varepsilon}\right)\frac{\partial \varepsilon}{\partial z}\right) = \frac{\varepsilon}{E_k}(C_{1\varepsilon}P - C_{2\varepsilon}\varepsilon + C_{3\varepsilon}B),$$

where $P = -\overline{u_i'w'}\frac{\partial u_i}{\partial z}$ is TKE shear production, $B = \beta\overline{w'\theta'}$ is TKE buoyancy production (or consumption), σ_k, σ_ε, $C_{1\varepsilon}$, $C_{2\varepsilon}$, $C_{3\varepsilon}$ are closure parameters, which were set according to Mortikov et al. (2019): $\sigma_k = 1.0$, $\sigma_\varepsilon = 1.3$, $C_{1\varepsilon} = 1.44$, $C_{2\varepsilon} = 1.92$, $C_{3\varepsilon\mathrm{stable}} = -0.4$ and $C_{3\varepsilon\mathrm{unstable}} = -1.14$. These parameters allow for correct simulation of SBL main properties (Mortikov et al. 2019), as well as CBL growth rate (Burchard 2002).

3 Experiment Setup

All experiments where conducted on a uniform grid of 512 cells, with the vertical domain of 2314 m. The simulations were run for 12 h, where for the first 6 h a CBL was formed with the surface heat flux set to $F_{s(\mathrm{conv})} = 0.15$ K m s^{-1}, after which the surface heat flux changed to $F_{s0} = 0$ either abruptly or gradually, and the experiment continued for another 6 h. The experiment setup of CBL formation partially corresponds to the one in Debolskiy et al. (2019). Figure 1 shows the change

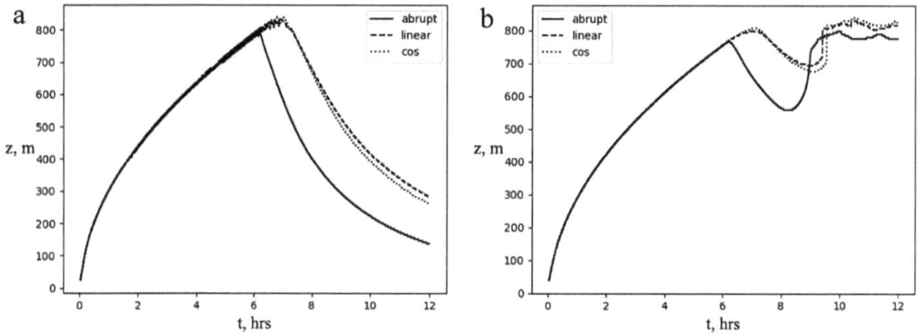

Fig. 1 Boundary layer height, $U_{geo} = 0$ (**a**) and $U_{geo} = 7.5$ m s^{-1} (**b**)

in boundary layer height, which was defined as the height where the heat flux reaches minimum value, from $F_{s(\mathrm{conv})}$ to $F_{s\,0}$ for two types of experiments.

For normalization of TKE and time, Deardorff velocity scale (Deardorff 1970) was used: $w_{*0} = (F_b h_{CBL})^{1/3}$ m s^{-1}, where h_{CBL} is CBL height. Surface buoyancy flux is $F_b = g\theta_0^{-1} F_s = 0.0061$ m^2 s^{-3}, where $g = 9.81$ m s^{-2} is gravitational acceleration, $\theta_0 = 241$ K is reference temperature, $\alpha_{th} = g/\theta_0 = 0.0061$ 1/K is the thermal expansion coefficient. The turbulence turnover time scale is $t_* = h_{\mathrm{CBL}}/w_{*0}$. Thus, the normalized TKE is $E_n = E/{w_{*0}}^2$ and the normalized time is $t_n = (t - \tau)/t_*$, where τ is the time it takes for heat flux to get from $F_{s(\mathrm{conv})}$ to F_{s0}.

All experiments varied by values of external parameters. The influence of heat flux decrease function, heat flux decrease rate, geostrophic wind speed, aerodynamic roughness and subsidence rate. When studying the influence of heat flux decrease function/rate and subsidence rate, cases without geostrophic wind ($U_{\mathrm{geo}} = 0$) and with it ($\mathrm{U_{geo}} = 7.5$ m s^{-1}) were considered.

Three functions describing the change of heat flux in time were used for experiments: abrupt change, when after 6 h $F_{s(\mathrm{conv})}$ changes to F_{s0} instantaneously; linear function, where heat flux is $F_s = F_{s(\mathrm{conv})} - \left(\frac{t-t_0}{\tau}\right)\left(F_{s(\mathrm{conv})} - F_{s0}\right)$; and sinusoidal (cos) dependence, where heat flux is $F_s = F_{s(\mathrm{conv})}\cos\left(\frac{\pi(t-t_0)}{2\tau}\right)$, where $t_0 = 21600$ s $= 6$ h is the moment when heat flux starts to change. In linear and cos experiments $\tau = 3600$ s.

In total, the results of 37 RANS experiments were studied, where h_{CBL} ranged from 664.4 to 964.9 m. Therefore, w_{*0} ranged from 1.6 to 1.8 m s^{-1}, t_* from 417.9 to 533.1 s, and τ/t_* from 8.6 to 6.8. Note that the time scale ratio τ/t_* is smaller than those considered in Nadeau et al. (2011).

4 Results

Figures 2, 3, 4, 5, and 6, display hours 6–12 of each numerical simulation (the TKE decay period).

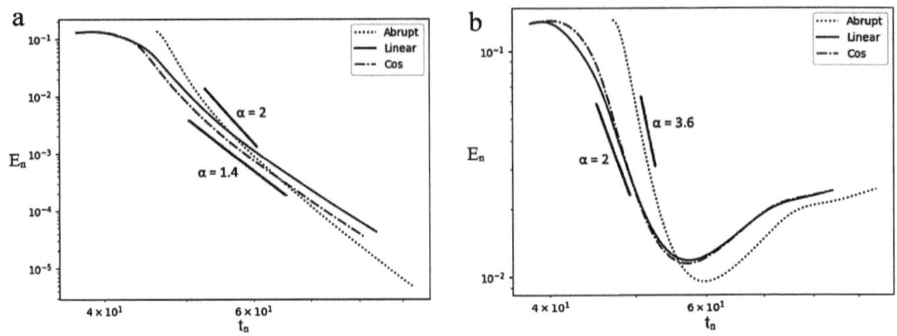

Fig. 2 TKE decay with different heat flux change functions, $U_{\text{geo}} = 0$ (**a**) and $U_{\text{geo}} = 7.5$ m s^{-1} (**b**)

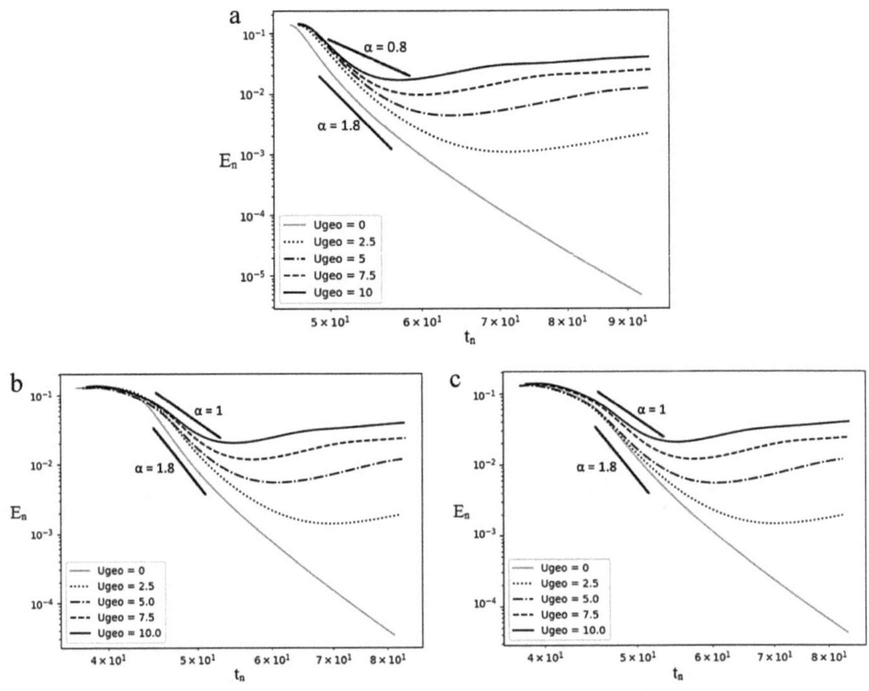

Fig. 3 TKE decay with different values of U_{geo} (m s^{-1}) for abrupt (**a**), linear (**b**) and cos (**c**) heat flux change

Figure 2 shows the results of simulations with different heat flux change functions. It should be noted that in experiments with no geostrophic wind (Fig. 2a) the TKE decay dynamics is in agreement with what was obtained in Saffman (1967)—in the experiment with the abrupt decrease the decay commences almost instantly and picks up the speed in a short amount of time, while there is a short delay period in both

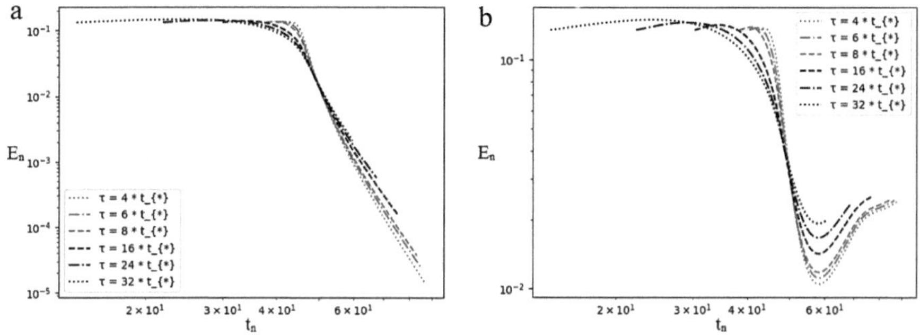

Fig. 4 TKE decay with different values of τ, cos heat flux change, $U_{\mathrm{geo}} = 0$ (**a**) and $U_{\mathrm{geo}} = 7.5$ m s^{-1} (**b**)

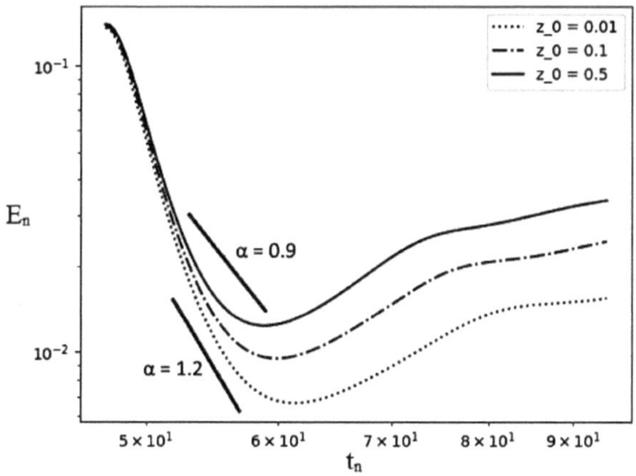

Fig. 5 TKE decay with different values of z_0, $U_{\mathrm{geo}} = 7.5$ m s^{-1}, abrupt heat flux change

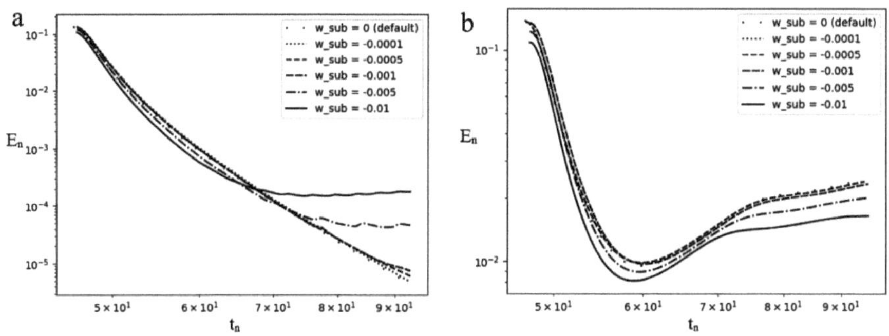

Fig. 6 TKE decay with different values of w_{sub}, abrupt heat flux change, $U_{\mathrm{geo}} = 0$ (**a**) and $U_{\mathrm{geo}} = 7.5$ m s^{-1} (**b**)

experiments with gradual change. However, the difference with LES experiments of Saffman (1967) is observed in the decay rate—here, the decay is faster for the experiment with the abrupt decrease, and in Saffman (1967), is was faster for the experiment with gradual change.

Experiments with different geostrophic wind speed (Fig. 3) also have shown the agreement of the TKE decay dynamics with results reported in Nieuwstadt and Brost (1986), Nadeau et al. (2011). The TKE decay rate is noticeably higher in experiments with no wind shear. Moreover, at a certain point the TKE decay ceases and the TKE even starts to grow, similarly to experiments in Nadeau et al. (2011). However, values of α derived here show that the TKE decay rate differs from LES results, even if by a small margin.

Figure 4 shows the results of experiments with different durations of heat flux decrease τ for cos experiments. In absence of wind shear (Fig. 4a) TKE decay dynamics are relatively similar for all experiments. Still, there is a noticeable divergence at the stage of the experiment where the decay rate increases. When wind shear is present (Fig. 4b), this divergence is even more prominent. The conclusion that may be derived here is that shear scale, induced by geostrophic wind, also affects TKE decay dynamics.

Experiments with various values of surface roughness z_0 (Fig. 5) also show similar TKE decay dynamics, with a difference arising in the TKE integral value when the TKE reaches the near-stationary state (closer to the end of the experiment). Therefore, the value of surface roughness likewise has an effect on the shear scale and vertical momentum flux distribution.

Finally, results of experiments with different values of large-scale subsidence rate w_{sub} (Fig. 6) demonstrate a level of diversity in TKE dynamics during last hours of experiments. For experiments with $U_{geo} = 0$, the smaller w_{sub} gets, the sooner TKE ceases to decay. A similar pattern can be observed in experiments with $U_{geo} = 7.5$ m s^{-1}, albeit not as prominent.

5 Conclusions

In this study the influence of changes in surface heat flux F_s, geostrophic wind speed U_{geo}, surface aerodynamic roughness z_0 and large-scale subsidence rate w_{sub} on evening transition dynamics was investigated. It has been shown that parameters in question have different effect on the TKE decay dynamics and may speed up and slow down the decay of TKE, thus affecting the duration of the evening transition.

The assumption that external parameters influence TKE decay dynamics was confirmed for all parameters considered, in one way or another. This influence mainly manifests itself in wind shear dependence (for different values of U_{geo} and z_0) and changes in TKE decay dynamics in general (for different functions of F_s and different values of U_{geo}, z_0, w_{sub}).

The TKE dynamics have shown a high level of similarity to those obtained through LES experiments. This constitutes a possibility of ABL evening transition modeling using two-parameter turbulence closures, potentially achieving the LES models accuracy. Still, results obtained in this study differ from LES results in TKE decay rates and its stages. Also, it should be stressed that idealized settings, that do not take into account the influence of surface heterogeneity and radiation transfer, were used for the experiments conducted here.

Further research will focus on confirming the results, acquired over the course of this study, with LES experiments in order to improve existing turbulent closures. This in turn will allow for better simulation of transitional periods and non-stationary dynamics of ABL in climate and weather prediction models.

Acknowledgements This study was funded by Russian Foundation of Basic Research within the project N 20-05-00776 and the grant of the RF President within the MK-1867.2020.5 project.

References

Burchard H (2002) Applied turbulence modelling in marine waters. Springer, Berlin, Germany, pp 57–59
Deardorff JW (1970) Convective velocity and temperature scales for the unstable planetary boundary layer and for Rayleigh convection. J Atmos Sci 27(8):1211–1213
Debolskiy A, Stepanenko V, Glazunov A, Zilitinkevich S (2019) Bulk models of sheared boundary layer convection. Izv Atmos Oceanic Phys 55(2):139–151
El Guernaoui O, Maronga B, Reuder J, Esau I, Wolf T (2019) Scaling the decay of turbulence kinetic energy in the free-convective boundary layer. Bound-Layer Meteorol 173:79–97
Holtslag AAM, Svensson G, Baas P, Basu S, Beare B, Beljaars ACM, Bosveld FC, Cuxart J, Lindvall J, Steeneveld GJ et al (2013) Stable atmospheric boundary layers and diurnal cycles—challenges for weather and climate models. Bull Am Meteorol Soc 94:1691–1706
Kolmogorov AN (1941) On degeneration of isotropic turbulence in an incompressible viscous liquid. Dokl Akad Nauk SSSR 31:538–540 [in Russian]
Mortikov EV, Glazunov AV, Debolskiy AV, Lykosov VN, Zilitinkevich SS (2019) Modeling of the dissipation rate of turbulent kinetic energy. Dokl Earth Sci 489(2):1440–1443
Nadeau DF, Pardyjak ER, Higgins CW, Fernando HJS, Parlange MB (2011) A simple model for the afternoon and early evening decay of the convective turbulence over different land surfaces. Bound-Layer Meteorol 141:301–324
Nieuwstadt F, Brost RA (1986) The decay of convective turbulence. J Atmos Sci 43(6):532–546
Oberlack M (2002) On the decay exponent of isotropic turbulence. In: Proceedings in applied mathematics and mechanics, vol 1, pp 294–297
Park S-B, Baik J-J, Han B-S (2020) Role of wind shear in the decay of convective boundary layers. Atmosphere 11:622
Pino D, Jonker HJJ, de Arellano JVG, Dosio A (2006) Role of shear and the inversion strength during sunset turbulence over land: characteristic length scales. Bound-Layer Meteorol 121:537–556
Rizza U, Miglietta MM, Degrazia GA, Acevedo OC, Marques Filho EP (2013) Sunset decay of the convective turbulence with Large-Eddy simulation under realistic conditions. Phys A 392:4481–4490
Saffman PG (1967b) Note on decay of homogeneous turbulence. Phys Fluids 10:1349
Sorbjan Z (1997) Decay of convective turbulence revisited. Bound-Layer Meteorol 82:501–515
Stull RB (1988) An introduction to boundary layer meteorology. Kluwer Academic Publishers, Dordrecht, The Netherlands

Svensson G et al (2011) Evaluation of the diurnal cycle in the atmospheric boundary layer over land as represented by a variety of single-column models: the second Gabls experiment. Bound-Layer Meteorol 140:177–206

Hydrogeological Responses of Fluid-Saturated Collectors to Remote Earthquakes

E. M. Gorbunova, I. V. Batukhtin, A. N. Besedina, and S. M. Petukhova

Abstract Reaction of confined and unconfined aquifers to propagation of seismic waves from distant earthquakes registered on the territory of geophysical observatory IDG RAS "Mikhnevo" is considered as dynamic deformation regime indicator. The reference point of measurements is located to the South of Moscow at the 80 km distance within the East–European platform. Based on the results of precision monitoring of groundwater levels, carried out from 2010 to 2018, database of seismic and hydrogeological data is formed, which includes registration of responses to 61 earthquakes with M_W 6.3–9.1 at epicentral distances in the range between 1863 and 16,507 km. On the basis of spectral analysis of a sample of experimental data series with duration of 6 h (3 h before and 3 h after the earthquake) the principal features of fluid–saturated reservoir reaction to seismic waves propagation and types of hydrogeological effects have been determined. In some cases the relative deformation of fluid saturated reservoir reaches (1.59–5.95)·10^{-7} at maximum amplitude of ground velocity 1.01–3.78 mm/s and exceeds by one or two orders of magnitude calculated poroelastic deformation of reservoir from tidal waves.

Keywords Confined and unconfined aquifers · Hydrogeological effects · Deformation mode · Remote earthquakes · Spectral analysis · Geophysical observatory

1 Introduction

Hydrogeological effects associated with propagation of seismic waves from distant earthquakes have been widely studied in seismic active regions and are considered as indicators of dynamic deformation of fluid-saturated reservoirs (Kissin 2015; Kopylova and Boldina 2019; Wang and Manga 2010). Based on the results of long–term observations of groundwater reaction on earthquakes the basic types of hydrogeological effects have been identified: precursors, coseismic, and postseismic changes of level.

E. M. Gorbunova · I. V. Batukhtin · A. N. Besedina · S. M. Petukhova (✉)
Sadovsky Institute of Geospheres Dynamics of RAS, Moscow, Russia

T. Chaplina (ed.), *Processes in GeoMedia—Volume IV*, Springer Geology,
https://doi.org/10.1007/978-3-030-76328-2_22

Increase of hydrogeological responses database to remote earthquakes at epicentral distances exceeding the length of seismogenic rupture with use of modern precision sensors of level variation with sampling frequency from 1 Hz up to 100 Hz allowed us to specify additional gradation of coseismic and postseismic effects. Coseismic dynamic variations of level—oscillations, which can accompany gradual and step–like (abrupt) changes of level caused by propagation of seismic waves from earthquakes, have been noted (Sun et al. 2015). Among postseismic hydrogeological effects two categories are considered for duration of manifestation: stable lasting for several weeks or months (less often more), and transient—within several hours or days.

For level oscillations caused by the propagation of seismic waves in open wells, analytical solutions have been proposed, which describe both the increase in response due to resonance and its weakening in the high–frequency part (Cooper et al. 1965; Liu et al. 1989). Variation of water level in a well at propagation of seismic waves from distant earthquakes can be caused by dynamic deformation of mineral skeleton and change of pore pressure. In both cases poroelastic reaction of fluid–saturated reservoir depends on ground velocity and is estimated using amplitude factor, which is the ratio of level spectra to ground velocity spectra at frequencies corresponding to synchronization of peak values (Brodsky et al. 2003). It is noted that for regional M_W 6.7–7.2 events and oscillation periods of less than 20 s, the amplitude factor is frequency–dependent and reaches 280 mm/(mm/s) at 0.06 Hz with a standard deviation of 49 mm/(mm/s). For teleseismic earthquakes with M_W 7.7–8.3 and more than 20 s oscillation periods such relation is absent.

For 21 local events in Liu et al. (2006), which have happened in the vicinity of Taiwan Island, and for two regional earthquakes (26.12.2004 and 28.03.2005) linear dependence between peak fluctuations of water level, ground velocity and three components of ground displacement is revealed, however the connection with ground acceleration is not traced. The amplitude–frequency spectra of ground velocity and water level from regional earthquakes coincide in the range of frequencies 0.01–0.1 Hz, as from local events spectra overlap in the frequency range 0.03–0.5 Hz.

At propagation of seismic waves from the earthquake 27.02.2010 in Chile M_W 8.7 in closed wells in Japan oscillations of water pressure are detected in the range of 0.002–0.1 Hz, which are frequency–dependent and inversely proportional to the volumetric stress (Kitagawa et al. 2011).

In six measurement points, located on the Taiwan Island and equipped with the broadband seismometers and observation wells, in 780–900 s after the catastrophic earthquake Tohoku 11.03.2011 M_W 9.1 ground vibrations and level variations of pressure horizons with duration from 800 to 1500 s have been registered (Shin et al. 2011). Spectral analysis of experimental data made it possible to identify the dominant frequency range of 0.03–0.05 Hz, within which high coherence of level changes with ground displacement along vertical and radial components is traced.

Filtration of hydrogeological and seismic data obtained in one of the observation wells in China during the propagation of seismic waves from Tohoku earthquake in the range up to 60 s and above has allowed to distinguish a step-like coseismic level rise in the low frequency region, which is absent in the corresponding seismogram (Sun et al. 2015).

The amplitude factor has been calculated for responses within the territory of GPO "Mikhnevo" to remote earthquakes during the period of observations 2010–2017 (Gorbunova et al. 2018a). The difference of applied approach to parameter determination lies in comparison of normalized seismic and hydrogeological parameters. Distribution of amplitude factor of confined and unconfined aquifers for the considered period of observations is a frequency independent function.

In this work on the basis of the data of long–term synchronized seismic and hydrogeological monitoring carried out on the territory of GPO "Mikhnevo" the reaction of fluid–saturated reservoir to remote earthquakes are studied and estimates of relative deformation have are obtained.

2 Brief Description of the Research Subject

On the territory of GPO "Mikhnevo", two aquifers are predominantly distributed. The upper unconfined aquifer is opened in the range of 44.0–56.2 m, the lower confined aquifer—in range of 92–115 m. Groundwater levels are located at the depth of 46 m and 68 m respectively. Amplitudes of annual variations of unconfined aquifer levels are from 0.4 to 1.9 m, of confined-from 0.5 to 2.7 m (Petukhova 2020). According to the pumping data, the water supply capacity of the unconfined aquifer is 15 m^2/day, that of the unconfined aquifer is 4 m^2/day. Isolation of the considered aquifers and dissociation of levels is due to the presence of regional aquiclide—dense clays with subordinate marl interlayers. The resonance frequency of observation wells is estimated by relation (Cooper et al. 1965):

$$f_0 \approx \frac{1}{2\pi}\sqrt{\frac{g}{H_e}} \tag{1}$$

where $H_e = H + 3d/8$, g—acceleration due to gravity, H—height of water column in the well casing, d—interval of the open part of the well. Resonance frequency of the well, where the confined aquifer is opened, is 0.09 Hz, for the well, drilled in the unconfined aquifer—0.19 Hz.

The object of study is a carbonate collector of pore-fractured type, irregularly cracked, silicified, represented by limestone fine oolite with cavities of leaching, subordinate layers of dolomite, marl and clay. Petrographic analysis of grindings made from samples taken in the interval of 39–53.5 m indicates a decrease in rocks permeability with depth. Fine grained calcite particle size varies from 5 to 2–3 microns per 85% of the grinding area. The sample contains relicts of microorganisms (echinoderms, sponges) partially made with chalcedony, which occupy about 5–7% of the grinding area. The size of pores and caverns of irregular elongated-oval shape varies from 0.05 to 0.5 mm. The total visible porosity of the grindings decreases from 15–20 to 5–10%. Gas permeability of samples varies widely from 0.7 to 13.9 mD, according to laboratory measurements.

3 Methods of Measurement and Data Processing

To clarify the engineering and geological structure and hydrogeological conditions within the territory of GPO "Mikhnevo", wells with a depth of 115 m and 60 m were drilled in December 2006 and March 2013 respectively. Upon end of drilling, a set of geophysical studies and test pumping were carried out in wells. The wells have been equipped with the precision LMP308i level sensors (Germany) the first since February 2008, the second—since July 2013. Accuracy of level registration with frequency of 1 Hz is 1.7 mm.

Seismic data are registered by the broadband seismometers STS–2 and CM-3-E installed in the mine at a depth of 20 m, with a sampling rate of 100 Hz (Adushkin et al. 2016). For signals with period more than 10 s the phase shift introduced by seismometer becomes significant and requires additional correction. To compensate this effect, the frequency characteristics of the sensors are extended to the periods of 1200 s by synthesis of inverse filter with consideration of seismometers zeros and poles. To remove possible artifacts, appearing at restoration of frequency response, filtering of high frequencies of seismic records with boundary frequency 0.0025 Hz is performed. The obtained seismic data are preliminary decimated by 100 times for comparison with hydrogeological data. Water level data are cleared from long-period trend above 400 s for correct comparison with seismic records.

Primary data processing consisted in sampling and systematization of hydrogeological and seismic (by vertical component) records in intervals corresponding to the time of waves propagation from distant earthquakes with M_W more than 6. The earthquakes are selected from the lower threshold of sensitivity of considered aquifers (double amplitude of velocity more than 0.05 mm/s) for remote earthquakes (Besedina et al. 2016). For each earthquake the record is formed with duration of 6 hours (3 h before and 3 h after the arrival of wave from the earthquake to the station). In accordance with the data of ground velocity, level of confined and unconfined aquifers the maximum values of amplitudes between consecutive maximum and minimum in the group of surface waves are determined. Measurements are made in a sliding window of 72 s duration with 50% overlap.

In accordance with seismic and hydrogeological data for each earthquake amplitude spectra normalized with respect to microseismic noise and variations of water levels 3 h before the wave arrival from the earthquake have been calculated, which allowed to eliminate the influence of local features of station location.

Basic parameters of considered earthquakes have been taken from the catalog of Federal Research Center of Unified Geophysical Survey RAS (FRC UGS

RAS (http://www.ceme.gsras.ru)): time of the earthquake in the origin, depth, coordinates and time of longitudinal wave arrival to "Obninsk" station (OBN), located at a distance of ~100 km North-West of Mikhnevo. Magnitude M_W is given by Global CMT Catalog (http://www.globalcmt/org/CMTsearch.html).

4 Results and Discussion

During the period of observations from 2010 to 2018 on the territory of GPO "Mikhnevo" hydrogeological responses to propagation of seismic waves from 61 distant earthquakes M_W 6.3–9.1 at epicentral distances from 1863 to 16,507 km have been registered (Fig. 1).

Most of the earthquakes are finely focused with the origin depth up to 70 km. In three cases, responses to medium-focus earthquakes in the depth range from 130 km (in the South Alaska region on January 24, 2016) to 215 km (in the Hindu Kush district, Afghanistan on October 26, 2015) are traced. The response of fluid saturated reservoirs is recorded at 4 deep-focus earthquakes in the depth interval of 560–680 km within the Western Pacific belt (in the Sea of Okhotsk and along the Western ridge of Bonin, Tonga and Fiji Islands in the Pacific Ocean).

At ground velocity (by vertical component here and further) from 0.06 to 3.78 mm/s the amplitude of hydrogeological responses of unconfined aquifer varies from 2.2 to 110.6 mm, confined—from 2.0 to 41.5 mm (Fig. 2). At ground velocity from 0.08 to 0.14 mm/s, amplitudes of confined and unconfined aquifer responses are close. In the range of registered velocities 0.6–1.3 mm/s the difference in amplitudes of hydrogeological responses confined and unconfined aquifers increases by an order of magnitude during passage of seismic waves from the same earthquake.

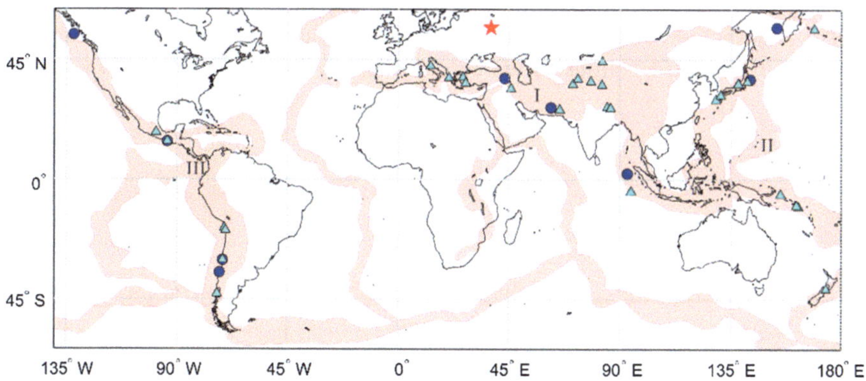

Fig. 1 Scheme of epicenter location of the earthquakes, which are reflected in variations of confined (circles) and unconfined (triangles) aquifers within the territory of GPO "Mikhnevo" (MHV—star) and Earth's seismic belts (dark outline): I—Mediterranean-Trans-Asian, II—West Pacific, III—East Pacific

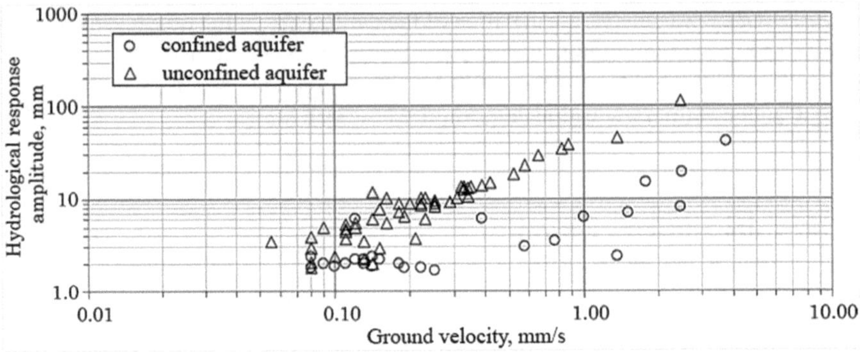

Fig. 2 Dependence of changes in the aquifers level in the GPO "Mikhnevo" (MHV) from the velocity during propagation of seismic waves from earthquakes

For each aquifer there is a degree of dependence between the maximum values of the ground velocity and the amplitude of hydrogeological response. Variations of ground water level are mostly synchronous with changes of ground velocity in vertical component from distant earthquakes. The extremums in level variations of confined and unconfined aquifer are manifested both simultaneously and with delay in seismic wave arrival. Reaction of ground waters to the arrival of longitudinal wave is not traced (Fig. 3).

Comparison of amplitude spectra based on seismic and hydrogeological data before and after the arrival of seismic waves from earthquakes is aimed to allocate events, in which the level variations after the earthquake exceed the background values before the earthquake. For further analysis normalized spectra of ground velocity and levels of confined and unconfined aquifers are used, calculated for each parameter separately as spectra relations in intervals of 3 h after the arrival of seismic waves from distant earthquakes to background parameters before the event.

The general view of the diagrams under consideration differs. In particular, the response of confined aquifer to the earthquake, which has happened in Chile on

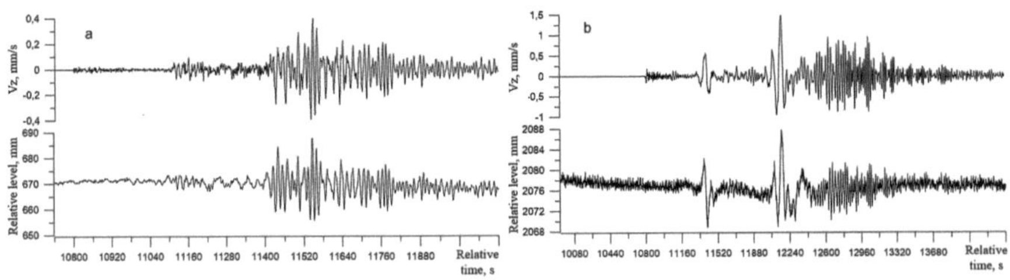

Fig. 3 Seismograms of *Z*-component (upper row) and level variations (lower row) of unconfined aquifer in case of the earthquake in Tajikistan M_W 7.2 December 07, 2015 (**a**) and confined aquifer at the earthquake near North Sumatra M_W 8.6 April 11, 2012 (**b**)

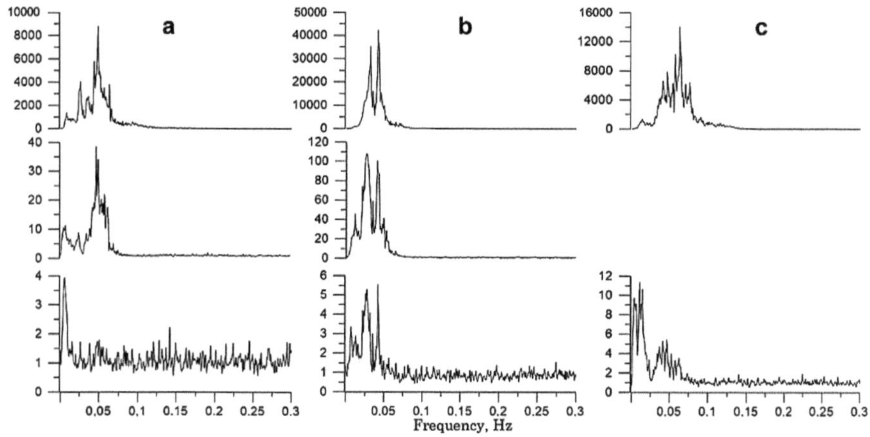

Fig. 4 Normalized velocity spectra (upper row), water levels of unconfined (middle row) and confined (lower row) aquifers from the earthquakes which have happened in Chile on September 16, 2015 M_W 8.3 (**a**), in Mexico on September 08, 2017 M_W 8.2 (**b**) and near North Sumatra on April 11, 2012 M_W 8.6 (**c**)

September 16, 2015 M_W 8.3, is registered in a narrow frequency range of 0.01–0.02 Hz as a single peak (Fig. 4a). In unconfined aquifer extremums in the vicinity of frequencies 0.01–0.02 Hz and 0.04–0.06 Hz are distinguished, which correspond to a wide band of seismic vibrations in the range of 0.01–0.06 Hz with a ground velocity of 0.58 mm/s.

Confined horizon response to the earthquake in Mexico on September 08, 2017 M_W 8.2 at the amplitude of ground velocity 2.46 mm/s appears in the frequency band 0.01–0.05 Hz and is characterized by presence of several peaks in the normalized spectrum (Fig. 4b). Basic variations of unconfined horizon level and seismic vibrations are traced in the range 0.01–0.075 Hz. Extremes of ground velocity and levels of confined and non–confined horizons are marked in the region of high frequencies and coincide at 0.03 and 0.04 Hz. In the low-frequency region in the 0.01–0.02 Hz range, additional extremums occur in variations of the levels, which are absent in the ground velocity.

Normalized spectra of seismic vibrations and pressure level variations for the earthquake in North Sumatra on April 11, 2012 *MW* 8.6 differ (Fig. 4c). Seismic vibrations with the amplitude of ground velocity 2.51 mm/s are traced in the frequency range 0.01–0.12 Hz, the extremum is clearly marked out at the frequency 0.06 Hz. In contrast, confined horizon level extremum are located in low-frequency range of 0.01–0.02 Hz. In the high-frequency area in the vicinity of 0.04 Hz, the extremum is poorly pronounced. Resonating asymmetric view of confined horizon level spectra with shifting of maximum level amplitudes into low frequency region is inherent to spectra of confined horizon response to the earthquakes of February 27, 2010 in Chile M_W 8.8 and March 11, 2011 in Tohoku M_W 9.1 at ground velocity

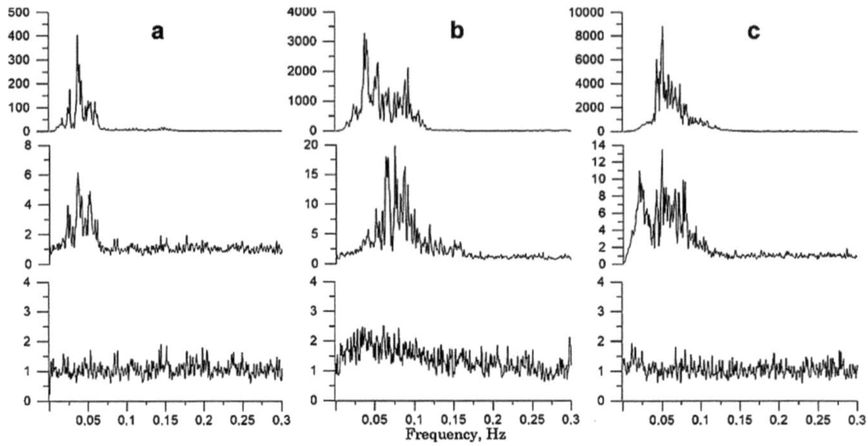

Fig. 5 Normalized velocity spectra (upper row), levels of unconfined (middle row) and confined (lower row) aquifers from the earthquakes that occurred near the Solomon Islands on December 08, 2016 M_W 7.8 (**a**), in Tajikistan on December 07, 2015. M_W 7.2 (**b**) and in the Commander Islands region on July 17, 2017 M_W 7.7 (**c**)

of 1.79 and 3.78 mm/s respectively. Extreme velocity on spectra of three above mentioned earthquakes are distinguished at the frequencies 0.06 and 0.09 Hz.

Normalized spectra of ground velocity and unconfined horizon level mainly coincide with the values of ground velocity 0.09–0.58 mm/s (Fig. 5a). Extreme values are shown in the frequency range 0.04–0.12 Hz from the earthquakes in Mediterranean-Trans-Asian seismic zone, 0.03–0.06 Hz—from those in West and East Pacific zones.

Normalized spectra of ground velocity and unconfined aquifer level differ at the values of velocity 0.31–1.37 mm/s from the earthquakes in Mediterranean-Trans-Asian seismic belt and 0.12–0.33 s—from those in West and East Pacific belts. Maximum amplitudes of response of unconfined aquifer are traced in the vicinity of 0.04–0.08 Hz frequencies and are shifted to the region of high frequencies by 0.02–0.04 Hz with respect to the extremum, determined in the normalized spectrum of ground velocity (Fig. 5b), less often—the shift to the region of low frequencies by 0.02 Hz is noted. In some cases, for the earthquakes in Afghanistan on October 26, 2015, in the region of Commander Islands on July 17, 2017 etc. in the spectrogram low-frequency region of 0.01–0.02 Hz the trend of increase in response intensity of unconfined aquifer relative to the background by 1.5 and more times is observed (Fig. 5c).

The extremums of normalized spectra of confined aquifer levels have been traced in the low-frequency region 0.01–0.02 Hz from the earthquakes with M_W 7.8–8.3, which occurred within the Western and Eastern Pacific zones, in the extended frequency range 0.01–0.08 Hz from the earthquakes with M_W 7.1–7.7, which occurred within the Mediterranean-Trans-Asian belt.

The noted differences in the normalized spectra confirm the correctness of the previously used approach to the analysis of coseismic variations and the isolation of postseismic effects from the data of precision groundwater level monitoring conducted on the territory of IDG RAS GPO "Mikhnevo" (Gorbunova et al. 2018b). Typization of registered hydrogeological responses from remote earthquakes is based on comparison of spectrograms of seismic and hydrogeological data from remote earthquakes filtered with bandpass filter 1–60 s (Sun et al. 2015). Coseismic variations of ground water level differ by intensity in high and low frequency ranges. The first type (I) includes hydrogeological effects, which are limited only in high or low-frequency areas (Fig. 4). The second type (II) is traced in a wide range of frequencies and is synchronous with seismic vibrations (Fig. 5). The third type (III) is represented by post-seismic level changes in the form of abrupt or smooth rise or fall (Fig. 6). Such changes are absent in corresponding seismic recordings from distant earthquakes.

The established synchronization and differences between normalized spectra of ground velocity and hydrogeological responses to remote earthquakes, probably, indicate the difference in deformation modes of fluid-saturated reservoirs. Relative deformation of the reservoir can be calculated using the ratio:

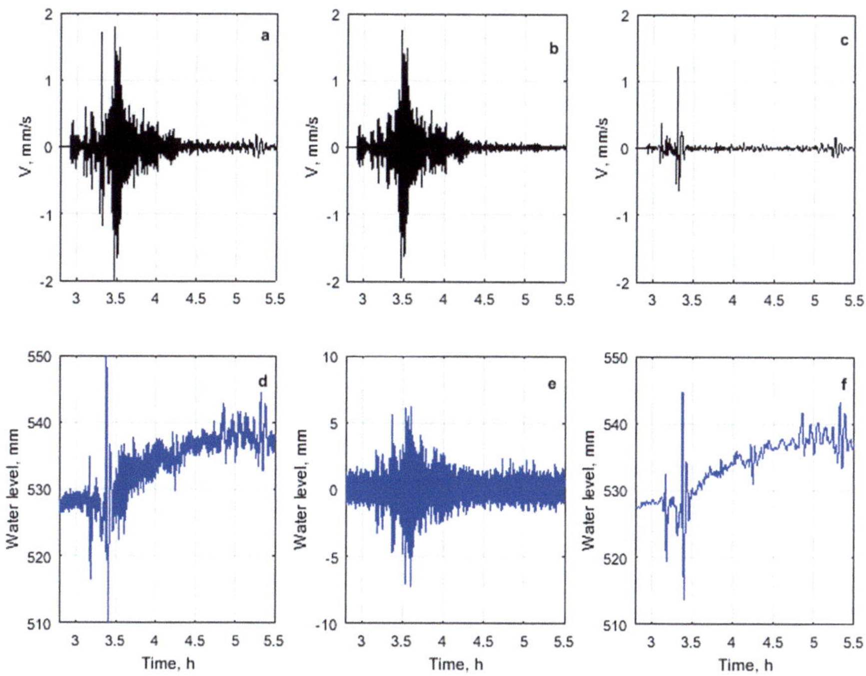

Fig. 6 Comparison of water level for confined aquifer (WLC) and vertical component of seismic data recorded by MHV station on March 11, 2011. Fist column (**a, d**)—initial data, second column (**b, e**)—after applying 60 s high-pass filter, third column (**c, f**)—after applying 60 s low-pass filter. Three hour (3 h) at the time axis corresponds to wave arrival to the station

$$\varepsilon = \sigma_d / E, \quad (2)$$

where E—Young's modulus, GPa; σ_d dynamic stresses, which are evaluated as (Kocharyan et al. 2011):

$$\sigma_d = \rho \cdot C \cdot V_m, \quad (3)$$

where ρ—density, kg/m^3; C—velocity of longitudinal waves propagation, m/s; V_m—maximum value of ground velocity at propagation of seismic waves from earthquakes, m/s.

At the average values of the basic parameters for dense limestones ρ ~ 2680 kg/m^3; C ~ 4700 m/s, E ~ 80 GPa (Terminological dictionary-reference of engineering geology 2011) and the range of values of ground velocities V_m from 0.08 to 3.78 mm/s, registered at the territory of the GPO "Mikhnevo", the relative deformation of fluid saturated collector ε varies within (0.12–5.95)10^{-7}.

The maximum average monthly values of volume deformation amplitudes for fluid-saturated collector calculated using ETERNA 3.0 software package for the coordinates of the measuring point—GPO "Mikhnevo", in the vicinity of half-day tidal waves are (7.7–9.2)·10^{-9}, that in range of daily tidal waves—reach (1.4–1.5)·10^{-8} (Kabychenko et al. 2020) and correspond to the background threshold parameters of the collector. At ground velocity 0.08–0.2 mm/s, registered during the propagation of seismic waves from distant earthquakes, relative and volume deformation in the range of daily tidal waves has similar order of values. Average values of amplitudes of confined aquifer responses do not exceed 2.4 mm, of unconfined aquifer—4.9 mm. When ground velocity increases up to 1.0 mm/s, relative deformation of fluid-saturated reservoir is one order of magnitude higher than values calculated from tidal waves, average values of amplitudes of confined and unconfined horizon response increase up to 3.3 and 15.3 mm respectively.

Principal factors determining the degree of influence on fluid-saturated collector include earthquake magnitude (M_W) and epicentral distance (r, km) to the measuring point, which can be used for estimation of seismic energy density (e, J/m^3) in accordance with relation proposed in Wang (2007):

$$M_W = 2.7 + 0.69 \lg e + 2.1 \lg r \quad (4)$$

When comparing the obtained values power-law dependence of relative deformation of fluid-saturated collector on seismic energy density is observed in the range from $1 \cdot 10^{-6}$ to $3 \cdot 10^{-3}$ J/m^3 (Fig. 7). The minimum density of seismic energy, at which stable hydrogeological effects on earthquakes in different regions of the world are registered, equals 10^{-3} J/m^3 (Shi et al. 2015). At the same time single coseismic oscillations are registered at the density of seismic energy ~10^{-6} J/m^3, for example, responses in Devils Hole, Nevada, confined to karst-exposed limestones (Weingarten and Ge 2014).

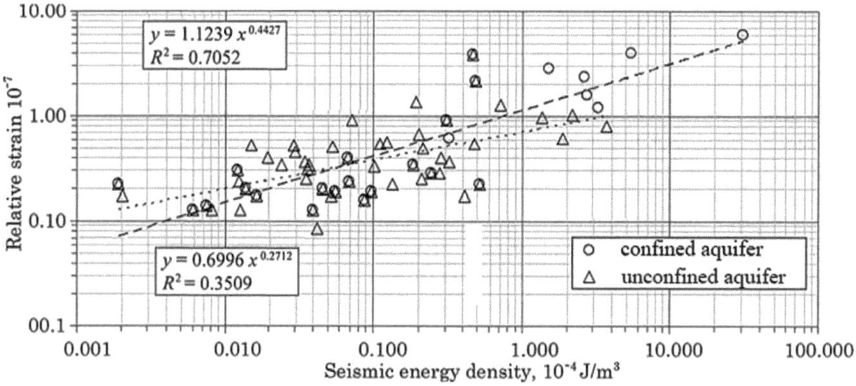

Fig. 7 Dependence of relative deformation of the collector on the density of seismic energy of distant earthquakes, registered by on the territory of GPO Mikhnevo

Experimental data obtained on GPO "Mikhnevo" territory supplement the worldwide database of hydrogeological effects in the far zone of the earthquakes at the epicentral distances of 1863–16,507 km at M_W 6.3–9.1 and indicate dynamic deformation of fluid-saturated reservoirs at seismic energy density in the range of 10^{-6}–10^{-3} J/m^3.

5 Conclusion

In the process of experimental data processing general functions of fluid-saturated reservoir reaction to seismic waves propagation from distant earthquakes are identified. The variations of the level are manifested both in narrow and wide frequency ranges. Synchronization of normalized spectra of ground water levels and ground velocities confirms poroelastic deformation of fluid-saturated collector.

Distinction in normalized spectra of groundwater levels and ground velocity, the shift of level extremums to the high frequency region, and less frequently to the low frequency region, relative to the maximum amplitude of the ground velocity, may occur due to a skin-effect onset at the dynamic impact caused by cracks colmatation/decolmatation in the near-wellbore space.

In some cases, at higher amplitude values of ground velocity greater than 2.82 mm/s and M_W, higher than 8.5 the maximum values of the ratio of spectrum modules in the low-frequency area may be larger than the extremums in the high-frequency area and differ from the velocity spectra.

Acknowledgements The research was carried out on the subject of State task No. AAAA–A17–117112350020–9 and with the financial support of the Russian Foundation for Basic Research under scientific project No. 20–35–90016.

References

Adushkin VV, Ovchinnikov VM, Sanina IA, Riznichenko OYu (2016) Mikhnevo: from seismic station no. 1 to a modern geophysical observatory. Izv Phys Solid Earth 1:108–119

Besedina A, Vinogradov E, Gorbunova E, Svintsov I (2016) Chilean earthquakes: aquifer responses at the Russian platform. Pure Appl Geophys 173(2):321–330

Brodsky EE, Roeloffs E, Woodcock D, Gall I, Manga M (2003) A mechanism for sustained groundwater pressure changes induced by distant earthquakes. J Geophys Res 108(B8 2390):7-1–7-10

Cooper HH, Bredehoeft JD, Papadopulos IS, Bennett RR (1965) The response of well-aquifer systems to seismic waves. J Geophys Res 70(16):3915–3926

Gorbunova EM, Besedina AN, Vinogradov EA (2018a) Dynamics of fluid-saturated reservoir deformation according to precision hydrogeological monitoring. Dynamic proc in geospheres Is. 10. IDG RAS, Moscow, pp 74–83 [in Russian]

Gorbunova EM, Besedina AN, Vinogradov EA (2018b) Reaction of the fluid saturated collector during passage of seismic waves. AIP Conference Proceedings, vol 2051, p 020100. https://doi.org/10.1063/1.5083343

http://www.ceme.gsras.ru

http://www.globalcmt/org/CMTsearch.html

Kabychenko NV, Gorbunova EM, Besedina AN (2020) Study of amplitude-frequency characteristics of water-saturated collector. AIP Conference Proceedings

Kissin IG (2015) Fluids in the Earth's crust. Geophysical and tectonic aspects. Nauka, Moscow, 328 p [in Russian]

Kitagawa Y, Itaba S, Matsumoto N, Koizumi N (2011) Frequency characteristics of the response of water pressure in a closed well to volumetric strain in the high frequency domain. J Geophys Res 116(B08 301):1–12

Kocharyan GG, Vinogradov EA, Gorbunova EM, Markov VK, Markov DV, Pernik LM (2011) Hydrologic response of underground reservoirs to seismic vibrations. Izv Phys Solid Earth 47(12):1071–1082. https://doi.org/10.1134/S1069351311120068

Kopylova GN, Boldina SV (2019) Hydrogeoseismic variations of the water level in the wells of Kamchatka. Kamchatpress, Petropavlovsk-Kamchatsky, 144 p [in Russian]

Liu L-B, Roeloffs E, Zheng X-Y (1989) Seismically induced water level fluctuations in the Wali well, Beijing, China. J Geophys Res 94:9453–9462

Liu C, Huang M-W, Tsai Y-B (2006) Water level fluctuations induced by ground motions of local and teleseismic earthquakes at two wells in Hualien, Eastern Taiwan. TAO 17(2):371–389

Petukhova SM (2020) Influence of exogenous factors on hydrogeological situation (by the example of GPO Mikhnevo). Processy v Geosredah 2(24):710–717 [in Russian]

Shi Z, Wang G, Manga M, Wang C-Y (2015) Mechanism of co-seismic water level change following four great earthquakes—insights from co-seismic responses throughout the Chinese mainland. Earth Planet Sci Lett 430:66–74

Shin DC-F, Wu Y-M, Chang C-H (2013) Significant coherence for groundwater and Rayleigh waves: Evidence in spectral response of groundwater level in Taiwan using 2011 Tohoku earthquake, Japan. J Hydrol 486:57–70

Sun X, Wang G, Yang X (2015) Coseismic response of water level in Changping well, China, to the Mw 9.0 Tohoku earthquake. J Hydrol 531:1028–1039

Terminological dictionary-reference of engineering geology KDU (2011) Moscow 952 p [in Russian]

Wang C-Y (2007) Liquefaction beyond the near field. Seismol Res Lett 78(5):512–517

Wang C-Y, Manga M (2010) Earthquakes and water. Springer-Verlag, Berlin Heidelberg, p 228

Weingarten M, Ge S (2014) Insights into water level response to seismic waves: A 24 year high-fidelity record of global seismicity at devils hole. Geophys Res Lett 41:1–7

Deformation Processes Modelling Throughout Underground Construction Within Megapolis Limits

D. L. Neguritsa, G. V. Alekseev, A. A. Tereshin, E. A. Medvedev, and K. M. Slobodin

Abstract The results of finite element modeling by applying geomechanical soil models in the PLAXIS 2D software package and with 3D deformation processes during renovation of transport and transfer hub are presented. On the basis of the analysis of the received values the conclusions about plastic flow of deformation process are made, at normalization of construction conditions and decay of deformation processes the return of position of building structures to the designed condition is predicted. Recommendations on deformation process monitoring during construction and operation of underground complex and objects and structures falling into the zone of influence of construction are developed.

Keywords Modeling · Deformation processes · Model of strengthening soil · Soil and rock mass · Building facilities · Surrounding buildings · Geomonitoring

D. L. Neguritsa (✉)
Sergo Ordzhonikidze Russian State University for Geological Prospecting, 23 Miklukho-Maklaya str, Moscow, Russia 117997
e-mail: neguritsadl@mgri.ru

G. V. Alekseev · A. A. Tereshin · E. A. Medvedev · K. M. Slobodin
National Research Moscow State University of Civil Engineering, 26 Yaroslavskoe Highway, Moscow, Russia 129337
e-mail: alexeev_german@bk.ru

A. A. Tereshin
e-mail: TereshinAA@mgsu.ru

E. A. Medvedev
e-mail: MedvedevEA@mgsu.ru

K. M. Slobodin
e-mail: SlobodinKM@mgsu.ru

T. Chaplina (ed.), *Processes in GeoMedia—Volume IV*, Springer Geology,
https://doi.org/10.1007/978-3-030-76328-2_23

1 Introduction

Prospects for development and renovation of the transport structure of megacities causes the necessity to build complex engineering structures in the form of multi-functional centers in areas with already developed infrastructure to optimize traffic and people flows, to create urban infrastructure for business purposes to provide a new level and range of services (Trubetskoy and Iofis 2007; Merkin and Konyukhov 2017). Construction of underground part is carried out in technogenically disturbed soil and rock massif. Complex conditions of construction of facilities with underground sections in Moscow city require to secure the engineering part of construction, surrounding buildings and supplying systems falling into zone of construction influence (Kartoziya 2015; Iofis and Neguritsa 2006; Kulikova 2018; Glozman 2016). This problem is solved comprehensively at the stage of engineering surveys, design and construction and at the beginning of construction operation before deformation processes stabilization (Neguritsa 2018; Broere 2012). To assess the impact of construction and set monitoring criteria calculations of displacements and deformations of soil and rock mass and structures falling into the construction zone are performed, in accordance with the chosen planning scheme of the facility.

The calculations performed by the finite element method are aimed to predict the impact of the new construction of the multifunctional center on the soil and rock mass, the surrounding area and engineering networks. Modeling of deformation processes includes the following stages:

- Study of available technical documentation of objects;
- Construction of a finite element geomechanical model of the soil and rock mass, which includes the object under construction and the surrounding buildings;
- Determination of soil and rock mass parameters to calculate on the basis of engineering and geological surveys;
- Calculation of stress–strain state (SSS) of base rocks in 3D and 2D problem formulation taking into account interaction with underground part of designed building;
- Analysis of SSS calculation taking into account stage-by-stage construction process;
- Plotting of SSS components isolines, including isolines of stress and displacement components in the soil and rock mass;
- Overall analysis of the obtained results.

Calculations are performed for 2D and 3D problem formulation in the PLAXIS 2D and in the PLAXIS 3D software packages respectively.

2 Analysis of Mining and Geological Conditions of Construction

The geological structure of the construction field of the multifunctional center facility with a bus station is based on the results of engineering and geological surveys carried out prior to design and construction.

In geomorphological terms, the site is located within a watershed water-glacial plain. Converted during the construction period, the relief of the site is flat, with a slight slope to the East. Construction field is located on an asphalted territory, which includes an old building of bus station. The underground space is obstructed by wide variety of engineering networks. Tracks of subway station reverse dead-end are located along the Western border of the site. Engineering and geological conditions of the construction site are of II (medium) complexity category. In terms of lithological composition and physical and mechanical properties, the soil and rock mass falling into the zone of influence of the projected facility is divided into 14 engineering-geological elements (hereinafter referred to as EGE).

When the underground part of the building is located 25 m below the ground level, the bottom of the foundation slab will be in water-saturated sands of morainic water-glacial debris (22 m below the first and 11 m below the second aquifer levels). Organization of the excavation is possible only under the "wall in the ground" protection. Hydrostatic pressure on foundations have to be taken into account when designing the excavation. Underground part of the building must have a reliable waterproofing. In terms of waterlogging the projected building area is considered as water logged in natural conditions. Organization of the excavation will be carried out in technogenic soils (IV category in terms of developability—15% of the total withdrawal amount), in the sand (I category—35%), loams with gravel and pebbles (III category—50%). Aggregate depth of seasonal freezing is 1.6 m. The construction site is under risk of karst-suffosion processes. There are no other negative processes besides water-logging and seasonal freezing.

The designed building has 6 aboveground and 5 underground floors and one technical floor. Maximum height mark of the building is +35 m. Designed underground part of the building in plan has rectangular shape with dimensions (in axes) 217.6 × 68.4 m. Spacing of bearing monolithic reinforced concrete constructions is 9.0 m.

3 Principal Technological Solutions

The constructive and technical decisions made in the project documentation take into account the following conditions of the construction field. Ground water level of the second aquifer is 9–10 m above the excavation bottom mark. Depth of the designed excavation from the Earth surface is 23–24.5 m. The fencing of the excavation is made in the form of 800 mm thick monolith trench wall in the soil.

The wall in the soil during the construction period (when the underground part is erected) and during the period of operation perceives horizontal loads from the ground, as well as loads from the columns of the above-ground part and five slabs resting on the wall along the entire perimeter of the underground part of the building, being a permanent load-bearing structure. The wall in the ground is designed for vertical load from the weight of the floors and the construction load from the materials and small appliances stored over—during the building period, as well as the constant load from the floors as the calculated (operating) load—during the functioning of the facility. The wall depth in the ground is 53 m upon condition that the wall is deepened in the waterproof layer of clay for at least 1.5 m.

To minimize the impact on the surrounding area, existing buildings and communications, construction of the underground part of the projected facility is carried out by the "top-down" method. When constructing the underground part of the building using the "top-down" method, the discs of slabs of the underground floors of the building under construction are used as spacing constructions ensuring strength and stability of the wall in the ground at excavation organization and development. The slabs are staged as the soil is withdrawn from the excavation and are based on temporary columns of rolled metal. Temporary columns are made according to the grid of designed columns of the facility underground part before the beginning of soil development in the foundation pit. The column is a rack made of rolled metal, which is inserted by the bottom end into the footing part in baretta form (slotted foundation) with the dimensions of 2.8×0.8 m, performed by a flat grapple technology "wall in the ground". After soil withdrawal and concreting the floor slabs, temporary columns are concreted and serve as part of permanent columns of the building during the operational period.

Such scheme of underground space construction of the projected building provides minimal horizontal deformations of the wall in the ground and, as a result, minimal additional subsidence of the surrounding territory, structures and communications located in the influence zone.

To reduce deformation, force and magnitude of wall deepening in the ground below the bottom of the excavation by pre-fixing the soil thickness in the zone of passive pressure (pinching the wall in the ground), an artificial geomassive is created from the ground-concrete elements with increased strength and deformation characteristics with the use of "jet-grouting", which is formed from the ground surface before organization of the excavation.

4 Modeling of Geomechanical Processes Promotion

The principle factors determining the possible impact of construction on the existing buildings and structures are changes in the stress–strain state of the soil base of existing facilities caused by the excavation organizing near them, changes in hydrogeological conditions (Gattinoni et al. 2014), impact of anthropogenic factors, such as dynamic effects caused by construction of deep foundations and enclosing structures of the excavation, the special types of "jet-grouting" works, injections and negative processes in the soil mass, associated with mining—suffusion processes, quicksands formation, thixotropic decompaction, etc.

Estimation of the impact of construction on the state of the surrounding buildings is performed by mathematical simulation of changes in the stress–strain state of the soil mass by the finite element method using geomechanical models of the soil in the PLAXIS 2D and 3D software packages. In PLAXIS 3D software package, which implements the finite element method to the volumetric problem, geomechanical model is created to determine the impact of construction on the surrounding buildings and structures.

Since PLAXIS 3D software package has limitations when modeling lenses in the soil-rock massif the three-dimensional geomechanical mathematical model is constituted on the basis of geomechanical elements (GME) of the soil-rock massif. Projected characteristics for GME are determined with the following assumptions: GME is a set of separate engineering-geological elements (EGE), combined into one element with relatively little difference in physical and mechanical characteristics of these EGE, the calculated characteristics of the GME are determined by the weakest EGE included in specific GME.

Seven GMEs are identified in the massif, combined by the affinity of physical and mechanical characteristics and the period of formation. Deposits having low power are not reflected in the geomechanical model, as they do not affect the stress–strain state of the system "basis-building".

The association of engineering-geological elements in geomechanical model is presented on engineering-geological section (Fig. 1). Comparison of measured characteristics values for individual EGEs as parts of GMEs is presented in Table 1.

In accordance with design features the underground part of the projected complex is modeled in full, the wall in the ground and floors as flat elements with corresponding strength characteristics. Absolutely flexible load from the building, transferred to the geomechanical elements at the bottom of the foundations, is set, and rigidity of the upper structure is not taken into account in calculations.

Pile foundations—pile bushes under the supports are modeled as massive bulk elements of reduced stiffness. The stiffness characteristics of elements are determined by the ratio of the total volume of the foundation piles to the total volume of pile and ground massif. At creation of geomechanical model underground communications as objects are not specified, but their deformations are determined on the basis of deformations of those areas of geomechanical model (clusters), in which underground communications are directly located.

The obtained three-dimensional model with dimensions 390 × 225 × 60 m (x, y, z) is divided into volumetric finite elements of tetrahedron shape. When constructing the finite element model, the presumed distribution of stress and strain fields is taken into account, which results in unevenness of the finite element grid in the calculated area. Boundary conditions in displacements—bonds are set on the boundaries of the finite element model. At the vertical boundaries of the computational domain, the

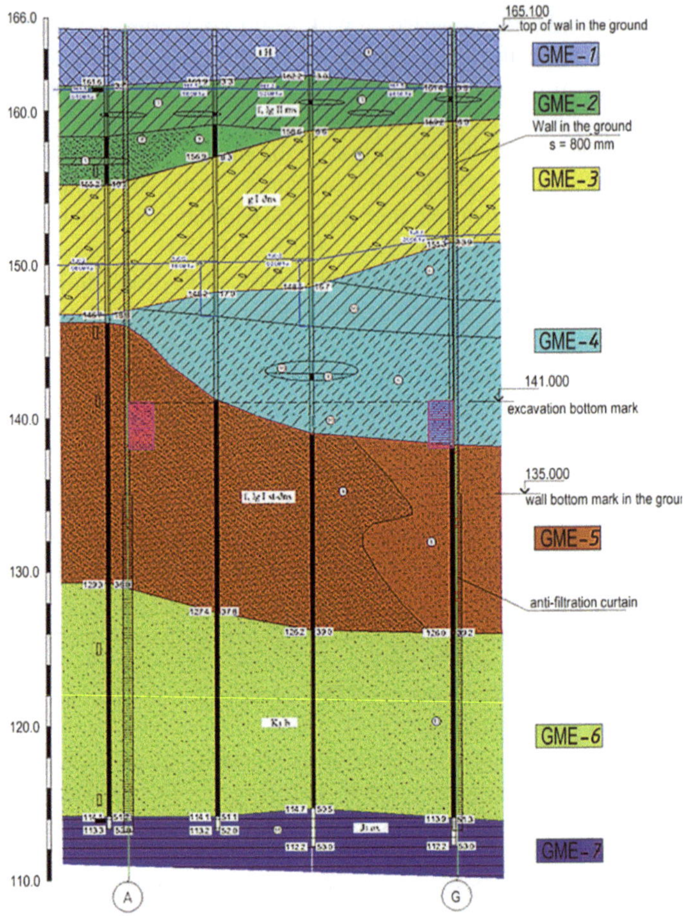

Fig. 1 Scheme of combining the geomechanical elements of the model

bonds are superimposed in the direction normal to the edges. At the lower boundary, the bonds are set in three directions. No bonds are set on the surface of model.

The weight of buildings and structures included in the geomechanical model in the calculations is transferred to the corresponding area of footing structures as uniformly distributed load. The following prerequisites are applied to calculations:

- Soil conditions are modeled using geomechanical soil elements compiled according to the data of engineering and geological surveys,
- Values of physical and mechanical characteristics of engineering and geological elements are calculated as for the determination of deformations,
- Constructions of the underground part, spacer system are erected by "top-down" method under protection of slabs in the form of reinforced concrete disks with technological openings,

Table 1 Characteristics of engineering and geological elements in the geomechanical model

GME No.	IEGs in GME	Stratigraphic Index	Density, g/cm^3		Specific clutch, MPa		Internal friction angle, grad		Deformation module, MPa	
			EHE	GME	EHE	GME	EHE	GME	EHE	GME
1	EHE-1	*tH*	1.96	1.96	10	10	20	20	10	10
2	EGE-2	*f,lg IIms*	2.08	2.08	0.032	0.032	15	15	20	20
	EGE-3		2.10		0.006		35		30	
	EGE-4		2.10		0.003		37		45	
3	EGE-7	*gIdns*	2.10	2.1	0.037	0.037	17	17	18	18
4	EGE-6	*gIdns*	1.99	1.99	0.023	0.023	18	18	14	14
5	EGE-8	*f,lgI st-dns*	2.11	2.10	0.006	0.008	37	36	45	42
	EGE-9		2.10		0.008		36		42	
6	EGE-12	*K1b*	2.11	2.11	0.007	0.007	35	35	39	39
	EGE-13		2.11		0.005		38		50	
7	EGE-14	*J3ox, J2k*	1.84	1.84	0.110	0.110	14	14	120	120

- Mathematical modeling of construction is carried out stage-by-stage in accordance with the construction stages,
- Only static loads were taken into account during modeling. Shock, dynamic, vibratory and other technological loads and impacts associated with the production are not taken into account in the calculations.

For modeling of mechanical behavior of excavation fencing structures, spacer structures and engineering-geological elements when changing their stress–strain state, the elastic model and the model of strengthening soil are used. The elastic model is used to simulate the effect of structure materials (concrete, metal) of the excavation fencing elements. Characteristics of the trestle is adopted on the basis of project documentation materials: structure scheme is single-span, distance between pillars amounts 42 m and more, the foundations of the facility—drilling piles, combined by reinforced concrete raft. Results of modeling are shown in Figs. 2, 3, 4, 5, and 6.

Analysis of obtained results at modeling of geomechanical processes evolution during construction of underground part of multifunctional center with transport and transfer hub (TTH) showed that zone of influence of underground part erection of the building on adjacent territory is 32–68 m. Maximum wall deflection in the ground reaches 40 mm. The maximum additional draught of the trestle pillar according to the results of calculations will be 0.8 mm, which is less than the permissible value of 1.0 mm determining the limit of construction process impact. The obtained values of additional pillar deformations will not affect the trestle operational reliability. The

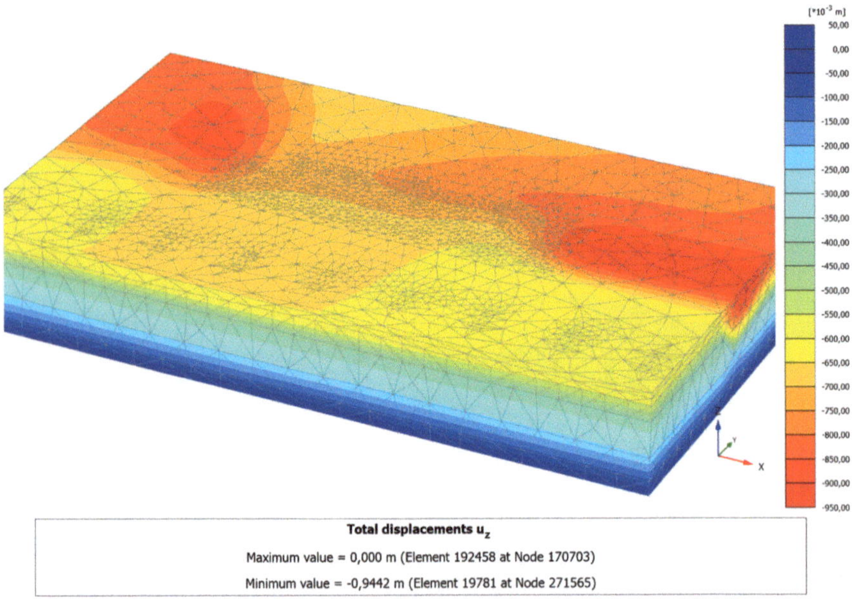

Fig. 2 Sediment isopole of the soil mass at the initial stage of modeling

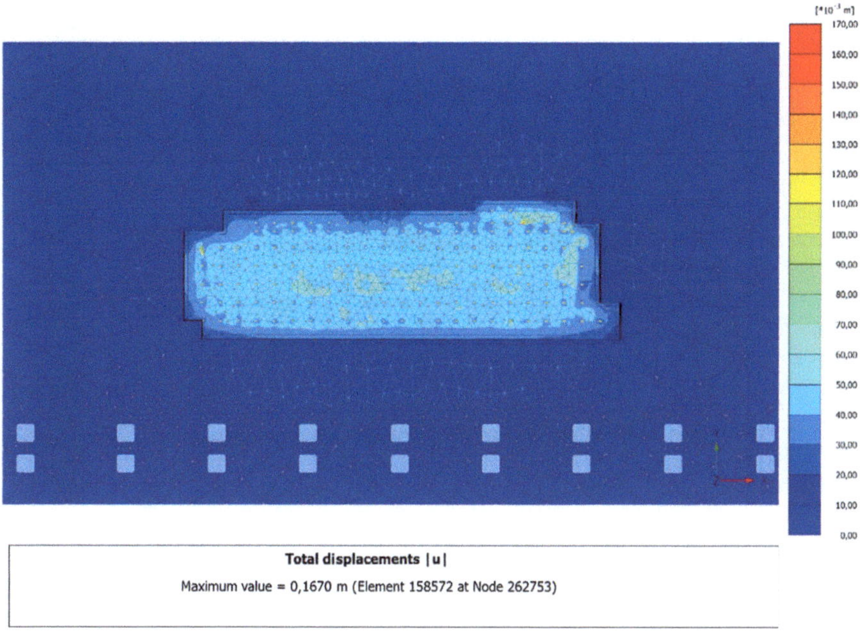

Fig. 3 Isopole of additional deformations of the soil mass at the last stage of calculations resource

obtained values of additional deformations of the surrounding territory structures do not exceed permissible values.

In general, the construction of the projected building, while the technology of construction works and the order of their execution are preserved, will not lead to excess of the maximum permissible values of additional foundation lowering of buildings of the surrounding facilities, located in the influence zone. The construction of the bus station building will not have a negative impact on the existing subway facilities.

5 Safety Recommendations for Construction and Operation

In order to ensure the safety of construction it is necessary to carry out measures to protect the environment during the new construction and after its completion. Geotechnical monitoring of the building under construction and existing buildings movement and communications displacement should be carried out in the zone of influence of new construction (Neguritsa et al. 2020; Iofis et al. 2020; Tereshin et al. 2017; Setan and Singh 2001).

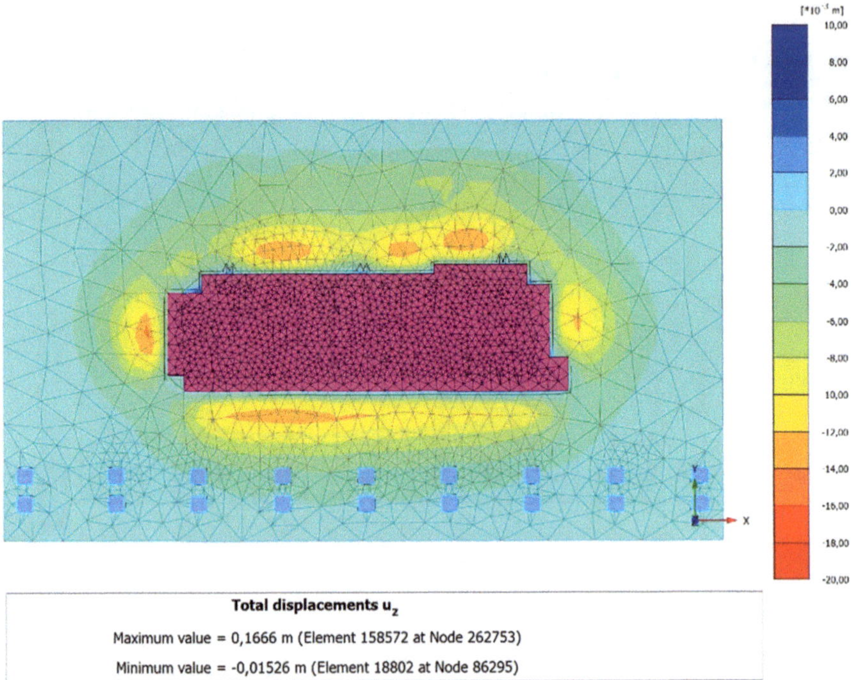

Fig. 4 Isopole of additional sediments of the surrounding soil mass at the last stage of calculations

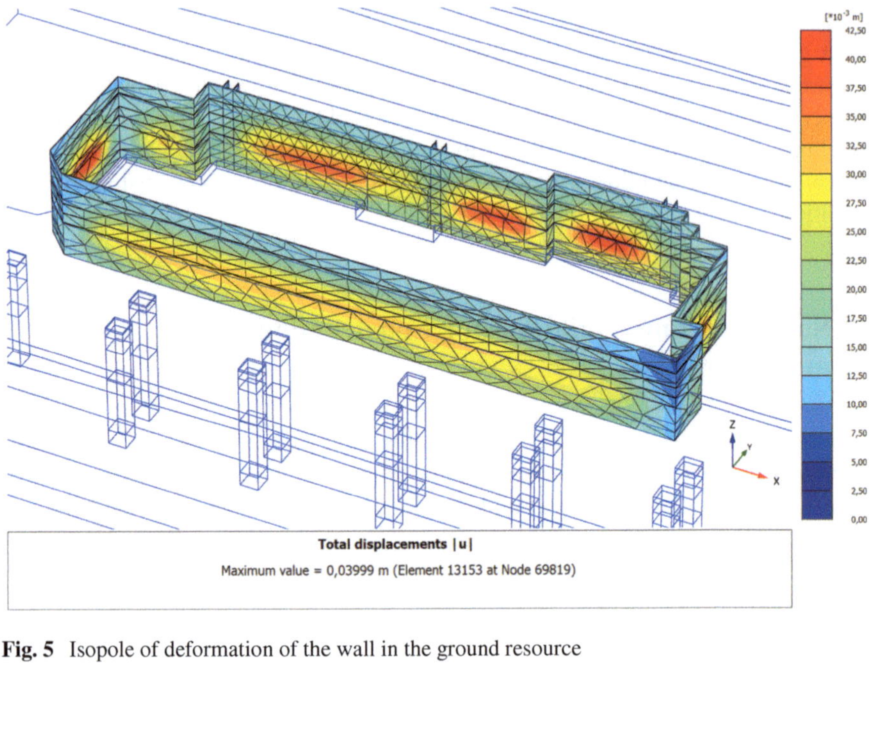

Fig. 5 Isopole of deformation of the wall in the ground resource

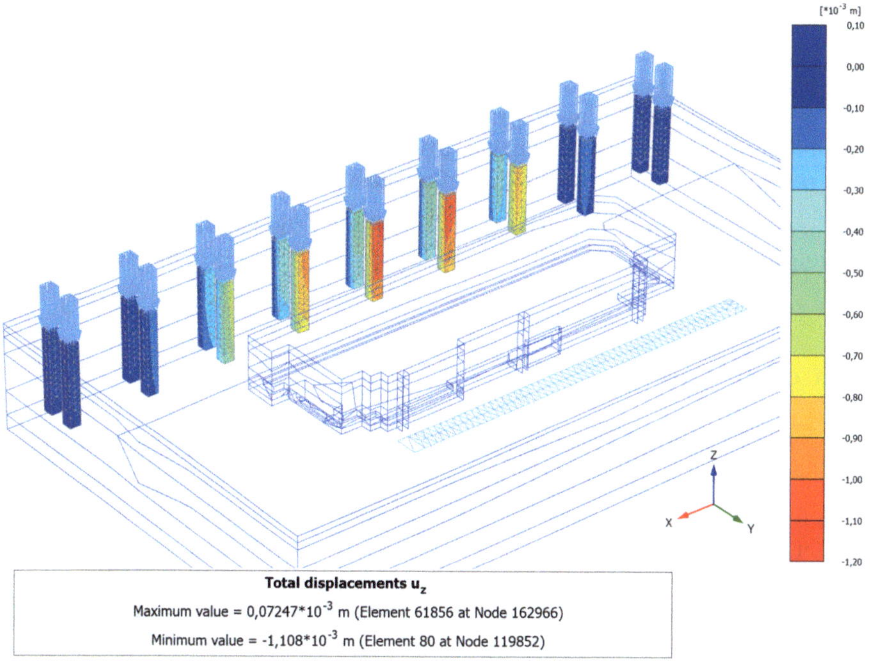

Fig. 6 Isopole of additional sediment from the overpass supports

When the underground part of TTH building is constructed under the protection of the wall in the ground, additional "technological" deformations of the wall and lowering of the ground surface caused by both deviations and errors in the construction work, and serious violations of regulatory and technological requirements.

In order to avoid additional deformations of the wall in the ground and, as a result, greater impact on the surrounding area, the following requirements and recommendations have to be met:

1. The soil shall be developed only when the wall in the ground is properly executed (without water leaks, non-concreted areas, gaps at the junction of concrete blocks, etc.). At detection of the wall defects during the organization of the excavation the defects must be eliminated in order to prevent soil deposition into the excavation, groundwater leaks and, respectively, the decompaction of the surrounding area. The defects repair should be performed by a specialized organization with an individual project.
2. When arranging the slab disks, it is necessary to ensure their tight fit to the wall in the ground.
3. Carry out control over existing communications in the area of construction influence. Do not allow leaks from the communication pipelines. If leaks occur, take measures to eliminate as soon as possible.

4. Run geodetic monitoring of horizontal deformations of the wall in the ground at each stage of the excavation organization and construction of the underground part, as well as monitoring of the surrounding territory in accordance with the geotechnical monitoring program.
5. Filling the wall trenches in the ground with bentonite solution should be performed only under the control of a special construction laboratory.

To reduce the deformation of the wall in the ground, the surrounding buildings and existing communications need to strengthen the soil in the area of the wall patching.

6 Conclusions

The analysis of dynamics of stress–strain state of soil and rock massif at construction of underground part of multifunctional center with TTH testifies to plastic flow of deformation process, at normalization of construction conditions and attenuation of deformation processes it is forecasted to return position of building structures to the designed condition. Recommendations are developed to follow deformation processes in the course of construction. During the construction of walls in the ground and erection of the projected building structures it is necessary to predict changes in deformation of the soil and rock massif, to carry out geomonitoring and geodetic monitoring of deformations of the building under construction, as well as existing buildings and communications located in the zone of influence of new facility. Geomonitoring should be carried out from the beginning of construction, during the entire construction period and continue for at least two years after completion.

References

Broere W (2012) Urban problems—underground solutions. In: Proceedings of the 13th world conference ACUUS, 2012, pp 1528–1539

Gattinoni P, Pizzarotti E, Scesi L (2014) Engineering geology for underground works. Engineering geology for underground works, vol 9789400778504. Published by Springer Netherlands, pp 1–305. https://doi.org/10.1007/978-94-007-7850-4_1

Glozman OS (2016) Underground planning of Moscow. Hous Constr 11:14–19 [in Russian]

Iofis MA, Neguritsa DL (2006) Monitoring of the state of structures when their bases are deformed. Mining Informational Anal Bull 10:138–143 [in Russian]

Iofis MA, Neguritsa DL, Esina EN (2020) Movement of rocks in the development of the earth's interior: monograph. RUDN, Moscow, 287 p [in Russian]

Kartoziya BA (2015) Development of underground space in large cities. New trends. Min Informational Anal Bull Sci Tech J S1:615–630 [in Russian]

Kulikova EY (2018) Defects of urban underground structure and their prediction. In: IOP conference series: materials science and engineering, p 012108. https://doi.org/10.1088/1757-899X/451/1/012108

Merkin VE, Konyukhov DS (2017) Main problems, tasks and prospects of development of underground space in Moscow. Metro Tunnels 1–2:18–23 [in Russian]
Neguritsa D (2018) The problems of monitoring the deformation processes in the integrated development of the underground space of metropolitan cities. In: E3S web of conferences, vol 56, p 02027. https://doi.org/10.1051/e3sconf/20185602027
Neguritsa DL, Tereshin AA, Medvedev EA (2020) Features of Geomonitoring of an underground complex during freezing of a soil-rock massif in the conditions of a megapolis. Processy v geosredah 3(25):831–838 [in Russian]
Setan H, Singh R (2001) Deformation analysis of a geodetic monitoring network. Geomatica 55(3):333–346
Tereshin AA, Neguritsa DL, Kirkov AE (2017) Restoration of reference points of observation stations during deformation monitoring. Herald Russia University Peoples' Friendship. Ser Eng Res 18(1):14–19 [in Russian]
Trubetskoy KN, Iofis MA (2007) State and problems of development of the underground space of the city of Moscow. Surveyor's Bull 4(62):27–30 [in Russian]

Analytical and Experimental Modelling of the Hydrocarbon's Spot Form and Its Spreading on the Water Surface

A. V. Kistovich, T. O. Chaplina, and E. V. Stepanova

Abstract The effect of different physical and chemical properties on the hydrocarbons spreading process on the water surface is described, as well as the comparison of analytically acquired relations for the form of the boundary of the hydrocarbon spot with observed in experiments for different parameters of experiments. The equations, which define quasi-stationary form of the oil patch on water surface, are numerically solved. The experiments on the compact oil spot area growth under different physical conditions are made. The defining conditions for hydrocarbon spot spreading rate on the initial volume, viscosity and surface tension of oil are pointed. It is shown that oil spill patch area growth with time is greater in fresh water than in saline, and also depend on oil volume and properties.

Keywords Petroleum · Analytical model · Surface tension · Hydrocarbon spreading

1 Introduction

Petroleum (oil), gas, and coal remain the main natural sources that cover humanity's energy necessities. Currently, about 10–15% of the reserves of explored coal deposits, and about 65–70% of the reserves of oil fields are exploited. Emergency oil releases are possible at all stages of its transportation from the well outfall to the oil processing plants. Accidents with oil leaks occur during the production, collection and storage of oil, during discharge, the release of petroleum products to consumers, during transportation through pipelines, etc. The amount of leaks reaches from 5 to 17% of the extracted volume according to various sources (Bazhenova et al. 2000).

Leakage is not only loss of valuable resources, but also a cause of significant damage to the environment. The causes for the hydrocarbons release into environment, the characteristics of sources, and the pollution amount estimation at oil and gas production facilities are widely considered (Zatsepa et al. 2018; Gritsenko et al.

A. V. Kistovich · T. O. Chaplina (✉) · E. V. Stepanova
Ishlinsky Institute for Problems in Mechanics of the Russian Academy of Science, Moscow, Russia
e-mail: chaplina_to@inbox.ru

T. Chaplina (ed.), *Processes in GeoMedia—Volume IV*, Springer Geology,
https://doi.org/10.1007/978-3-030-76328-2_24

1997; Davydova and Tarasov 2004; Zyryanov and Shurganova 2015). Once released into the environment, petroleum hydrocarbons cause disruption of the biological balance of ecosystems for a long time.

Ocean pollution with oil products issues and, in particular, the dynamics of oil patches on the water surface areas began to be actively engaged in the sixties of the twentieth century. According to many researchers, the oil film is one of the most common forms of existence of oil as an ocean pollutant.

When planning and carrying out works to deal with accidental spills in the Sea, it becomes necessary to predict the spreading of oil. Such prognosis allow to warn about possible oil pollution of the coastal zone, about the intersection with intensive economic activity areas, ship courses, etc. Spreading of oil in the sea in case of emergency spills is a complex process, which requires taking into account large number of various factors. The lifetime of an oil slick depends non-linearly on the wind velocity and volume of the initial spill. The results of simulations show that when the wind velocity changes from 2.5 to 25 m/s, the spot lifetime decreases by 16 times, and when the wind velocity changes from 10 to 25 m/s, it only decreases by 3 times (Lobkovsky et al. 2005; Ovsienko 2005).

In recent years, various States have made great efforts to improve systems for preventing and eliminating the consequences of accidental oil and oil product spills, but the problem still remains relevant. In order to reduce possible negative consequences, special attention is required to study the methods of localization, elimination of oil spills and development of an additional set of measures for collection and disposal of hydrocarbons that have fallen into the external environment.

This paper presents the results of a study of the process of spreading hydrocarbons with different physical and chemical properties on the water surface, as well as a comparison of analytical expressions that describe the established form of the surface of the hydrocarbon slick on the water with observed in experiments for different values of the experiment parameters.

2 Evolution of the Oil Patch on the Water Surface

Oil is a natural oily combustible liquid, a mixture of hydrocarbons. The physical and chemical properties of oil can vary over a wide range depending on its composition (Driatzkaya et al. 1975; https://petrodigest.ru/).When oil hits the water surface, it immediately begins to undergo changes due to the influence of various physical and chemical factors. This interaction of oil with the environment can be divided into several stages: spreading, evaporation, dissolution, emulsification, dispersion, oxidation, and dispersion.

Different processes take different times and involve different fractions of oil—some occur with light components, others with heavier ones, and the composition of petroleum products on the surface changes under their influence. Spreading and evaporation predominate on hours time scales, emulsification and dispersion play the most significant role on scales of several days, while deposition and biodegradation—on scales of more than a week. After a week, the light fractions evaporate, and the remaining hydrocarbons undergo biodegradation. Oil has the ability to tighten solid particles from water, after which the formed aggregates acquire negative buoyancy—solid sediment falls to the bottom (Zatsepa et al. 2018). The process of degradation of an oil slick on the water surface can take up to several years.

Oil spreading in open water occurs under the influence of wind, surface currents, surface tension and gravity forces, forming areas covered with thin oil films. In the absence of wind and currents 1 m^3 of crude oil can spread over an area with a radius of 50 m in 1.5 h (Gritsenko et al. 1997).

The physical model of petroleum spreading in absence of wind and currents is based on the regarding only of surface tension and gravitational forces. So in the vicinity of contact line of these three media (water, petroleum, air) the surface tension is greater than gravitational and inertial forces in 10^9 times, the angles of the mutual sloping of the media on the contact line are defined only by the action of surface tension and are invariable for any sizes and forms of petroleum spot.

For this reason the quasi-static model based on the equation of the equilibrium of contact line is used for the description of the petroleum spot form. The scheme of the contact line is shown in Fig. 1, where σ_{wa}, σ_{wp}, σ_{pa} are the surface tension forces acting normally to the contact line along the boundary surfaces "water-air", "water-petroleum" and "petroleum-air" respectively.

The provision of the contact line equilibrium has the form, expressed by the angles shown in Fig. 1:

$$\sigma_{wa} + \sigma_{wp} + \sigma_{pa} = 0 \tag{1}$$

The relation (1) is divided into two equations

$$\sigma_{wa} = \sigma_{pa} \cos \alpha_{wa} + \sigma_{wp} \cos \alpha_{wp}, \sigma_{pa} \sin \alpha_{wa} = \sigma_{wp} \sin \alpha_{wp} \tag{2}$$

The presumed form of the petroleum (oil) spot on the water surface is presented in Fig. 2, where ζ, η, θ are the water-air, petroleum-air and water-petroleum boundaries respectively, R_p is the spot's radius.

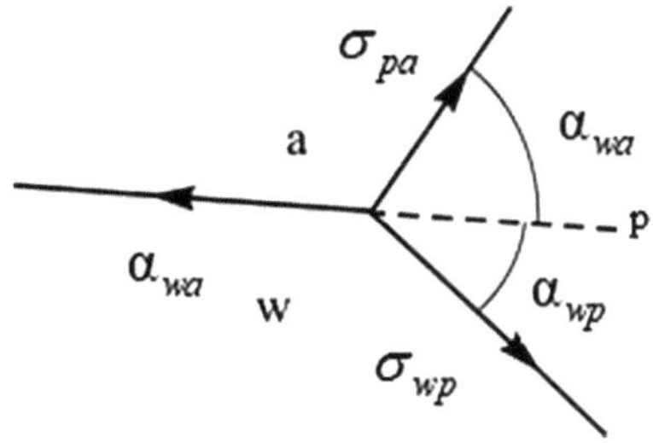

Fig. 1 The surface tension forces on the contact line of the three media

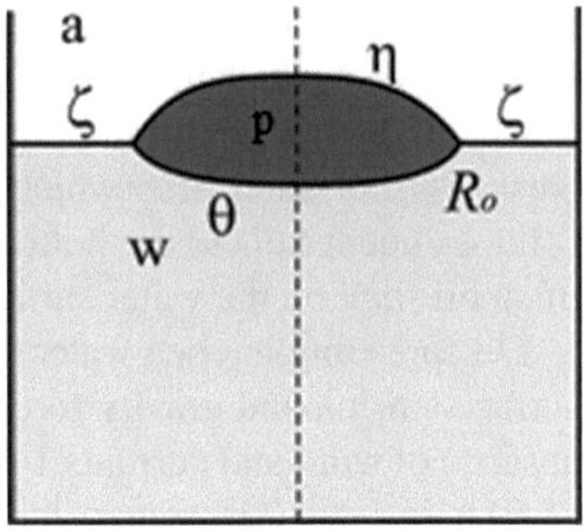

Fig. 2 The form of the oil spot

In the model, the water and the petroleum are considered as incompressible media enclosed into vertical cylindrical container of height H and radius R and the atmospheric pressure is constant. The pressures in the water and petroleum are presented in the forms.

$$p_w^{(1)} = p_a + \rho_w g(\zeta - z) + q_w^{(1)}(r, z), \quad \text{when} \quad R_0 \le r \le R \tag{3}$$

$$p_w^{(2)} = p_a + \rho_p g(\eta - \theta) + \rho_w g(\theta - z) + q_w^{(2)}(r, z), \quad \text{when} \quad 0 \le r \le R_0 \tag{4}$$

$$p_p = p_a + \rho_p g(\eta - z) + q_p(r, z), \quad \text{when} \quad 0 \le r \le R_0 \tag{5}$$

Here $q_w^{(1)}(r, z), q_w^{(2)}(r, z)$ and $q_p(r, z)$ are the parts of pressure fields due to surface tension. For the description of the boundary conditions the shapes of boundaries between the media are defined by the relations.

$$S_{wa} : z = \zeta(r); \quad S_{pa} : z = \eta(r); \quad S_{wp} : z = \theta(r) \tag{6}$$

Then the dynamic boundary conditions have the following forms.

$$\begin{aligned} &p_w^{(1)} - \sigma_{wa} \mathrm{K}_{wa}\big|_{z=\zeta(r)} = p_a, \quad p_p - \sigma_{pa} \mathrm{K}_{pa}\big|_{z=\eta(r)} = p_a, \\ &p_w^{(2)} - p_p - \sigma_{wp} \mathrm{K}_{wp}\big|_{z=\theta(r)} = 0 \end{aligned} \tag{7}$$

where K_i are the curvatures of the interfacial areas which are defined by the formula

$$\mathrm{K}(f) = -\frac{f''_{rr} + \left(1 + f_r'^2\right) f'/r}{\left(1 + f_r'^2\right)^{3/2}} \tag{8}$$

in that the f function denotes one of the corresponding boundaries $\zeta(r)$, $\eta(r)$ or $\theta(r)$.

On the contact line (when $r = R_p$) the condition of the continuity for all the physical fields is relevant. In the model the evaporation is not taken into account then the conditions of conservation of water and petroleum volumes are used (V_p is petroleum volume).

$$\int_{R_0}^{R} \zeta(r) r\,dr + \int_{0}^{R_0} \theta(r) r\,dr = \frac{HR^2}{2} \tag{9}$$

$$2\pi \int_{0}^{R_0} (\eta(r) - \theta(r)) r\,dr = V_p \tag{10}$$

On the rigid cylindrical sidewall the water surface has the fixed angle α_w, which depends on the physical properties of the cylinder material and the boundary condition.

$$\zeta_r^{'}|_{r=R} = \operatorname{ctg}\alpha_w \tag{11}$$

From the boundary conditions (7) the relation

$$\sigma_{wa}K_{wa} = \sigma_{pa}K_{pa} + \sigma_{wp}K_{wp} \tag{12}$$

is followed also as equations

$$\begin{aligned} &\rho_w g\zeta + \sigma_{wa}K_{wa}|_{z=\zeta(r)} = a, \quad \rho_p g\eta + \sigma_{pa}K_{pa}\big|_{z=\eta(r)} = b, \\ &(\rho_w - \rho_p) g\theta + \sigma_{wp}K_{wp}\big|_{z=\theta(r)} = a - b \end{aligned} \tag{13}$$

where a and b are some constants.

The relations (13) describe the "lens" of oil on the water surface as it is shown in Fig. 2. In the explicit form the first boundary condition of (13) is specified by the relation.

$$\zeta - \frac{\sigma_{wa}}{\rho_w g} \frac{\zeta_{rr}^{''} + \left(1 + \zeta_r^{'2}\right)\zeta_r^{'}/r}{\left(1 + up\zeta_r^{'2}\right)^{3/2}} = \frac{a}{\rho_w g} \tag{14}$$

The introduction of the dimensionless radial coordinate $\overline{r} = k_{wa} r$ and surface deviation $\overline{\zeta} = k_{wa}(\zeta - a/(\rho_w)g)$, where $k_{wa} = \sqrt{\rho_w g / \sigma_{wa}}$ reforms (14) into equation.

$$\overline{\zeta} - \frac{\overline{\zeta}_{\overline{r}\overline{r}}^{''} + \left(1 + \overline{\zeta}_{\overline{r}}^{'2}\right)\overline{\zeta}_{\overline{r}}^{'}/\overline{r}}{\left(1 + \overline{\zeta}_{\overline{r}}^{'2}\right)^{3/2}} = 0 \tag{15}$$

From the second boundary condition (13) the equation of the same form for dimensionless variables $\overline{r} = k_{pa}\, r$ and $\overline{\eta} = k_{pa}(\eta - b/(\rho_p g))$, $kp_{pa} = \sqrt{\rho_p g/\sigma_{pa}}$ follows.

Analogously, from the third condition (13) the equation also in the same form is followed for the dimensionless variables $\overline{r} = k_{wp}\, r$ and $\overline{\theta} = k_{wp}(\theta - (a-b)/(\rho_w - \rho_p)g)$, where $k_{wp} = \sqrt{(\rho_w - \rho_p)g/\sigma_{wp}}$.

Thus the all introduced dimensionless surfaces are subjects of the unique equation.

$$F - \frac{F''_{\xi\xi} + \left(1 + F'^2_{\xi}\right) F'_{\xi}/\xi}{\left(1 + F'^2_{\xi}\right)^{3/2}} = 0, \tag{16}$$

where F and ξ denote the corresponding dimensionless variables.

Despite the fact that different regions are scaled differently one important property arises as a result of the performed transformations which is necessary for solving (16).

$$F'_{\xi} = \begin{Bmatrix} \zeta'_r \\ \eta'_r \\ \theta'_r \end{Bmatrix} \tag{17}$$

Here and later symbols F_{ζ}, F_{η} and F_{θ} denote the dimensionless surfaces corresponding to the surfaces ζ, η and θ, and symbols F'_{ζ}, F'_{η} and F'_{θ} denote the derivatives with respect to the corresponding dimensionless radial coordinate.

From relation (1) on the contact line follows

$$\begin{aligned} \operatorname{tg}\alpha_{wp} &= \frac{\sqrt{4\sigma^2_{wa}\sigma^2_{wp} - \left(\sigma^2_{wa} + \sigma^2_{wp} - \sigma^2_{pa}\right)^2}}{\sigma^2_{wa} + \sigma^2_{wp} - \sigma^2_{pa}} = \left.\frac{F'_{\theta} - F'_{\zeta}}{1 + F'_{\theta}F'_{\zeta}}\right|_{r=R_0} \\ \operatorname{tg}\alpha_{wa} &= \frac{\sqrt{4\sigma^2_{wa}\sigma^2_{wp} - \left(\sigma^2_{wa} + \sigma^2_{wp} - \sigma^2_{pa}\right)^2}}{\sigma^2_{wa} + \sigma^2_{wp} - \sigma^2_{pa}} = \left.\frac{F'_{\zeta} - F'_{\eta}}{1 - F'_{\zeta}F'_{\eta}}\right|_{r=R_0} \end{aligned} \tag{18}$$

The equations $\zeta(R_0) = \eta(R_0) = \theta(R_0)$ on the contact line have the form.

$$\frac{a}{\rho_w g} + \left.\frac{F_{\zeta}}{k_{wa}}\right|_{\xi = k_{wa}R_0} = \frac{b}{\rho_p g} + \left.\frac{F_{\eta}}{k_{pa}}\right|_{\xi = k_{pa}R_0} = \frac{a-b}{(\rho_w - \rho_p)g} + \left.\frac{F_{\theta}}{k_{wp}}\right|_{\xi = k_{wp}R_0} \tag{19}$$

From the conservation laws the equation follow

$$k_{wa}^{-3}\int\limits_{k_{wa}R_0}^{k_{wa}R} F_\zeta(\xi)\xi\,d\xi + k_{wp}^{-3}\int\limits_0^{k_{wp}R_0} F_\theta(\xi)\xi\,d\xi = \frac{HR^2}{2} - \frac{a\left(R^2 - R_0^2\right)}{2\rho_w g} - \frac{(a-b)R_0^2}{2(\rho_w - \rho_p)g}$$
$$k_{pa}^{-3}\int\limits_0^{k_{pa}R_0} F_\eta(\xi)\xi\,d\xi - k_{wo}^{-3}\int\limits_0^{k_{wp}R_0} F_\theta(\xi)\xi\,d\xi = \frac{V_p}{2\pi} - \frac{bR_0^2}{2\rho_p g} + \frac{(a-b)R_0^2}{2(\rho_w - \rho_p)g} \tag{20}$$

Now to get the final results it is necessary to solve (16). This equation is nonlinear hence its solutions in the general case are inaccessible (except the trivial case of the flat surface $F = 0$ but in this case the petroleum patch is absent).

The implementation of priorly stated information of the quasi-flat water surface on the boundary "water-air" permits to use the approximation $\left|F_\zeta' << 1\right|$. Then on the base of (18) for the real data for surface tension coefficients $\sigma_{wa} \approx 0.073$, $\sigma_{pa} \approx 0.033$, $\sigma_{wp} \approx 0.050$ N/m it follows that on the "petroleum-air" boundary the estimation is valid

$$\max\left(\left|F_\eta'\right|\right) = \max\left(\left|\eta_r'\right|\right) = \operatorname{tg}\alpha_{wa} \approx 0.72, \tag{21}$$

and on the "water-petroleum" boundary the result has place

$$\max\left(\left|F_\theta'\right|\right) = \max\left(\left|\theta_r'\right|\right) = \operatorname{tg}\alpha_{wp} \approx 0.42. \tag{22}$$

The presented estimations (21, 22) permit to rewrite the Eq. (16) in the approximate form for the whole diapason of the ξ variation

$$F - F_{\xi\xi}''\left(1 - \frac{3}{2}F_\xi'^2\left(1 - \frac{5}{4}F_\xi'^2 + \frac{35}{24}F_\xi'^4\right)\right)$$
$$- \frac{1}{\xi}F_\xi'\left(1 - \frac{1}{2}F_\xi'^2\left(1 - \frac{1}{4}F_\xi'^2 + \frac{5}{8}F_\xi'^4\right)\right) \approx 0, \tag{23}$$

the solution of that satisfies the condition $F_\xi'\Big|_{\xi=0} = 0$ (due to the cylindrical symmetry of the spot) has the form

$$F = AI_0(\xi) + \frac{A^3\xi^4}{64}\left(1 + \frac{5\left(2 + A^2\right)\xi^2}{36} + \frac{\left(82 + \frac{677}{4}A^2 + \frac{261}{4}A^4\right)\xi^4}{2304} + \ldots\right), \tag{24}$$

where I_0 is the modified Bessel function.

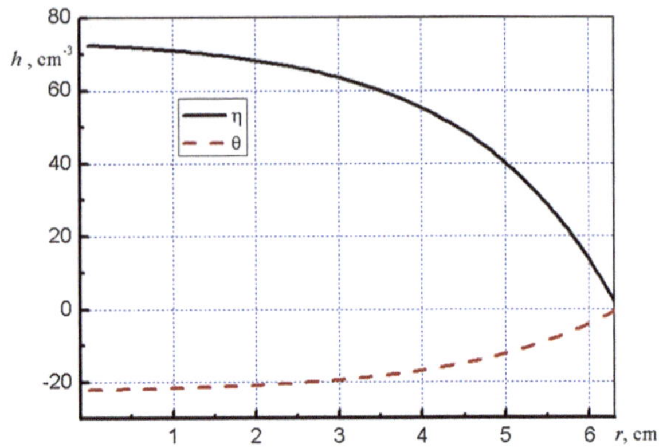

Fig. 3 The calculated boundaries of petroleum spot on the water surface

To present in the explicit form the form of the spot it is necessary to find six values A_ζ, A_η, A_θ, R_0, a and b for the definition of that there are six equations and six unknown dimensionless coefficients. The numerical solution is obtained by means of "Maxima" software product (http://maxima.sourceforge.net/) with implementation of the built-in sets of functions. The graphic representation of the solution is shown in Fig. 3.

3 Experimental Study of Hydrocarbon Spreading Over the Water Surface

The experimental study of the dynamics of spreading of a hydrocarbon spill-spot on the water surface under various controlled conditions is performed. Experiments are carried out in fresh and saline water with different volumes of admixture (various organic, mineral and synthetic oils, petroleum itself). The wide range of the defining parameters variability allows to study the basic properties of flows, taking into account the essential similarity criteria and modelling conditions. The initial spot area and spreading process are monitored using video recording. A large number of experiments are carried that allow to reliably identify the laws of the area growth of the hydrocarbon spill-spot on the water surface at rest.

Experiments have shown that oil spill patch area growth rate with time is greater in fresh water than in saline, and also depend on the admixture volume and properties. The area growth dynamics is presented in Fig. 4 for machine oil patch. Also, observations have shown that the area of spreading in fresh water is sufficiently larger than in saline water.

It is necessary to note that in the majority of the hydrocarbons spreading models the condition is accepted that the interpenetration of oil into water and vice-versa is absent. Actually such interpenetration has place almost all the time when the portion of oil is spilled on the water surface. In this case in the water in the vicinity of the oil

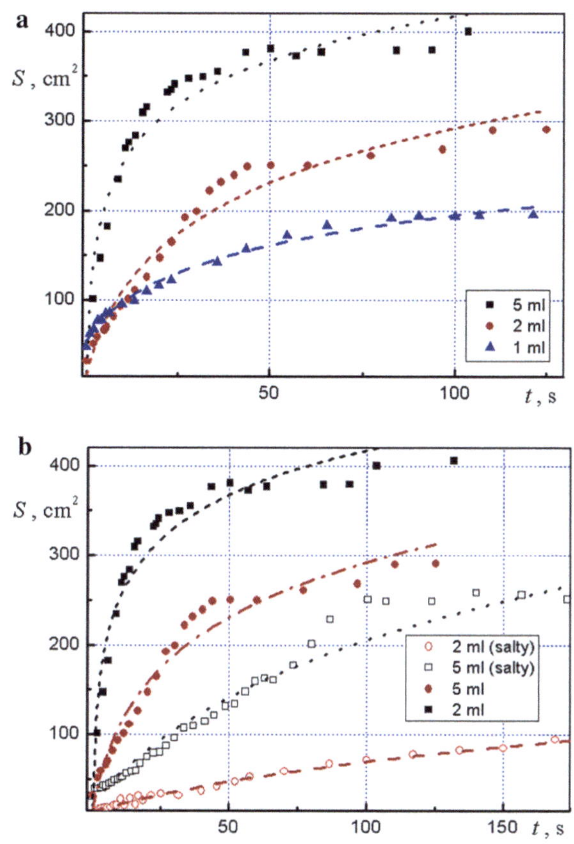

Fig. 4 Spill area evolution for machine oil: **a** different oil spill volumes; **b** comparison of oil patches for fresh and saline water with respect to admixture amount

drop surface the emulsion "oil in water" is formed consisting of the small oil spheres surrounded by water. Herein the greater the surface tension coefficient is the bigger are these spheres (oil drops). In the saline water due to complex physical–chemical processes the very fine-dispersed emulsion is formed. For the correct description of the energy balance it is necessary to consider the intake for the emulsion's formation both in fresh and saline water. The surface area of the emulsion in the saline water is sufficiently greater than in the fresh one. Thus on the surface of saline water the drop of oil will spread slower than on the surface of fresh water in the exact agreement to the experimental results.

4 Conclusions

The differential equations, which define the "oil body" form, and its solutions for the quasi-stationary form of the hydrocarbon spot on the water surface are obtained. The experimental study of the process of the compact spot area growth on the water surface under different physical conditions (temperature and salinity of water) is performed. On the saline water surface the drop of oil spread slower than on the surface of fresh water. The conditionality for hydrocarbon spot spreading rate on the initial volume, viscosity and surface tension of oil is revealed.

The work was carried out with the financial support of the project of the Russian Federation represented by the Ministry of Education and Science of Russia No. 075-15-2020-802.

References

Bazhenova OK, Burlin Yu K, Sokolov BA, Hain VE (2000) Geology and geochemistry of oil and gas. Moscow state University publishing House, Moscow, 384 p (in Russian)

Davydova SL, Tarasov VI (2004) Oil and oil products in the environment. Russian Peoples' Friendship University Publishing House, Moscow, 163 p (in Russian)

Driatzkaya ZV, Mician MA, Pomace NM (1975) Oil of the USSR. Chemistry, Moscow, 81 p (in Russian)

Gritsenko AI, Akopova GS, Maksimov VM (1997) Ecology. Oil and gas. Nauka, Moscow, 597 p (in Russian)

https://petrodigest.ru/

Lobkovsky LI, Levchenko DG, Leonov AV, Ambrosimov AK (2005) Geo-ecological monitoring of marine oil and gas areas. Nauka, Moscow, 326 p (in Russian)

Maxima http://maxima.sourceforge.net/

Ovsienko SN (2005) Modeling of oil spills and assessment of environmental impact risk. In: Proceedings of GOIN 209, pp 248–271 (in Russian)

Zatsepa SN, Ivchenko AA, Solbakov VV, Stanova VV (2018) Forecasting the spread of petroleum products in the event of an emergency spill in marine areas. AO "Finpol", Moscow, 140 p (in Russian)

Zyryanov VN, Shurganova SV (2015) Spreading of oil spots on the sea surface taking into account weathering. Processy v geosredah 1:30–36 (in Russian)

Fluid Conductivity of Natural Shear Fractures in Vicinity of a Production Well During Directional Unloading

N. V. Dubinya

Abstract The paper presents a theoretical study of spatial orientations of hydraulically conductive natural shear fractures existing in well surrounding rock masses. Critically stressed fractures concept is used to analyze all possible spatial orientations of shear fractures existing in a stress field altered by presence of stress concentration caused by well. Evolution of these spatial orientations is studied under conditions of directional unloading — a procedure aimed at enhancing filtration properties of rock masses near the well. It is shown that changes in stress state near the wellbore caused by directional unloading lead to alteration of spatial orientations of hydraulically conductive fractures. This alteration results into involvement of the larger number and variety of natural shear fractures into filtration paths leading to an overall increase of filtration properties of well surrounding rock masses.

Keywords Directional unloading · Critically stressed fractures · Reservoir geomechanics

1 Introduction

Problems of geomechanics frequently occur at different stages of hydrocarbon reservoir exploration and development. One has to deal with mechanical behavior of fluid-saturated rocks, study stresses and strains, emergence and development of fractures. Current paper focuses on mechanical behavior of natural shear fractures existing in fluid-saturated rock masses subjected to varied stress state.

Natural fractures play a major role in fluid filtration in rocks with low matrix permeability. In fact, in some rocks almost all filtration paths for fluid are provided by open natural fractures. These filtration paths are dependent on stress state of the rock: processes of natural fractures opening and closure may be caused by changes in stress state of the rock (Rogers 2002; Mattila and Follin 2019). But current stress state is not the only factor influencing permeability of natural fractures: other factors

N. V. Dubinya (✉)
Moscow Institute of Physics and Technology, Dolgoprudny, Moscow Region, Russia

Schmidt Institute of Physics of the Earth, Moscow, Russia

T. Chaplina (ed.), *Processes in GeoMedia—Volume IV*, Springer Geology,
https://doi.org/10.1007/978-3-030-76328-2_25

include geological factors and temperature (Fisher et al. 2003), apertures of fractures (Bisdom et al. 2016) and stress history (Sathar et al. 2012).

One of the basic concepts used to study relationship between stress state of a naturally fractured rock mass and its filtration properties is the concept of critically stressed fractures (Barton et al. 1995). According to this concept, a fracture is considered to be critically stressed if shear stress acting on its surface is large enough to allow a whenever differential stress is slightly increased. It was shown that critically stressed fractures are closely related to hydraulically conductive fractures: fractures which are critically stressed under current stress state tend to be hydraulically conductive and vice versa. A number of studies (Rogers 2002; Mattila and Follin 2019; Townend and Zoback 2000; Ligtenberg 2005; Massiot et al. 2017) were carried out to verify this hypothesis at different objects. That provided a basis to develop approaches of using stress fields obtained from numerical geomechanical modeling of reservoir behavior to predict positions and properties of zones containing hydraulically conductive fractures, analyze their evolution with time, and incorporate the obtained results into hydrodynamic models of reservoir development (Schwab et al. 2017; Bayuk et al. 2019).

There is a variety of specific problems related to usage of this concept to predict behavior and filtration properties of natural fractures under specific conditions, including but not limited to: patterns of hydraulically conductive fractures in vicinity of large faults, at different depths and stress regimes, near wells, during reservoir development, considered in our previous studies (Bayuk et al. 2019; Dubinya 2019, 2020). The current paper is devoted to analysis of behavior of natural fractures existing in vicinity of a wellbore subjected to directional unloading technique (Karev et al. 2020) from the perspective of critically stressed fractures concept. The following problems are considered: once fluid pressure at well is varied leading to changes in the stress state of the rock masses surrounding the well, what happens to critically stressed fractures? How does variety of critically stressed and therefore hydraulically conductive fractures change? Does the preferable direction of fluid filtration through paths provided by hydraulically conductive fractures change when pressure at well changes and what is degree of this change? These questions may be answered with help of critically stressed fractures concept.

2 Theoretical Approach

2.1 Critically Stressed Fractures Concept

Once again, critically stressed fractures were introduced as natural shear fractures with shear stresses acting on their surfaces being large enough for a slip to occur alongside the surface whenever differential stress is slightly increased. In terms of mechanics this condition may be written as:

$$\tau_n \geq \mu\sigma_n \quad (1)$$

Here τ_n is shear stress acting on a fracture plane, σ_n is normal stress acting on its plane, μ is internal friction coefficient. In the current study any fracture is considered as a plane, its spatial orientation is determined through three direction cosines of the normal vector (pole) to the fracture plane. All stresses considered from now on are effective meaning that fluid pressure is taken into account through Biot's coefficient.

Internal friction coefficient is not the main point of interest in the current study, so following many similar studies we consider it to have a typical value of 0.6 following the well-known experimental results (Byerlee 1978) and keeping in mind the opportunity to change this coefficient while applying the same method to particular rock masses.

In order to check whether a certain fracture is critically stressed, one has to calculate normal and shear stresses acting on its surface. These stresses depend on stress state of the rock mass containing the fracture and its spatial orientation. There are two ways to study current stress state: one may use the full stress tensor σ_{ij} (i, j = 1, 2, 3 for three-dimensional space) or principal stresses σ_1, σ_2, and σ_3 alongside with their directions. A following numeration rule will be used throughout the paper: $\sigma_1 \geq \sigma_2 \geq \sigma_3$, compressive stresses are positive. Any fracture is considered as a plane, its spatial orientation is determined through three direction cosines of the normal vector (pole) to the fracture plane. Once six independent parameters (six stress tensor components or three principal stresses and three direction cosines) are known alongside with spatial orientation of the fracture plane, condition (1) may be checked for the considered fracture after solving the equations:

$$\sigma_n = \sigma_2 + l^2(\sigma_1 - \sigma_2) + n^2(\sigma_3 - \sigma_2), \quad (2)$$

$$\tau_n^2 = (\sigma_2 - \sigma_3)^2/4 + l^2(\sigma_1 - \sigma_2)(\sigma_1 - \sigma_3) - (\sigma_n - 0.5(\sigma_2 + \sigma_3))^2 \quad (3)$$

Here l is cosine of the angle between vector normal to the fracture plane and direction of maximum principal stress σ_1, n is cosine of the angle between normal to the fracture plane and direction of minimum principal stress σ_3.

Although these equations are generally used to check, whether any particular fracture is critically stressed under the given stresses, the inverse problem may be solved as well. One may obtain all possible spatial orientations of fractures being critically stressed for the given stress tensor. Direction cosines l and n of all critically stressed fractures provide the variety normal (2) and shear stresses (3) that satisfy criterion (1). It was shown (Dubinya 2019, 2020) that these cosines may be found for an arbitrary stress tensor. Moreover, although being very complicated in mathematical terms, this solution is analytical and explicit. One may refer to paper (Dubinya 2020) for a detailed description of this solution.

The analytical form of the described solution provides an opportunity to easily apply it to solve various problems of reservoir geomechanics. In fact, once stress state of naturally fractured rock mass is calculated for any problem, one may consequently obtain all possible spatial orientations of critically stressed fractures that also tend to be hydraulically conductive for any representative volume considered in the problem. It is to be mentioned, that technically one may consider only spatial orientations of critically stressed fractures but not their number: a few information may be obtained regarding relative quantity of these fractures and even less can be said about quantitative analysis of permeability of these fractures due to the complicated relationship between fractures mechanical and conductive properties (Rogers 2002; Mattila and Follin 2019; Fisher et al. 2003; Bisdom et al. 2016; Ligtenberg 2005).

The current paper is mainly focused at patterns of critically stressed fractures in vicinity of a production well. It is well-known that any well drilled in rock mass acts as a stress concentrator, providing a complicated stress field in its vicinity. This leads to an disturbance of critically stressed fractures spatial orientations in vicinity of the well. This disturbance is analyzed in the current paper.

2.2 *Stress State in Vicinity of a Vertical Well*

Solution to the stress field in vicinity of a well drilled in a rock mass originates from the well-known Kirsch's problem of stress concentration around a circular hole (Kirsch 1898). In the current paper we will consider a vertical well of a circular section drilled in fluid-saturated rock mass. We will use the modified Kirsch's solution taking into account vertical stress, two-directional loading in horizontal plane, pressure at well, and pressure of fluid saturating the rock mass (Jaeger et al. 2007; Zoback et al. 1985). All stresses are considered as effective, Biot's coefficient is taken to be equal to one.

$$\begin{cases} \sigma_r = \dfrac{\sigma_H + \sigma_h}{2}\left(1 - \dfrac{R^2}{r^2}\right) + \left(P_{well} - P_{por}\right)\dfrac{R^2}{r^2} \\ \qquad + \dfrac{\sigma_H - \sigma_h}{2}\left(1 - 4\dfrac{R^2}{r^2} + 3\dfrac{R^4}{r^4}\right)\cos 2\theta, \\ \sigma_\theta = \dfrac{\sigma_H + \sigma_h}{2}\left(1 + \dfrac{R^2}{r^2}\right) - \left(P_{well} - P_{por}\right)\dfrac{R^2}{r^2} \\ \qquad - \dfrac{\sigma_H - \sigma_h}{2}\left(1 + 3\dfrac{R^4}{r^4}\right)\cos 2\theta, \\ \sigma_z = \sigma_V - 2\upsilon(\sigma_H - \sigma_h)\dfrac{R^2}{r^2}\cos 2\theta, \\ \tau_{r\theta} = -\dfrac{\sigma_H - \sigma_h}{2}\left(1 + 2\dfrac{R^2}{r^2} - 3\dfrac{R^4}{r^4}\right)\sin 2\theta. \end{cases} \tag{4}$$

The equations are written in a cylindrical coordinate system with its axis coinciding with well axis (Fig. 1). σ_V is vertical effective stress acting at the considered

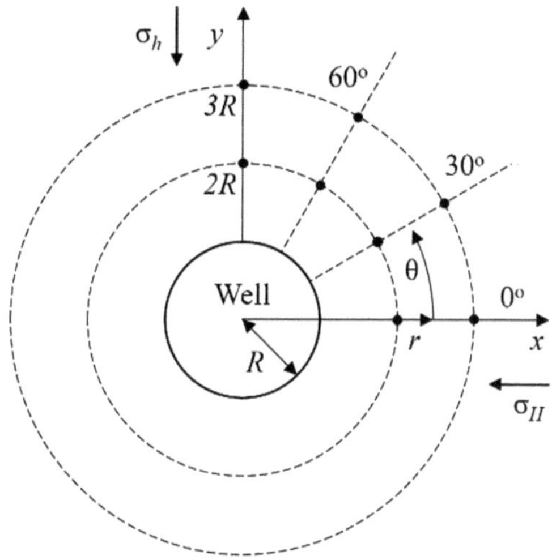

Fig. 1 Schematic representation of the considered problem

depth, σ_H and σ_h are effective maximum and minimum horizontal stresses respectively. θ is polar angle measured from the direction of σ_H, r is distance from wellbore center. R is the radius of the well, P_{well} is pressure of the fluid at the well, P_{por} is fluid pressure at the considered point. v is Poission's ratio—the third equation of system (4) is only valid for elastic medium (Jaeger et al. 2007). Tangential stresses τ_{rz} and $\tau_{\theta z}$ are equal to zero providing that one of the principal stresses is always vertical.

Any point in the vicinity of the well may be characterized by six components of stress tensor obtained from system (4). One may calculate corresponding magnitudes and spatial orientations of principal stresses as well after solving the eigenvector problem. Moreover, it is possible obtain the stress tensor components for Cartesian coordinate system with x axis aimed at direction of maximum horizontal stress and y aimed at σ_h direction (see Fig. 1).

All these parameters may be obtained for any valid set of input parameters: far-field stresses σ_h, σ_H and σ_V, and well pressure P_{well}. If some of these parameters are altered, stress state changes in well vicinity.

The current study focuses on changes in stress field near the wellbore and corresponding changes in spatial orientations of hydraulically conductive fractures immediately after change in pressure at well—analogous to direct unloading method of enhancing filtration properties of well surrounding rock masses (Karev et al. 2020). We calculate stress state of the rock mass using system (4) for two conditions: before unloading (for initial values of input data parameters that will be listed below) and immediately after the drop of well pressure value. Several points (shown as black dots at Fig. 1) are considered as follows: all components of effective stress tensor are calculated using system (4); the problem of eigenvectors is solved at each point providing principal stresses magnitudes and directions; principal stresses are used in Eqs. (2) and (3) to obtain all the direction cosines l and n corresponding to all fractures that are critically stressed under current conditions for the considered point.

It is worth mentioning that these cosines are related to current directions of principal stresses. That means that if alteration of well pressure leads to rotation of principal stresses directions, spatial orientations of critically stressed fractures are altered as well.

2.3 Visualization of Results

In the current paper lower hemisphere Lambert azimuthal equal-area projections will be used to visualize spatial orientations of critically stressed fractures. In the current case, construction of these projections consists in transforming the set of possible pairs of direction cosines l and n into a set of strike and dip angles of critically stressed fractures (Allmendinger et al. 2012) (the procedure is discussed in detail in Dubinya (2020)). The following approach is used to show any direction on lower hemisphere Lambert equal area stereonet. A sphere of unit radius is constructed from the center of coordinate system; latitude and longitude of the point of intersection between this sphere and any vector are calculated. This is a tool to visualize both principal stresses directions and spatial orientation of any plane (through direction of normal vector to this plane). Figure 2 represents a typical solution of the problem of finding spatial orientations of critically stressed fractures for an arbitrary stress tensor: diagram of direction cosines l–n is shown to the left, and lower hemisphere projection of critically stressed fractures domain is shown to the right. Solid blue line borders the domain of critically stressed fractures spatial orientations on both figures.

Red, yellow, and green circles at the right figure represent directions of maximum σ_1, intermediate σ_2, and minimum σ_3 principal stresses—the same notation will be used hereafter. It is worth mentioning that there are four typical areas on the stereonet corresponding to spatial orientations of critically stressed fractures. In this case one of these areas is the ellipse-like area in the North-Western part of the figure; the other one is located in the Eastern side. Small area near the Western direction is in fact the continuation of the second area—its presence is caused by consideration of only lower hemisphere.

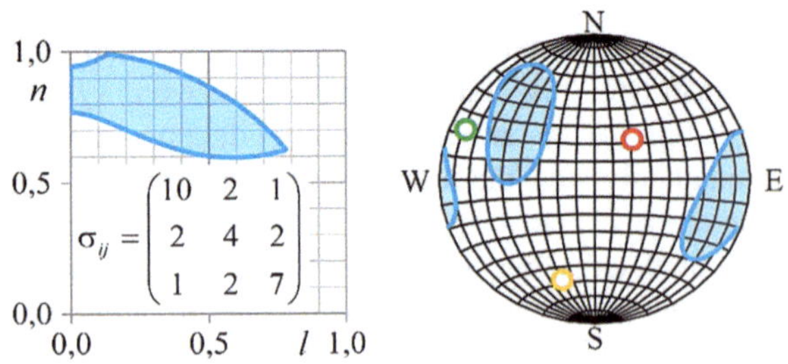

Fig. 2 Typical solution of critically stressed fractures spatial orientations problem

Although l–n diagram has a clearer mathematical meaning, stereonet contains information on both critically stressed fractures orientations and principal stresses direction in a global coordinate system being more informative. As a result, only stereonets will be used hereafter.

It is also worth mentioning, that all stresses are in linear relationship in condition (1) providing an opportunity to normalize stress tensor components: if all components are divided by the same positive value, spatial orientations of critically stressed fractures do not change. As a result, we normalize all stress tensor components in a global coordinate system as we divide them by the smallest stress tensor component.

The variety of critically stressed fractures spatial orientations may be qualitatively represented by the area of the zone shown in the left figure. Strictly speaking this area is equal to the number of fractures that are critically stressed under given conditions normalized by the total amount of fractures in the representative volume in an 'idealized' medium: a fractured medium with spatial orientations of natural fractures being distributed homogeneously. As far as this assumption is clear to be wrong, this area may only be considered as the qualitative representation of number of critically stressed fractures. To make a quantitative estimation one has to consider a function dependent on this area and accumulated inelastic strain in the medium (Bayuk et al. 2019).

The problem considered in the paper may be stated as follows: how do the domains of critically stressed fractures at stereonets change during directional unloading in well surrounding rock masses?

3 Results and Discussion

Eight different points in vicinity of a well were considered from the perspective of critically stressed fractures during directional unloading. These points are shown as black dots at Fig. 1: two different radii r ($r = 2R, 3R$) and four different polar angles ($\theta = 0°, 30°, 60°, 90°$) were considered. Well pressure changes have a smaller influence on stress state for greater radii, while greater polar angles reveal the same behavior of natural fractures due to symmetry of the considered problem. One specific set of input data was considered: effective vertical stress σ_V was assumed to be equal to 60 MPa, maximum horizontal stress σ_H was taken as 30 MPa, and minimum horizontal stress σ_h was considered to be equal to 15 MPa. With far-field stresses being in relationship $\sigma_V \geq \sigma_H \geq \sigma_h$ Normal Fault regime was considered. Initial pore pressure was considered to be equal to 20 MPa, initial pressure at well was 15 MPa. During directional unloading pressure at well was decreased by 10 MPa.

Stereonets with spatial orientations of critically stressed fractures and directions of principal stresses were constructed for all mentioned points before and right after directional unloading (Figs. 3, 4, 5 and 6). Blue solid line at these figures serves as borderline for the zone of critically stressed fractures before directional unloading, red line corresponds to the stress state immediately after unloading. In some cases, change at well pressure leads to considerable changes in principal stresses directions.

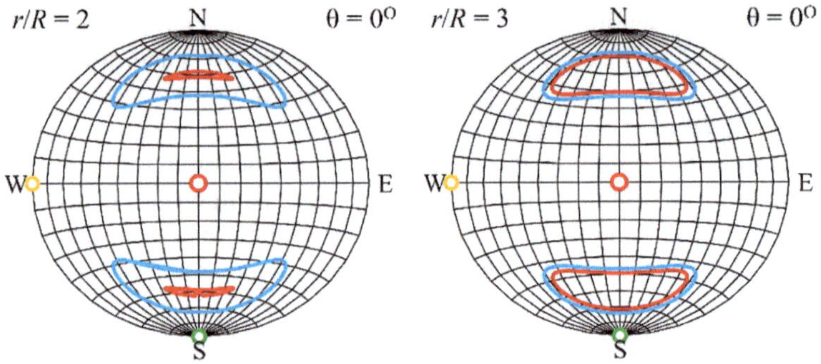

Fig. 3 Stereonets for σ_H direction ($\theta = 0°$)

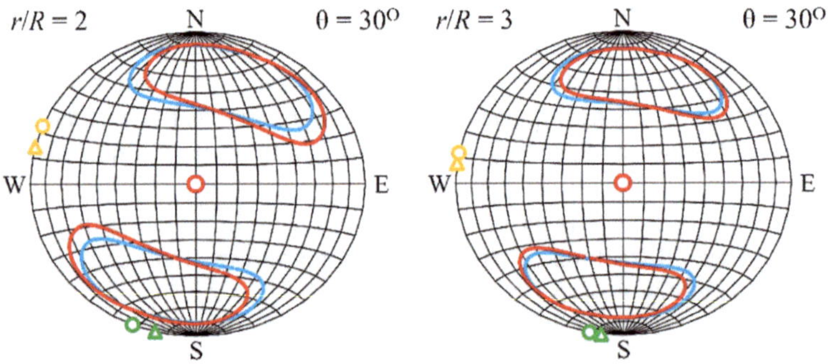

Fig. 4 Stereonets for the first intermediate direction ($\theta = 30°$)

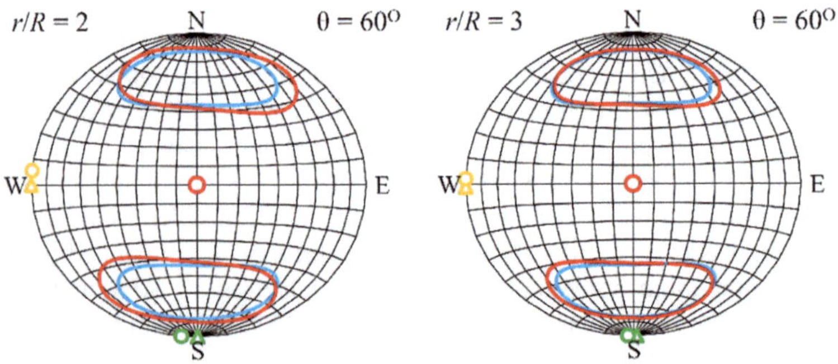

Fig. 5 Stereonets for the second intermediate direction ($\theta = 60°$)

In these cases, two directions of principal stresses are plotted at the same figure: circles represent directions of principal stresses after unloading, while triangles of the same color scheme represent their initial directions. Once again, red corresponds to maximum principal stress σ_1, yellow corresponds to intermediate principal stress

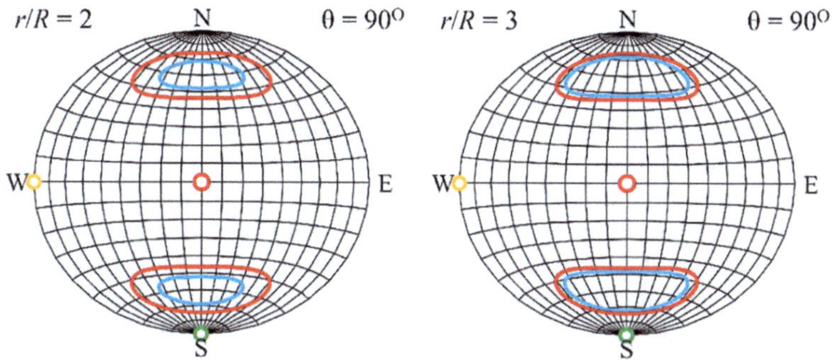

Fig. 6 Stereonets for σ_h direction ($\theta = 90°$)

σ_2, and green represents minimum principal stress σ_3. Maximum far-field horizontal stress σ_H is considered to act in EW direction.

Figure 3 represents evolution of spatial orientations of critically stressed fractures during directional unloading in the direction of maximum far-field horizontal stress σ_H ($\theta = 0°$). The left figure represents a point close to the well ($r = 2R$), the right figure is plotted for the point relatively far from the well ($r = 3R$).

We can draw several qualitative conclusions from this figure. First of all, in this particular direction the variety of critically stressed fractures (the area bordered by solid line) contracts. This effect is considerable near the well ($r = 2R$) but becomes less important with the growth of distance to the wellbore. That means that a limited number of natural fractures that were critically stressed before directional unloading close during this process, but this effect matters in a relatively small zone near the wellbore in direction of maximum horizontal stress σ_H. It is also worth mentioning, that principal stresses directions are not rotated during unloading for $\theta = 0°$.

The situation is different for another direction in horizontal plane. The case of $\theta = 30°$ is shown in Fig. 4. The left part, once again, represents a point closer to the well than the point represented by the right figure.

We observe a considerable rotation of both principal stresses directions and zones of critically stressed fractures for this case. The areas of the critically stressed fractures spatial orientation zones grow during directional unloading. That means that directional unloading leads to involvement of new natural fractures into filtration paths accompanied by closure of the previous filtration paths. One may conclude that the total amount of hydraulically conductive fractures remains relatively constant during unloading for this polar angle. At the same time, preferable direction of filtration paths is a little bit altered providing new paths for fluid filtration.

Figure 5 is plotted for the second polar angle between directions of σ_H and σ_h ($\theta = 60°$). All notations are maintained.

The positive effect of directional unloading is much more evident for this direction compared to the previous ones. Preferable direction of filtration paths is altered a little, but the number of fractures involved into these paths definitely grows after the drop of well pressure. This effect may be also seen at the distant point ($r = 3R$):

new natural fractures join the fluid conductive fractures network even for this point. Unloading does not lead to closure of the fractures that previously acted as filtration paths.

Finally, direction of minimum principal stress σ_h is analyzed with the help of Fig. 6 with maintained notations.

Here we can see the evident growth of the variety of natural fractures involved into filtration paths during directional unloading. This positive effect is comparable to the negative one shown in Fig. 3: directional unloading leads to simultaneous closure of hydraulically conductive fractures in the zones near the well at direction of maximum horizontal stress and opening of previously closed natural fractures in the zones near the well at direction of minimum horizontal stress. According to all the obtained results, one might expect that these two processes are in balance so that filtration properties of fractures do not change a lot if the whole zone around the well is considered. Nevertheless, there are two points to be mentioned. Firstly, not only the total number of fractures, but also their preferable spatial orientations matter. As it was shown in Figs. 4 and 5, directional unloading leads to rotation of the zones of critically stressed fractures around vertical axis. That means that totally new, not affected by previous filtration processes, fractures are involved into the conductive fracture network. One might expect that these fractures had better filtration properties enhancing filtration properties. Secondly, as it has already been mentioned, the area of the zone of critically stressed fractures spatial orientations is technically not the same as their number. One has to consider accumulation of inelastic strain to predict the number of fractures in medium (Bayuk et al. 2019) and it has been shown (Karev et al. 2020) that this accumulation really takes place during directional unloading.

Figures 7 and 8 represent the qualitative estimation of relative number of fractures that are hydraulically conductive during directional unloading. Figure 7 is plotted for the closer set of points ($r/R = 2$), Fig. 8 is for distant dots ($r/R = 3$). Horizontal axis represents the magnitude of change in well pressure during unloading, vertical

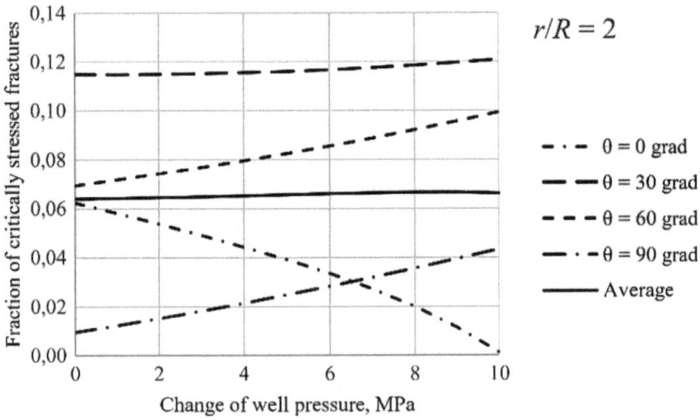

Fig. 7 Estimation of number of critically stressed fractures during unloading ($r/R = 2$)

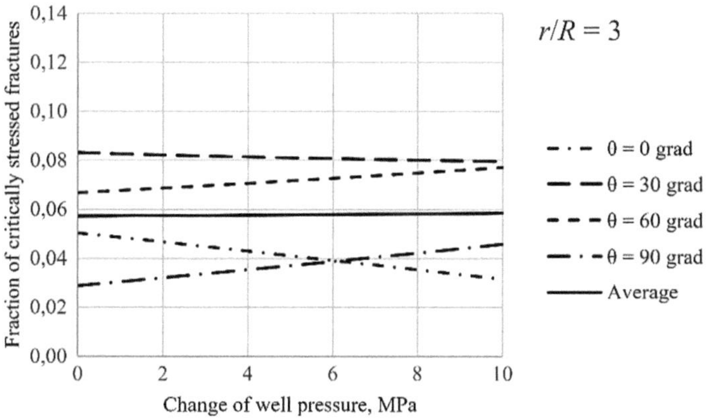

Fig. 8 Estimation of number of critically stressed fractures during unloading ($r/R = 3$)

axis represents that qualitative estimation of hydraulically conductive fractures—area of the zone at direction cosines diagram l–n (Fig. 2). Once again, this area only represents the total amount of hydraulically conductive fractures divided by the total amount of natural shear fractures in representative volume if all the fractures and their spatial orientations are distributed homogeneously. This value will be referred to as fraction of critically stressed fractures. Different dotted lines represent different points for which Figs. 3, 4, 5 and 6 were plotted. Solid line represents the average value of fraction of critically stressed fractures.

It can be clearly seen that the relationship between fraction of critically stressed fractures, change of well pressure, and polar angle is strongly nonlinear, while increase of radius leads to decrease of the effect, as expected. The average fraction of critically stressed fractures remains generally constant with stress state variations having a negligible effect on it. Interestingly enough, it is also nonlinear and has its maximum—it is achieved with well pressure $P_{well} = 7$ MPa for the considered case.

Once again, fraction of critically stressed fractures should be calculated with respect to real distribution of fractures spatial orientations to become more than a qualitative assessment (Bayuk et al. 2019).

4 Conclusions

The presented results are related to mechanical behavior of natural shear fractures existing in vicinity of a vertical well during directional unloading procedure. It was shown that changes of pressure at well leads to corresponding changes in patterns of critically stressed and therefore hydraulically conductive fractures. These changes are not always beneficial: immediately after directional unloading natural fractures existing in the zones located in the directions of maximum horizontal compression tend to close and can't serve as paths for fluid filtration any more. Nevertheless, this negative effect is countered: in other zones a number of initially closed fractures open and join fracture network for fluid flow. More importantly, preferable spatial orientation of hydraulically conductive natural fractures is altered: that means that hydraulically conductive fracture network becomes more diverse, enhancing the overall filtration properties of well surrounding rock masses.

Several important questions were not covered by this study. First of all, only one stress regime was considered, yet the algorithm to natural fracture characterization during directional unloading proposed in this paper may be easily applied to other stress conditions. Secondly, with the known solutions for stress concentration around horizontal or slant wells analogous results may be obtained for these cases as well.

Current study did not cover the problem of pore pressure field evolution. Although one could have used the known solution for pore pressure distribution around a production well, it would be wrong to use it without any modification. In fact, during directional unloading new fractures may be formed around the well which results from inelastic strain accumulation. These fractures will interact with natural fractures considered in this paper, some of them will become hydraulically conductive, and some will not. This will definitely lead to a complicated system of fluid filtration paths leading to anisotropy of rock mass permeability. It is expected that one has to carry out coupled hydrodynamic and geomechanical modeling to obtain valid results regarding stress state of well surrounding rock masses and, therefore, spatial orientations of hydraulically conductive fractures.

The reported study was funded by RFBR, project number № 20–05-00629.

References

Allmendinger RW, Cardozo N, Fisher DM (2012) Structural geology algorithms vectors and tensors. Cambridge University Press, Cambridge

Barton CA, Zoback MD, Moos D (1995) Fluid flow along potentially active faults in crystalline rocks. Geology 23(8):683–686

Bayuk IO, Dubinya NV, Garagash IA, Tikhotskiy SA, Tikhotskaya OA (2019) Multiscale rock-physics modeling of effective elastic properties of fractured reservoir rocks. Proceedings of the 53rd US rock mechanics/geomechanics symposium, New York, NY, USA, 23–26 June 2019:1–9

Bisdom K, Bertotti G, Nick HM (2016) The impact of different aperture distribution models and critical stress criteria on equivalent permeability in fractured rocks. J Geophys Res: Solid Earth 121(5):4045–4063

Byerlee J (1978) Friction of Rocks. Pure Appl Geophys 116:615–626

Dubinya NV (2019) Patterns of hydraulically conductive fractures' positions and spatial orientations near large faults. Proceedings of the SPE Russian petroleum technology conference, Moscow, Russia, October 22–24:1–21

Dubinya NV (2020) An analytical solution for spatial orientations of critically stressed fracture problem and its application for transitional stress regimes. Proceedings of the 53rd US rock mechanics/geomechanics symposium, Golden, Colorado, USA, 28 June-1 July 2020:1–9

Fisher QJ, Casey M, Harris SD, Knipe RJ (2003) Fluid-flow properties of faults in sandstone: the importance of temperature history. Geology 31(11):965–968

Jaeger JG, Cook NGW, Zimmerman RW (2007) Fundamentals of rock mechanics, 4th edn. Blackwell Publishers, Hoboken

Karev V, Kovalenko, Y, Ustinov K (2020) Directional unloading method is a new approach to enhancing oil and gas well productivity. Geomechanics of oil and gas wells. Adv Oil Gas Explor Prod:155–166

Kirsch EG (1898) Die Theorie der Elastizitat und die Bedurfnisse der Festigkeitslehre, Zeitschrift des Verlines Deutscher Ingenieure. Zeitshrift Des Vereines Deutscher Ingenieure 42:797–807

Ligtenberg JH (2005) Detection of fluid migration pathways in seismic data: implications for fault seal analysis. Basin Res 17:141–153

Massiot C, Nicol A, McNamara DD, Townend J (2017) Evidence for tectonic, lithologic, and thermal controls on fracture system geometries in andesitic high-temperature geothermal field. J Geophys Res: Solid Earth 122(8):6853–6874

Mattila J, Follin S (2019) Does in situ state of stress affect fracture flow in crystalline settings? J Geophys Res: Solid Earth 124(5):5241–5253

Rogers S (2002) Critical stress-related permeability in fractured rocks. Fracture and in situ stress characterization of hydrocarbon reservoirs. Geological Society of London, Special Publications 209:7–16

Sathar S, Reeves HJ, Cuss RJ, Harrington JF (2012) The role of stress history on the flow of fluids through fractures. Mineral Mag 76(8):3165–3177

Schwab DR, Bidgoli TS, Taylor MH (2017) Characterizing the potential for injection-induced fault reactivation through subsurface structural mapping and stress field analysis, Wellington Field, Sumner County, Kansas. J Geophys Res: Solid Earth 122:10132–10154

Townend J, Zoback MD (2000) How faulting keeps the crust strong. Geology 28(5):399–402

Zoback MD, Moos D, Mastin L, Anderson RN (1985) Well bore breakouts and in situ stress. J Geophys Res 90:5523–5530

Hypothesis of a Possible Cause of Warming in the Arctic Due to Energy Dissipation of Intensive Two-Phase Vortices in Earth's Mantle

S. Yu. Kasyanov

Abstract According to the considered hypothesis, the source of warming in the Arctic is the heat dissipation during the retardation of divercely directed jets of two intense two-phase (vapor–liquid) vortices in the Earth's mantle in the zone of their contact, located under Eastern Siberia. The substance displacement in the vortices occurs at large hydrostatic pressure drops. Therefore, evaporation of substance takes place on the ascending vortex branches. The resulting vapors can reach the permafrost layer and the Ocean floor. The heat release at vapors condensation can cause abnormal methane outflows from gas hydrates on the Eastern Arctic shelf and in coastal tundra, the formation of lava and mud volcanoes, the emergence of new islands (including sandy ones), as well as the formation of the Laptev and East Siberian Sea zone of increased heat fluxes from the lower to the upper layers of the Ocean and Great Siberian ice hole, which are located North of the Novosibirsk Islands and cover their environs.

Keywords Methane · Arctic warming · Ice hole · Global warming · Permafrost degradation · Gas hydrates · Shelf · Methane bomb · Vortices · Earth mantle

1 Introduction

The temperature increase occurring on Earth, which is most expressed in the Russian Arctic, is associated with an increase in atmospheric CO_2 concentration as a greenhouse gas. Climate models take into account the influence of water and CO_2 in the atmosphere on temperature and consider scenarios of the climate system change at various anthropogenic CO_2 emissions (see, for example, (Volodin et al. 2013)). The influence of methane, as a rule, is not taken into account due to the complex mechanism of methane oxidation in the atmosphere to CO_2 and water and its relatively small concentration. However, the process of methane yield under melting permafrost and gasification of gas hydrates on the shelf has increased dramatically

S. Yu. Kasyanov (✉)
N.N. Zubov State Oceanographic Institute, Roshydromet , 6, Kropotkinsky lane, 119034 Moscow, Russia

T. Chaplina (ed.), *Processes in GeoMedia—Volume IV*, Springer Geology,
https://doi.org/10.1007/978-3-030-76328-2_26

in recent years, which is associated with climate warming. Methane emissions create the effect of positive feedback in the climate system and can cause irreversible further warming, leading to increased methane yield due to accelerated permafrost melting and increased gasification of gas hydrates. A rapid temperature rise effect called a "methane bomb" may occur. Permafrost degradation leading to methane outcome occurs under the influence of heat flows both from the upper permafrost boundary, from water or air, and endogenous heat from the lithosphere and mantle. It is shown here that the endogenous heat generation on the seashelves of the Russian Eastern Arctic can be a determining factor making significant contribution to both regional and global warming and the geophysical mechanism for transporting intra-mantle heat to permafrost and Ocean water, controlled by astronomic factors is described.

2 The Zone of Increased Heat Flow to the Ocean Surface from the Underlying Water Layers

Novosibirsk Islands from the North are enclosed by the arcuate zone of increased heat flow to the Ocean surface from the underlying water layers (Fig. 1) (Mansurenkov

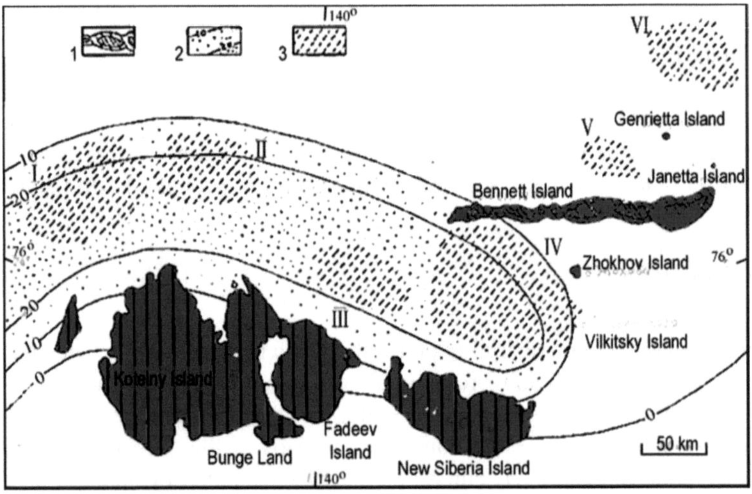

Fig. 1 Heat flux to the ocean surface from the underlying water layers, gas plume 18.02.1983 from an underwater source near Bennett Island, the places of "Sannikov Lands" observation and other anomalous phenomena according to different data: 1—gas plume; 2—zone of increased heat flux to the Ocean surface from the underlying water layers and its isolines in kcal/(cm^2 month) (Atlas of the oceans 1980), which practically coincides with the position of the Great Siberian ice hole in Fig. 3.; 3—alleged position of anomalous objects seen by different observers: I—Ya. Sannikov, II—E. Tollem, III—Ya. Sannikov, as well as Shparo and Shumilov (1982), IV—Ya. Sannikov, V—Google Earth, Data SIO, NOAA, US Navy, NGA, GEBCO (melted ice floe as if from a gas stream), VI—L. Mukhanov (Shparo and Shumilov 1982)

et al. 2012). To the East of this arc the basalt volcanism zone lies around the island. In the East of this arc lies the basalt volcanism zone around Bennett Island, where the bottom relief changes rapidly. In the period between 1974 and 1983, two elevations similar to basalt cones were formed off the Bennett Island coast. The position of these cones almost coincides with the position of gas plumes sources (76.5°N, 149°E). The neighboring Zhokhov and Vilkitsky Islands are also basalt volcanoes to the South (Mansurenkov et al. 2012). The rapid change of the lithosphere surface topography in this area could have led to the appearance above the water surface of uplifts summits with their subsequent vanishing and to the possibility of observing different "Sannikov lands" at different time intervals. Academician A.O. Gliko in the foreword to the monograph by Masurenkov et al. (2012), referring to the legendary "Sannikov Land", notes: "In this course of events, the periodic appearance of peaks of emerging one-act volcanoes above sea level may well provoke a real vision of land, and the powerful destructive activity of the Arctic Ocean quickly eliminates this fragile and short-lived reality".

In the proximate vicinity of the Bennett Island shore, intensive underwater sources of gas have been periodically observed since 1973 to form giant gas plumes (76.5°N, 149°E) (Fig. 2) (Mansurenkov et al. 2012; Bondur and Kuznetsova 2015).

Gas eruptions may continue for several days in a series of short (40–50 min) emissions, or they may be single ones. "Plumes occur above the water surface in open space or in the gaps between the ice. This gives ground to count the entire area around the Bennett island as capable to generate gases and water vapor. The energy base of the plume formation around Bennett Island is an endogenous fluid-magmatic flow, which can (and should) be connected to the day surface by a gas hydrate degassing process. Apparently, the share of volcanic and gas-hydrate components in the matter outcome can vary widely depending on the spatial and scale ratio of the endogenous flow of matter and energy with deposits of gas hydrates" (Mansurenkov et al. 2012).

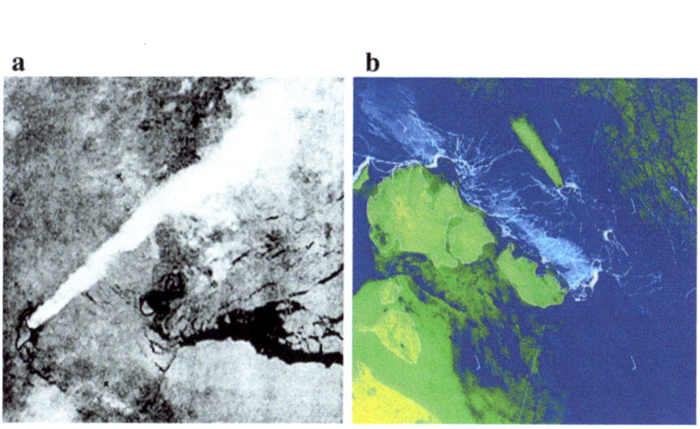

Fig. 2 Gas plumes at Bennett Island, obtained according to: **a**—NOAA_7 satellite, 18.02.1983; **b**—Suomi NPP satellite, 14.03.2013 (Bondur and Kuznetsova 2015)

The volcanic rocks of Bennett, Zhokhov, and Vilkitsky Islands (Mansurenkov et al. 2012) are related to the kimberlites of Yakutia by their mineralogical and chemical features. The Bennett fluid-magmatic system has its roots in the mantle depths and is characterized by processes similar to kimberlite formation at great depths. "The formation of magmatic melts is performed here at high water saturation with the day surface outcome of not only water-saturated basalt magma, but also with giant gas and steam plumes. This powerful fluid-magmatic system also forms a dome and ring structure, the center of which is characterized by the increased heat flux presence with the probable formation of areal volcanic region and a real impact on the thermal balance of the Ocean in the Great Siberian ice hole region" (Mansurenkov et al. 2012).

According to Bondur et al. (2014), "Great Siberian ice hole is a continuous strip of open water and young ice regularly forming behind the landfast outer edge at the region from Bolshoy Begichev Island in Laptev Sea to Bear Islands in East Siberian Sea" (Fig. 3). The existence of open water areas in the Arctic is associated both with the hydrodynamics of sea currents and atmospheric circulation, influence of heat from the North Atlantic Current spreading to the Laptev Sea, and also with methane fluids and modern volcanism (Bondur et al. 2014). However, the endogenous causes of the Great Siberian ice hole formation seem to be more preferable, since it is obvious that regions of warmer waters on the Ocean surface formed by the Earth's endogenous heat create low atmospheric pressure, which originates strong winds in this area.

The periodicity of the appearance of plumes at Bennett Island, as well as the periodicity of volcanic eruptions, is in certain relation with changes in the Earth's angular velocity (Mansurenkov et al. 2012), which, in turn, are controlled by astronomical factors through gravity anomalies created by the location of the planets in the Solar System (Barkin 2014).

In the West of the arcuate zone of increased heat flow to the Ocean surface from the underlying water layers (Fig. 1) in the Laptev Sea disappeared the Vasilievsky

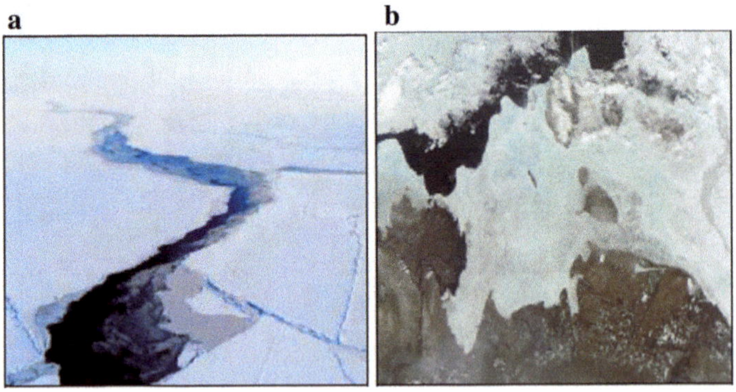

Fig. 3 Great Siberian ice hole from: **a** helicopter, **b** TERRA satellite/Modis equipment (Bondur et al. 2014)

Island (74.233 °N, 133.433 °E) by 1936, and the Semyonovsky Island (74.250 °N, 133.917 °E) between 1952 and the beginning of the 1960s . The Islands of Vasilyevsky and Semyonovsky were formed from frozen rocks and disappeared, melting under the influence of heat flows from the atmosphere, Ocean and lithosphere. In 2013, helicopter pilots near the vanished Vasilyevsky Island discovered newly formed sandy Island Yaya (73.99033 °N, 133.09114 °E), similar in shape to volcano crater (Fig. 4). Formation of Yaya Island is not explained yet.

On the East Siberian and Laptev Sea shelves there is an intensive emission of gas containing methane (Shakhova et al. 2010). At the same time, in the water column there are methane gas flares coming out of bottom sediments and boiling of the sea surface called "methane furnace tubes" (Figs. 5 and 6) (Bondur and Kuznetsova 2015). It seems that gasification of such strong methane flows from gas hydrates requires a significant heat flow from the lithosphere.

Thus, areas of high concentration of dissolved methane on the East Siberian Sea and the Laptev Sea shelf may indicate zones of increased heat flow from the lithosphere and mantle to the gas hydrate layer. The intensity of methane yields from

Fig. 4 Yaya Island formed on the place of the former Vasilevsky Island (http://arcticuniverse.com/set/000417.jpg)

Fig. 5 Methane pop-up bubbles (Bondur and Kuznetsova 2015)

Fig. 6 Methane bubbles under ice (Bondur and Kuznetsova 2015)

gas hydrates on the shelf of the Laptev Sea and East Siberian Sea can be seen from the maps of concentration of dissolved methane distribution in water (Shakhova et al. 2010) (Fig. 7).

Maps of methane concentration in the bottom water layer are the most indicative. Around the Bennett Island and to the South of it the area of high methane concentration is really visible. The maximum concentration is then observed near 166 meridian in the Laptev Sea near the settlement of Tiksi.

Areas of high concentration of dissolved methane in Fig. 7 are consistent with the location of high heat flux zones from the underlying Ocean layers to the surface in Fig. 1 and with the location of the Great Siberian ice hole (Fig. 3), above which, due

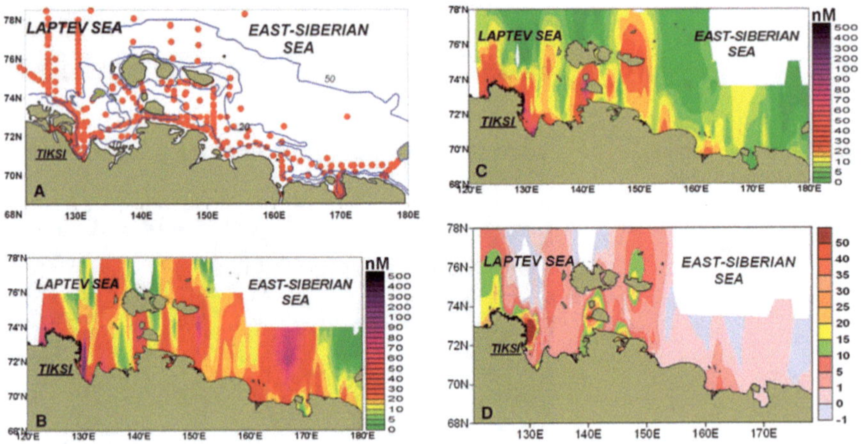

Fig. 7 Summer observations of dissolved methane (Shakhova et al. 2010): **a** position of oceanographic stations in the Eastern part of Laptev Sea and East Siberian Sea, **b** dissolved methane in bottom water, **c** dissolved methane in the surface water layer, **d** methane flow into the atmosphere by ESAS

to intensive heat fluxes from Ocean to atmosphere (from the surface of non-freezing clear water or through thin young ice), a zone of low pressure is formed, which has a significant impact on atmospheric circulation and climate not only directly in the Arctic, but also on global scale.

Thus, one can say that the existence of the Great Siberian ice hole is associated with the existence of a zone of increased heat emission from the lithosphere and mantle. The fact of existence of the arcurate zone of increased heat emission from the lithosphere and mantle is in turn determined by three main phenomena. First, basalt volcanism of Vilkitsky, Zhokhov, Bennett Islands and methane outflows around Bennett Island—in the East, and melted permafrost islands Semyonovsky and Vasilyevsky, as well as newly formed sand island Yaya—in the West. Second, active methane release from gas hydrates, the region of which covers Novosibirsk Islands, Bennett Island and the shelf of the Laptev and East Siberian Seas. Third, high seismicity of this area. Besides that, an important sign of the considered natural phenomenon is that time course of gas plume eruptions as well as time course of volcanoes eruptions (Barkin 2014) is connected with changes in angular velocity of Earth rotation.

3 Intense Hot Two-Phase Vortices in the Earth's Mantle

The combination of all the mentioned above facts and phenomena is unimodally explained on the basis of the model of three energy-bearing vortices in the Earth mantle (Kasyanov and Samsonov 2019), which is based on seismic data and data on relief and gravity anomalies. According to the model, at present time there is a system of three energy-bearing vortices in the Earth mantle (Fig. 8), which is the result of evolution of two-phase hot intensive vortex ring covering almost the whole mantle thickness, formed by tangential invasion of temporary Earth satellite with radius about 930 km (Kasyanov 2012).

The vortex ring disintegrated under the action of Earth rotation. Evolution of its parts led to the formation in the Earth mantle system of quasi-stationary intensive hot two-phase (gas–liquid) vortices, the total kinetic momentum vector of which is now almost collinear to the vector of angular velocity of Earth rotation. Three most significant intensive energy-bearing vortices can be distinguished in this system, shown in Fig. 8. The vortex (1) of Himalayas + Tien Shan − Kolyma (HTS-K), the vortex (2) of Kolyma–Yellowstone (K–Y) and the vortex (3) of Yellowstone–Hudson Bay–Eastern Mediterranean Sea (Y-HB-EMS).

Structure and geometry of vortices are determined on the base of mantle geometry, relief fields and gravitational anomalies taking into account seismology data. The description of vortex structure and geometry is given in Kasyanov and Samsonov 2019.

The substance movement in all three vortices is two-phase and quasi-stationary. The vortex geometrical parameters are given in Tables 1, 2 and 3.

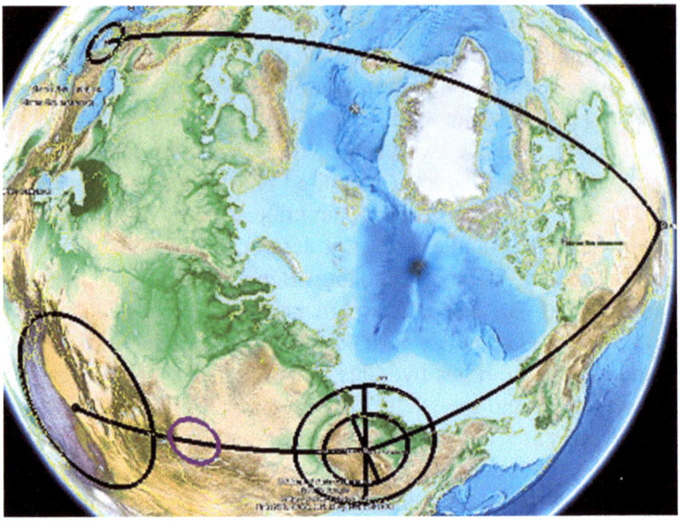

Fig. 8 Structure of the system of three connected intensive vortices in the Earth mantle (Kasyanov and Samsonov 2019)

Table 1 Vortex (1) geometrical parameters

Vortex (1): Himalayas + Tien Shan − Kolyma (HTS-K)		
Vortex outcome	Himalayas − Tian-Shan	Kolyma
Vortex outcome center	39.768°N 88.720°E	67.194°N 140.339°E
Vortex outcome depth, km	1200	1700
Vortex outcome radius in the mantle, km	1300	700
Vortex radius on the surface, km	1600	954
Direction of rotation in the vortex outcome (look from the Earth's surface)	Clockwise	Counterclockwise
Vortex axis of rotation angle to vertical	Not determined	71.8°
Vortex outcome deepening direction	North-East	

Vortex (1) (HTS-K) at its lower point of the central part comes into contact with the liquid core of the Earth. The contact of the vortex moving substance with the substance of the liquid core of the Earth generates stationary hydrodynamic pressure disturbances, at isostatic compensation of which the Baikal rift and the relief around Lake Baikal were formed on the day surface. The two-phase substance of the vortex moves in an inclined plane with large differences in depth and hydrostatic pressure, which leads to the appearance of a stationary condensation-evaporation front when the vortex substance moves. On the day surface the condensation-evaporation front manifests itself in the form of a wavy line inside the Himalayan-Tien Shan mountain ring, separating the areas of high and low pressure at the upper edge of the

Table 2 Vortex (2) geometrical parameters

Vortex (2): Kolyma – Yellowstone (K-Y)		
Vortex outcome	Kolyma	Yellowstone
Vortex outcome center	67.006°N 140.694°E	44.745°N 110.328°E
Vortex outcome depth, km	500	100
Vortex outcome radius in the mantle, km	525	30–50
Vortex radius on the surface, km	570	Not determined
Direction of rotation in the vortex outcome (look from the Earth's surface)	Counterclockwise	Quasihorizontal
Vortex axis of rotation angle to vertical	65.6°	Quasihorizontal
Vortex outcome deepening direction	South–West	

Table 3 Vortex (3) geometrical parameters

Vortex (3): Yellowstone – Hudson Bay – Eastern Mediterranean Sea (Y-HB-EMS)	
Vortex outcome	Yellowstone
Vortex outcome center	44.745°N 110.328°E
Vortex outcome depth, km	700
Vortex outcome radius in the mantle, km	350
Vortex radius on the surface, km	Not determined
Direction of rotation in the vortex outcome (look from the Earth's surface)	Counterclockwise
Vortex axis of rotation angle to vertical	30–50°
Vortex outcome deepening direction	South–East

vortex outcome, that result in the areas of high and low relief. In the field of gravity anomalies Buge built according to the WGM2012 (Bonvalot et al. 2012) data the de-densification in the vortex hot area, which is most clearly manifested at the vortex end-faces.

The **Kolyma node** (Fig. 9) is of special interest for the mantle vortex energy transport to the Ocean. Here, the vortex (1) outcome **(HTS-K)** is in contact with the overlying vortex (2) outcome. Due to the difference of velocities in vortices at mixing of their jets there is a dissipation of kinetic energy and heat generation. Since in the Kolyma node the vector of angular velocity in the vortex (1) outcome is directed to the North-East, and the vector of angular velocity in the vortex (2) outcome—to the North-West, both have positive axial (along the Earth's axis) component, the gyroscopic forces seeking to combine the vortex axes with the Earth's rotation axis, in the vertical direction, affect the vortex (1) and (2) outcomes in opposite directions. Therefore, as a result of gyroscopic forces, the area along the Lena River valley, under which the border of the lower vortex (1) passes, is lowered, and the area along the Verkhoyansky Ridge, under which the border of the upper vortex (2) located—raised.

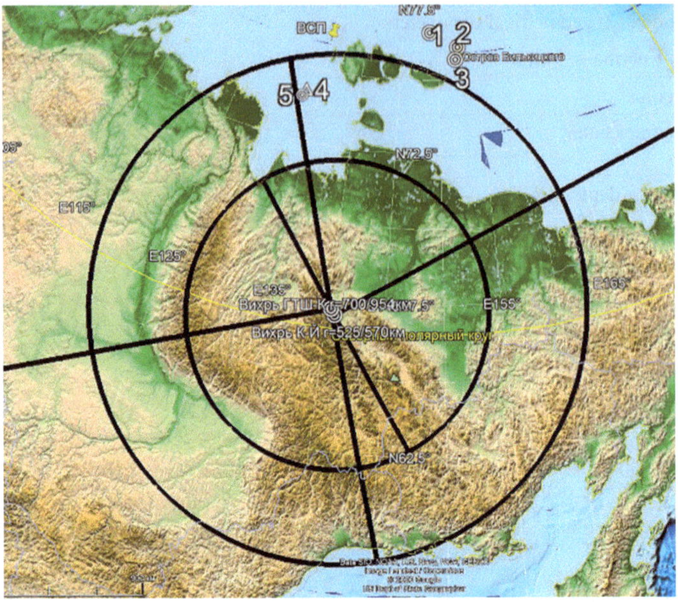

Fig. 9 Schematic representation of the Kolyma node. Numbers indicate the islands located on the ascending branch of the vortex (1): 1—Bennett, 2—Zhokhov, 3—Vilkitsky, 4—vanished islands Vasilyevsky (1936) and Semenovsky (after 1952), which were built of permafrost rocks, 5—new sandy island Yaya (discovered in 2013)

The vortex outcomes in Fig. 9 are schematically depicted in circles. In reality, due to the large angles of vortex axes inclination to the vertical, the projections of vortex outcomes to the sphere of the day surface are elipse shapes stretched in perpendicular to the vortex axes directions. The vortex outcomes contact each other in the South-Western sector so that the upper point of vortex (1) is in contact with the lower point of vortex (2).

Comparison of **vortex (2) (K-Y)** outcomes radii in Table 2 with taking into account the depth of their centers shows that the vortex has a conical shape and passes away from the liquid Earth core. Parameters of vortex outcomes in **Yellowstone node** in Tables 2 and 3 are determined from seismic tomography data (https://vimeo.com/125650792; Obrebski et al. 2010). Coordinates of the vortex outcomes are taken from the map as an estimate. In accordance with seismic tomography data (https://vimeo.com/125650792; Obrebski et al. 2010) three related vortex rings are located under Yellowstone (Fig. 10).

In Fig. 10 on the right (Obrebski et al. 2010)—area under small upper ring with reduced velocities of seismic waves in the upper mantle. The big lower ring of the region with reduced velocities of seismic waves contacts to the overlying small upper ring of the analogous region with reduced velocities of seismic waves located in the upper mantle and lithosphere (left part in Fig. 10). As links of chain big yellow lower ring of the region with reduced velocities of seismic waves it is connected with

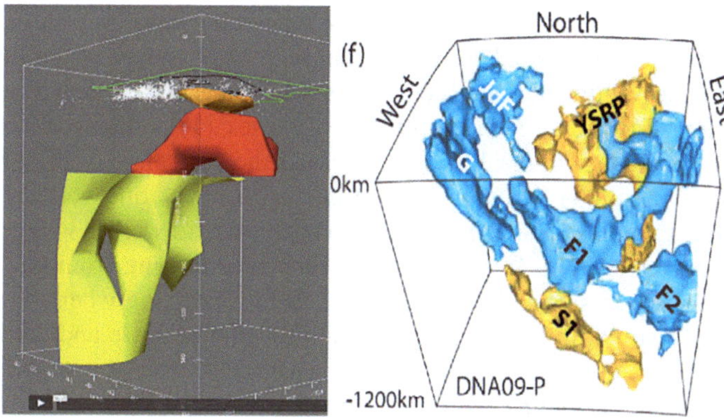

Fig. 10 Seismic tomography of the area under Yellowstone, where three bound vortex rings can be identified. The area on the right (Obrebski et al. 2010): from 37°N to 49°N and from −107°E to −124°E (1332 km along the meridian and 1380 km along the parallel 43°N); left—small upper ring with reduced rates of seismic waves in upper mantle and lithosphere (https://vimeo.com/125650792)

big blue ring of the region with increased velocities of seismic waves. Small upper yellow ring—the outcome of K-Y vortex, big yellow ring—outcome of Yellowstone—Greece (Mediterranean Sea) vortex. Blue ring is the tip of the previously existed vortex Chukchi Rise—Yellowstone, possibly a continuation of the vortex **HTS-K**. The modern vortex (2) (**K-Y**) was formed over this vortex as secondary to it and therefore rotates in the opposite direction.

The upper yellow ring of the vortex of small radius (30–50 km) has a quasi-horizontal axis and is the outcome of **K-Y** vortex. This ring at the vortex tip is characterized by low velocities of seismic waves and, consequently, hotter substance. The upper vortex is in contact with the underlying large hot ring. At contact of differently directed jets of upper and lower vortices these jets deceleration occur, as a result of which the substance in both vortices heats up and in this way two hot rings are formed, visible on seismic tomography (https://vimeo.com/125650792; Obrebski et al. 2010). The large blue ring—the region with high velocities of seismic waves—may represent the tip of earlier existed and stopped vortex (Chukchi Rise—Yellowstone), containing heavier substance at its periphery. This vortex may have previously been a continuation of vortex (1) (**HTS-K**). Vortex (2) (**K-Y**) was formed over this vortex as secondary to it and rotating in the opposite direction. The system of three vortices ends under the Mediterranean Sea.

4 The Dynamics of the Three Vortices System

The dynamics of the system of vortices is determined by astronomical factors. The movement of planets in the Solar System creates gravitational anomalies that shift the

Earth's inner core (Barkin 2014). The displacement of the inner core causes changes in the position of equipotential surfaces, including the boundary of the liquid outer core and mantle. Whilst, the liquid core substance under Baikal may flow into the vortex (1) area. In this case, the substance of vortex (1) is displaced into vortex (2) through the contact in the Kolyma node and the substance from vortex (2) is displaced into vortex (3). Thus, a chain of exchange of rotating vortices by mass, kinetic moment, impulse and energy between each other as well as with mantle and liquid core is created. The velocities of the substance flowing from one vortex to another differ both in size and direction. As a result of mixing when the substance flows from one vortex to another, deceleration of mixing flows and dissipation of mechanical energy into heat occurs, resulting in a thermal expansion of the substance in the vortexes, and changing the Earth' volume, which is reflected by changes in Ocean level, readout from the satellites orbits. Changes in the position of the Earth's center of mass, its kinetic momentum, and thus the vector of angular velocity of the Earth's mantle, heat is released, which is spent on gas hydrates gasification and heating of Ocean.

The strongest changes in geophysical parameters, characterized by the surges, occur at the gravitational resonances in the solar system (Barkin 2014). The surge of geophysical parameters in 1997–1998, in which the inner core and center of the Earth's masses shifted in the direction of Taimyr, was studied in most detail (Barkin 2014). For 1997–1998 surge the data (Barkin 2014) about change of the following geophysical parameters are obtained: position of Earth mass center (Fig. 11), position of Earth rotation pole (Fig. 12), increase of Ocean level by 8–9 mm, decrease of day duration by 0.7 ms and increase of World Ocean temperature by 0.16 °C.

A simple algebraic model of the system of three interrelated vortices in the Earth's mantle (Kasyanov and Samsonov 2019) is constructed. It allows to estimate the velocity in three vortices, the density of matter at the periphery of these vortices and volume of matter flowing from the liquid core of the Earth into vortex (1) in the contact zone of the vortex and liquid core near Baikal according to the first seven above known geophysical parameters of the 1997–1998 surge. The model includes seven nonlinear

Fig. 11 Change of Earth mass center position relative to the Earth center in mm (Barkin 2014)

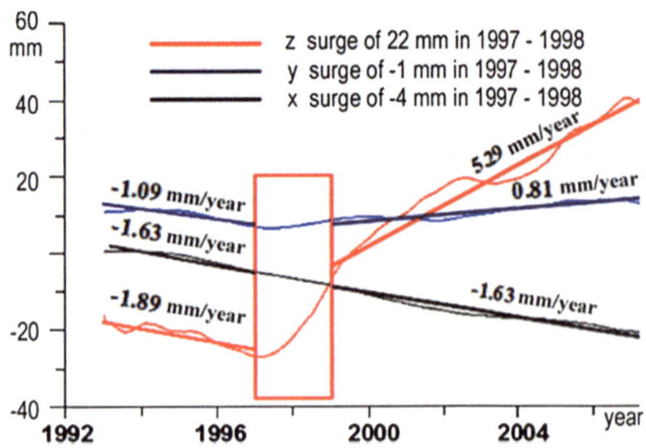

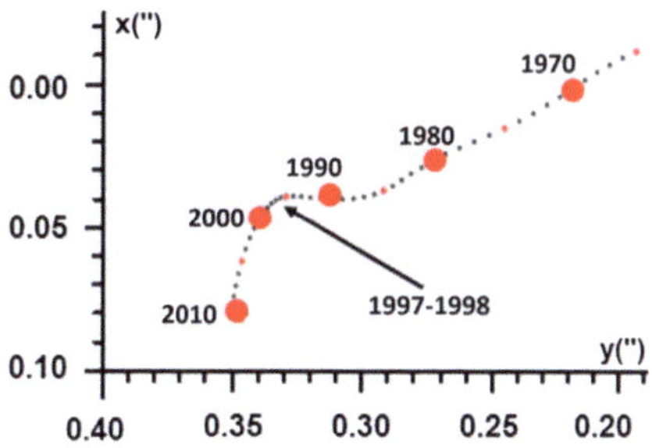

Fig. 12 Centennial drift of the Earth's axis of rotation in angular seconds (x increases toward the Greenwich meridian, y decreases toward 90°E) (Barkin 2014; Scalera 2011)

algebraic equations, which are solved numerically by iteration method. The model makes possible to obtain an estimate of the energy transmitted to the world Ocean and also to obtain an estimate of its temperature change. Since the data on changes in the World Ocean temperature is not used in the model, the obtained estimation can be used for independent control of the solution quality and adequacy of the built model to the natural phenomenon under study.

The numerical solution gives the following estimates of maximum linear velocities in vortices (from vortex (1) to (3)): 4130.76, −1244.69, 2940.59 m/s. The obtained negative value of linear velocity in vortex (2) here indicates that the vector of angular velocity of this vortex is directed from Yellowstone to Kolyma. The corresponding estimates of substance density at vortex periphery are: 11,800, 4600, 4180 kg/m^3. At these calculated velocities and densities of substance in vortices an estimation of increase in World Ocean temperature is found to be 0.16 °C.

It should be noted that the obtained value was roughly equal to the one observed in 1997–1998 for the value of the surge in the World Ocean temperature (0.16 °C). This fact serves as an independent confirmation of the adequacy of the proposed model (Kasyanov and Samsonov 2019) to the natural process under consideration. Another independent confirmation of the model adequacy is the fact that based on the modeling results it was found that the system of three mantle vortices in question forms the Earth's magnetic dynamo and explains the existence and basic structure of the Earth's magnetic field. Reynolds' magnetic numbers calculated by the electrical conductivity of the Earth's mantle are 6073.3, −985.4 and 1906.7, respectively. At the same time, as shown in Lowes and Wilkinson (1968) laboratory experiments, the magnetic dynamo in experimental facility with similar geometry of rotating rotors is excited at much lower critical values of Reynolds' magnetic numbers—from 200 to 1000. As can be seen, Reynolds' magnetic numbers in vortices (1) and (3) exceed critical values, so vortices (1) and (3) in the system of three vortices could excite magnetic dynamo. Thus, the movement of the conducting substance in vortices (1) and (3) explains the appearance of magnetic dynamo in the system of three intra-mantle vortices and the formation of the Earth's magnetic field. The location of vortices (1) and (3) explains the structure of the Earth's magnetic field maximums and the reasons for the formation of Canadian and Siberian global magnetic anomalies.

5 Mechanism of Energy Transfer from the System of Intra-mantle Vortices to the Ocean

The substance in the vortexes is in a saturated two-phase state. When the substance rises in the vortex hydrostatic pressure drops, which leads to evaporation of the most easily boiling fraction of the mantle, namely SiO_2 .

In the Kolyma node, at the very beginning of the ascent near the bottom point of the ascending branch of vortex (1) outcome, the movement is close to the horizontal, and therefore, although the hydrostatic pressure here is changing slowly and corresponndently a relatively small amount of SiO_2 vapor is formed, nevertheless the resulting SiO_2 vapors, seeking to rise upward, partly break through the mantle and lithosphere vertically to the layer of gas hydrates, heating them and causing gasification. In this case, vortex cords may appear, forming analogues of kimberlite tubes. The ascending branch of vortex (1) outcome in this section of the trajectory passes under the area of high concentration of dissolved methane in water at 166 meridian, then passes under the volcanic islands of Zhokhov, Vilkitsky and Bennett, then arcurates Novosibirsk islands from the North and then passes under the area of Yaya Island and the vanished islands of Vasilyevsky and Semenovsky. The presence of the ascending branch of vortex (1) outcome explains the possibility of large heat flows from the lithosphere and mantle into water and gas hydrates, which may lead to the existence of the Great Siberian ice hole and active gas hydrates gasification. Under the Laptev Sea shelf an ascending branch of the vortex (2) outcome is located.

Evaporation of liquid SiO_2 boiling when moving upwards in the ascending branch of the vortex (2) outcome, which steam jets in the form of vortex cords can penetrate into the gas hydrates layer, and can provide the heat fluxes necessary for intensive gas hydrates gasification, which is observed on the Laptev Sea shelf (Fig. 7). The highest concentration of dissolved methane in water is observed here, near the crumbling Mouostahi Island in the Buor-Haya Bay of the Laptev Sea (near Tiksi settlement).

Excreting vapors of SiO_2 and other substances can break through the mantle, forming volcanoes, both on the surface and underwater. If the SiO_2 vapor rise to the Ocean water, it forms finely dispersed suspension of solid SiO_2. This may form mud volcanoes containing this suspension. At high intensity and duration of excretion of gaseous SiO_2 in the water can form a sandy island, such as Yaya. When the gaseous SiO_2 condenses and cools down, a large amount of heat is released transferred to the surrounding material, and there are intense local heat sources at the bottom of the ocean, associated with volcanoes and faults. Thus, a process in which energy from a system of intense two-phase mantle vortices can be transported by SiO_2 evaporating on the ascending vortex branches and transferred to the waters of the seas of the Eastern Arctic and further on to the entire World Ocean is described.

6 Heat Release in the Arctic with Atlantic Multidecadal Oscillation and Quasi-Sixty-Year Cycle Relation

The generation of heat in the process of mass flows from one vortex to another in a system of three intense two-phase mantle vortices as well as volcanic eruptions and the above described gas plume emissions in the Arctic is controlled by gravity anomalies in the Solar System with periods, that are determined by the position of the planets. The kinetic energy of vortices transition to the thermal energy in mantle vortices leads to the thermal expansion of matter in them, changing the volume of the Earth and the heterogeneous expansion of the mantle and lithosphere. The latter leads both to earthquakes and volcanoes eruptions on Earth and to gas plume releases in Arctic.

The strongest gravitational influence on system of vortices appears at convergences of planets-giants Jupiter and Saturn which define a quasi-sixty-year cycle. It has been discovered that the North-Atlantic Multidecadal Oscillation (AMO), i.e. the time course of the Ocean surface temperature anomaly in the North Atlantic, numerically characterized by AMO index, is also subject to this cycle (Fedorov). The leap of global temperature in 1997–1998 is also sharply manifested in the AMO index (Fig. 13), which shifts from the "cold" phase to the "warm" phase in 1997–1998. Consequently, the parameters of our three vortex model were calculated from data directly related to the quasi-sixty-year AMO cycle. Therefore, there is a reason to assert that heat generation from the system of intensive two-phase mantle vortices, which is controlled by gravitational anomalies in the Solar System with rhythms

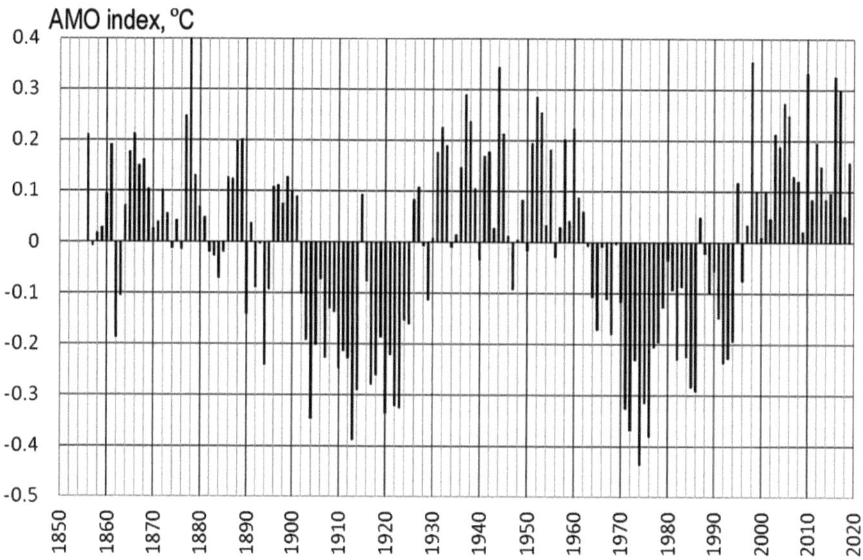

Fig. 13 Annual AMO index distribution in time (Fedorov)

determined by the position of the planets and including a quasi-sixty-year cycle of Jupiter and Saturn connexions, determines the existence of a quasi-sixty-year AMO cycle. It means that heat emission from the system of mantle vortices in the seas of the Russian Arctic influences the temperature of surface waters of the North Atlantic and can be the reason of manifestation of the quasi-sixty-year AMO cycle.

7 Conclusions

Inflow of heat from two-phase intense mantle vortices into the seas of the Russian Arctic can cause regional warming in the Arctic, which then spreads to the World's Ocean. This phenomenon is controlled by gravity anomalies created by the position of the planets of the Solar System with the inherent rhythms and can make a significant contribution to global warming, which also has (Barkin 2014) connection with the gravity anomalies from the position of the planets of the Solar System. Inflow of heat from two-phase intensive mantle vortices in the seas of the Russian Arctic may be the reason for the manifestation of a quasi-sixty-year AMO cycle.

The energy of mantle vortices, in particular, can support the existence of the Great Siberian ice hole. Lowering the pressure over warmer waters in the Great Siberian ice hole changes the atmospheric circulation, weakens the action of the first kind of heat machine (according to V.V. Shuleikin) and can lead to instability and anomalies of weather and climate.

References

Atlas of the oceans (1980) Arctic Ocean. USSR ministry of defense, p 188 [in Russian]

Barkin YV (2014) Geophysical effects of relative displacements and fluctuations of the earth's core and mantle. SIEPh RAS, Moscow. http://www.ifz.ru/fileadmin/,user_upload/subdivisions/506/OMTS/2014/16.09/Barkin_16_09_14.pdf [in Russian]

Bondur VG, Kuznetsova TV (2015) Gas sip detection in the Arctic seas water areas using remote sensing data. Earth exploration from space. No. 4. C.30–43 [in Russian]

Bondur VG, Kuznetsova TV, Vorobiev VE, Zamshin VV (2014) Detection of gas manifestations on the Russian shelf using space imagery data. Georesour, Geoenergy, Geopolitics. Episode. 1(9):1–23. http://oilgasjournal.ru/vol_9/bondur.html [in Russian]

Bonvalot S, Balmino G, Briais A, Kuhn M, Peyrefitte A, Vales N, Biancale R, Gabalda G, Reinquin F, Sarrailh M (eds) (2012) World gravity map. Commission for the geological map of the world. Paris, BGI-CGMW-CNES-IRD. (WGM2012)

Fedorov VM. North Atlantic oscillation and KWO http://www.solar-climate.com/geo/amo.htm, free access mode [in Russian]

http://www.esrl.noaa.gov/psd/data/correlation/amon.us.long.data

https://vimeo.com/125650792 [video]

Kasyanov SY (2012) Simulation of orbital motion of a large asteroid at decrease with deepening into lithosphere and mantle of the earth. Phys Probl Ecol (environ Phys) 18:151–164 [in Russian]

Kasyanov SY, Samsonov VA (2019) On the modern system of three energy-bearing vortices in the earth's mantle. "Physical and mathematical modeling of processes in GeoMedia" 5th international science conference-school of young scientists. Moscow, Ishlinsky Institute for Problems in Mechanics RAS, pp 77–79. [in Russian]

Lowes FJ, Wilkinson I (1968) Geomagnetic dynamo: an improved laboratory model. Nature 219:717–718

Mansurenkov YP, Sobisevich AL, Petrova VV, Slezin YB, Flyorov GB, Shuvalov RA, Kuzmin YD, Ovsyannikov AA (2012) Modern endogenous process activity on bennett Island (De Longa Archipelago, Arctic). SIEPh RAS, Moscow, p 160 [in Russian]

Obrebski M, Allen RM, Xue M, Hung S-H (2010) Slab-plume interaction beneath the Pacific NorthWest. Geophys Res Let 37:L14305

Scalera G (2011) South American volcanoes and great earthquakes. Volcano-seismic correlation versus secular PM. 37th interdisciplinary workshop international school geophysical Ext abstract book Sicily, 125–129

Shakhova N, Semiletov I, Salyuk A, Yusupov V, Kosmach D, Gustafsson Ö (5 Mar 2010) Extensive methane venting to the atmosphere from sediments of the East Siberian Arctic shelf. Science. 327(5970):1246–1250. https://doi.org/10.1126/science.1182221

Shparo D, Shumilov A (1982) Three Arctic riddles. Mysl', Moscow, p 142 [in Russian]

Volodin EM, Diansky NA, Gusev AB (2013) Reproduction and forecast of climatic changes in XIX–XXI centuries by means of the model of the earth climate system of INM RAS. Izv RAS Phys Atm Ocean 49(4):379–400

Change of Time Variations in Acoustic Vibrations in the Atmospheric Surface Layer in Moscow During Production Restrictions Due to Covid-19 Quarantine Measures in 2020

Svetlana Riabova, Alexander Spivak, and Yuri Rybnov

Abstract The presented research provides a comparative analysis of the features of the amplitude and spectral characteristics of acoustic noise in the conditions of Moscow during the quarantine period for COVID-19 in 2020 and during the period preceding the quarantine. As the initial data, we used the time series of instrumental observations of micropulsations of atmospheric pressure, carried out at the Center for Geophysical Monitoring in Moscow of Sadovsky Institute of Geosphere Dynamics of Russian Academy of Sciences from February to May 2020. Data plots were analyzed during periods characterized by calm meteorological conditions: the absence of strong atmospheric disturbances in the form of hurricanes and squalls, as well as atmospheric fronts.

Keywords Daily variation · Acoustic vibrations · COVID-19 · Spectral analysis

1 Introduction

The COVID-19 outbreak was declared a global health emergency by the World Health Organization in January 2020 and a pandemic in March 2020. In order to contain the spread of the virus, borders have closed, air traffic was limited, and production activities have significantly reduced. In addition, as a measure to combat the rapid spread of the virus, the governments of many states introduced a self-isolation regime for citizens, as a result, there was a significant restriction on the movement of citizens, and the number of vehicles on the roads decreased.

The radical changes in the daily life of the population during the period of self-isolation provided a unique opportunity to study how the environment reacts to such specific conditions. In particular, research results have become known, indicating a significant decrease in nitrogen oxide emissions into the atmosphere (Bauwens et al. 2020; Kaplan and Yigit Avdan 2020) and a decrease in the level of seismic noise (Lecocq et al. 2020; Poil et al. 2020; Xiao et al. 2020).

S. Riabova (✉) · A. Spivak · Y. Rybnov
Sadovsky Institute of Geospheres Dynamics of Russian Academy of Science, Leninsky Prospekt, 38, building 1, Moscow, Russia

T. Chaplina (ed.), *Processes in GeoMedia—Volume IV*, Springer Geology,
https://doi.org/10.1007/978-3-030-76328-2_27

In urban agglomerations, in particular in megalopolises (for example, Moscow), one of the factors that negatively affect the vital activity of living organisms and humans is acoustic noise (Adushkin and Spivak 1995). According to modern data, more than 30% of cities are located in the zone of increased acoustic noise (Spiridonova and Makarov 2013). In urban conditions, acoustic vibrations are characterized by a large variety of sources of technogenic origin associated with human activities (Garaeva 2010). The most common sources are the work of numerous manufacturing, construction and transport enterprises, power substations and traffic (Bocharov et al. 2012).

This paper examines the features of the amplitude and spectral characteristics of variations in microbaric vibrations in Moscow during the quarantine period for COVID-19 in 2020 and in the period preceding the quarantine. In Moscow, the self-isolation regime for all its residents was in effect from March 30 to June 9, 2020. In addition, in the period from April 13 to May 12, 2020, measures have introduced to reduce production activities. Comparative analysis has carried out using data from the Center for Geophysical Monitoring in Moscow of Sadovsky Institute of Geosphere Dynamics of Russian Academy of Sciences (IDG RAS).

2 Initial Data and Data Processing Methods

In this work, the processing and analysis of data of the microbaric variation registration carried out at the Center for Geophysical Monitoring in Moscow of IDG RAS (55.70°N, 37.57°E; Leninsky Prospect, Moscow; CGM) from February to May 2020. Data of instrumental observations of micropulsations of atmospheric pressure in the CGM, performed using MB-02 microbarometer in the frequency range 0.05–30 Hz and MB-03 microbarometer in the frequency range 0.001–10 Hz, are posted on the website of the IDG RAS in graphical and digital forms.

Areas of rapid changes in atmospheric pressure observed during the passage of the atmospheric front, during periods of precipitation, hurricanes and similar meteorological processes, have rejected.

To determine the spectral composition of the disturbances, the power spectrum was calculated (Krivosheev 2006). Outliers in the data have preliminarily identified and gaps have removed using the techniques described in Spivak et al. (2019). In addition to calculating the spectrum, the time base of the spectral harmonic variations at certain frequencies was calculated.

All calculations in this work have carried out using the codes of the Matlab software package.

3 Results

The processing and analysis of the data of long-term registration of atmospheric pressure micropulsations show the difference between the amplitude and spectral characteristics of microbaric vibrations in Moscow and in the territories located outside the zone of influence of the metropolis (Mikhnevo Geophysical Observatory of IDG RAS) (Rybnov et al. 2019; Riabova and Spivak 2019).

The mode of microbaric fluctuations in a megalopolis is influenced by technogenic sources (in particular, traffic flow).

The home regime of self-isolation of the majority of Moscow residents, introduced by the decree of the Mayor of Moscow from March 30, 2020, and, as a result, a decrease in the intensity of traffic flows, as well as the suspension from April 13 to May 12, 2020 of the activities of almost all enterprises and organizations, significantly reduced the man-made component of microbaric vibrations. This led to a decrease in the amplitude of acoustic vibrations by approximately 1.5–2.5 times.

The results of spectral analysis of instrumental observation data in the CGM demonstrate the difference in the nature of the spectral amplitudes in the spectra during the quarantine period and in the period preceding it. Examples of spectra are shown in Fig. 1. As seen from Fig. 1, the maxima of the spectral density of micro-vibrations of atmospheric pressure are observed in the frequency range 1–5 Hz. An analysis of the spectrum of variations registered in the period preceding the quarantine shows that eight fundamental spectral harmonics can be distinguished in the frequency range from 1.3 to 6.3 Hz and with spectral amplitudes from 1.4 to 5 $Pa/(Hz)^{1/2}$. The spectra during quarantine are characterized by the presence of five

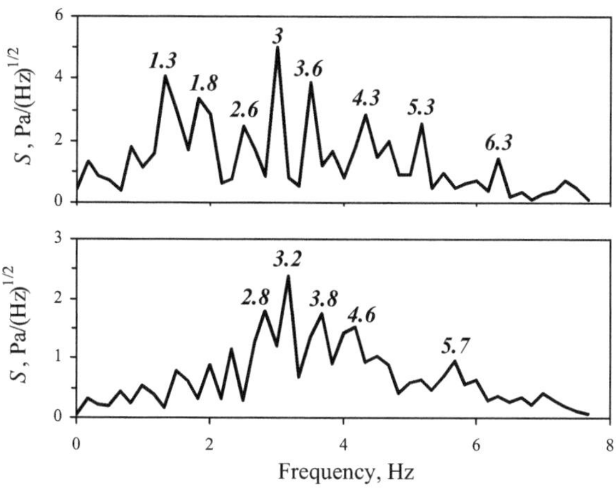

Fig. 1 Spectra of acoustic vibrations for the period from March 20, 2020 to March 24, 2020 (upper panel) and for the period from May 20, 2020 to May 24, 2020 (down panel); in the field of the figure are the values of the frequencies of the spectral maxima in Hz

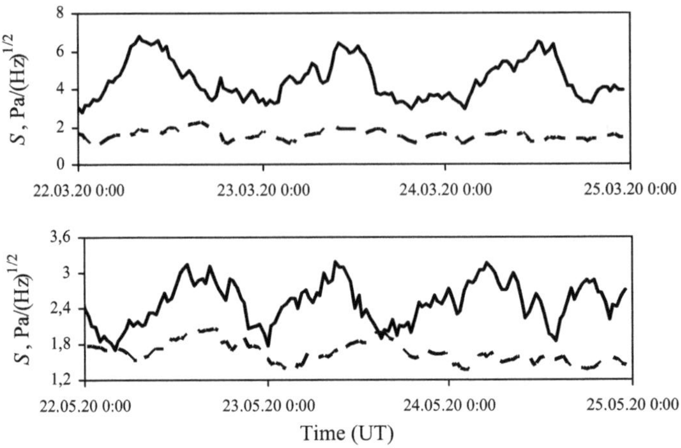

Fig. 2 Time variations of the amplitudes of the spectral density of acoustic vibrations from March 20, 2020 to March 24, 2020 (upper panel) and for the period from May 20, 2020 to May 24, 2020 (down panel) at frequencies of 3 Hz (dashed lines) and 5 Hz (solid lines)

spectral harmonics in the range 2.5–5.1 Hz and with spectral amplitudes from 1.0 to 2.4 Pa/(Hz)$^{1/2}$.

In addition to the analysis of the spectral composition, we studied the daily variation of the amplitudes of the spectral maxima during the quarantine period and in the period preceding it. Figure 2 shows examples of the time variation of the daily amplitude variation for frequencies of 3 and 5 Hz. From the data shown in Fig. 2, it follows that in the temporal variations of the amplitudes of the spectral maxima in the period before the introduction of quarantine, the daily periodicity is clearly manifested, while during the quarantine period the daily variation is almost constant. It should also be noted that the main spectral amplitudes of acoustic noise decreased during the quarantine period compared to the period preceding it.

4 Conclusion

Thus, it can be considered that the restrictive measures introduced due to the coronavirus pandemic have led to a 1.5–2.5-fold decrease in the amplitude of acoustic vibrations and to a narrowing of their frequency range. According to the results of the studies performed in this work, the daily periodicity in the variations in the amplitudes of the spectral maxima is manifested to a much lesser extent during the quarantine period than in the previous period. The manifestation of daily periodicity in microbaric variations, which is associated with the intensification of traffic flows in the city and the functioning of most industrial enterprises operating in one day shift, was practically not recorded during the quarantine period.

According to the authors, the data obtained in this work provide a basis for identifying the contribution of technogenic sources in urban agglomerations to acoustic vibrations based on a comparative analysis of their amplitude-frequency characteristics.

References

Adushkin VV, Spivak AA (1995) Megapolis: the problem of geophysical fields. Sci Russ 5:65–69 [In Russian]

Bauwens M, Compernolle S, Stavrakou T, Müller J-F, van Gent J, Eskes H, Levelt PF, van der Veefkind RAJP, Vlietinck J, Yu H, Zehner C (2020) Impact of coronavirus outbreak on NO_2 pollution assessed using TROPOMI and OMI observations. Geophys Res Lett 47(11), https://doi.org/10.1029/2020GL087978

Bocharov AA, Kolesnik AG, Soloviev AV (2012) Acoustic noises of urbanized areas on the example of Tomsk. Bull Tomsk Polytech Univ 321(1):191–196 [In Russian]

Garaeva KhR (2010) Acoustic discomfort in the urban environment (on the example of the cities of Baku and Astrakhan). Bull Astrakhan State Techn Univ 1(49):88–95 [In Russian]

Kaplan G, Yigit Avdan Z (2020) COVID-19: spaceborne nitrogen dioxide over Turkey. Eskişehir Techn Univ J Sci Technol A—Appl Sci Eng 21(2):251–255

Krivosheev VI (2006) Modern methods of digital signal processing (digital spectral analysis). UNN, Nizhniy Novgorod. [In Russian]

Lecocq T, Hicks SP, Noten KV, van Wijk K, Koelemeijer P, De Plaen RSM, Massin F, Hillers G et al (2020) Global quieting of high-frequency seismic noise due to COVID-19 pandemic lockdown measures. Science 369(6509):1338–1343

Poil P, Boaga J, Molinari I, Cascone V, Boschi L (2020) The 2020 coronavirus lockdown and seismic monitoring of anthropic activities in Northern Italy. Sci Rep 10:9404. https://doi.org/10.1038/s41598-020-66368-0

Riabova SA, Spivak AA (2019) Geomagnetic variations in the surface zone of the Earth. Grafiteks, Moscow [In Russian]

Rybnov YS, Spivak AA, Soloviev SP, Kharlamov VA (2019) Local sources of infrasonic noise in a metropolis. Optics of the atmosphere and ocean. Physics of the atmosphere: materials of the XXV international symposium. Novosibirsk, July 1–5, 2019. Publishing house of IAO SB RAS, Tomsk. D-27–D-30. [In Russian]

Spiridonova YuA, Makarov BA (2013) Noise and its effect on humans. Natl Priorities Russ 2(9):68–69 [In Russian]

Spivak AA, Rybnov YuS, Kharlamov VA (2019) Some characteristics of acoustic noise in Moscow. Geophys Process Biosph 18(1):118–124 [In Russian]

Xiao H, Eilon Z, Ji Ch, Tanimoto T (2020) COVID-19 societal response captured by seismic noise in China and Italy. Seismol Res Lett 91(5):2757–2768

Anisotropy of Sea Surface Roughness Formed by Waves of Different Scales

A. S. Zapevalov

Abstract The dependence of the variance of the upwind and crosswind components of the sea surface slope, as well as the anisotropy of the slope on the range of waves creating them is analyzed. We used radio and optical sounding data, as well as in situ measurements (measurements by laser inclinometers, string sensors, and wave buoys). Dependences of the integral characteristics of the slopes created by waves whose lengths are in the range with a changing boundary: from the longest waves to waves of a given length are constructed. It is shown that the estimation of the anisotropy coefficient obtained in the framework of the well-known Cox-Munk model, which is widely used in applications related to the reflection of electromagnetic waves, does not agree with the estimates of this coefficient calculated from in situ measurements.

Keywords Sea surface wave · Slope · Roughness · Slope anisotropy

1 Introduction

Specular reflection of radio waves from the sea surface is created by sea waves, the length of which is much longer than the length of the radio wave (Brekhovskikh 1952; Valenzuela 1978). Sea waves, whose length is comparable and less than the length of a radio wave, create diffuse scattering (Freilich and Dunbar 1999). The dependence of the physical mechanisms of interaction of radio waves with surface waves on the ratio of their lengths makes it necessary to determine the characteristics of the wave field in different wavelength ranges (Danilychev et al. 2009; Zapevalov 2020).

One of the main characteristics of the sea surface that determine roughness in electromagnetic wave scattering problems are slopes, which are the first spatial derivatives of elevations. In (Wilheit 1979; Danilytchev et al. 2009), the dependences of the total slope variance (the sum of the slope variances of its two orthogonal components) on the wavelength range were investigated on the basis of radiometric measurements.

A. S. Zapevalov (✉)
Marine Hydrophysical Institute RAS, Sevastopol, Russian Federation

T. Chaplina (ed.), *Processes in GeoMedia—Volume IV*, Springer Geology,
https://doi.org/10.1007/978-3-030-76328-2_28

Later, the same problem was solved based on the synthesis of data obtained using different types of equipment (Zapevalov et al. 2020). Use in situ measurement data (measurements by laser inclinometers (Hughes et al. 1977; Khristophorov et al. 1992), string sensors (Kalinin and LeikinI 1988), wave buoys (Longuett-Higgins et al. 1963)), as well as remote sensing data in the radio and optical ranges (Cox and Munk 1954; Bréon and Henriot 2006; Chen et al. 2015).

The purpose of this work is to analyze the vector characteristics of a two-dimensional wave field, the dispersion of the longitudinal and transverse relative to the wind direction of the sea surface slope components, as well as the anisotropy of the slope. The analysis is carried out in the framework of the previously developed approach (Zapevalov 2020) based on the synthesis of data from different types of measuring equipment.

2 Statement of the Problem

We will consider a wavelength range with a variable boundary: from the longest to the wavelength λ_0. It is customary to distinguish two components of slope (Cox and Munk 1954), oriented upwind and crosswind. Hereinafter, they are denoted by the indices *u* and *c*, respectively. If there is no subscript, this expression is valid for both slope components. The variance of components of slopes created by waves whose length is less than λ_0 is determined by the expression

$$\sigma^2(k_0) = \int_0^{k_0} \Psi(k)\, dk, \tag{1}$$

where k is the wave number of surface waves, $k_0 = 2\pi / \lambda_0$, $\Psi(k)$ is the spectrum of the slope component. The inferior limit of integration is determined by the maximum surface wave length $\lambda_{\max}$. Here the condition $\lambda_{\max} = \infty$ is accepted, which implies that the inferior limit of integration in (1) is equal to zero.

We will also analyze the anisotropy of the sea surface slopes. For this we introduce the dimensionless parameter,

$$\gamma(\lambda_0) = \sigma_c^2(\lambda_0) / \sigma_u^2(\lambda_0). \tag{2}$$

The dispersion of the sea surface slope from radiosonding data is determined within the framework of the Kirchhoff approximation. Specular reflection is created by surface waves satisfying the conditions (Brekhovskikh 1952; Valenzuela 1978)

$$\lambda \gg \Lambda, \ \ K r \cos^3 \theta \gg 1, \tag{3}$$

where Λ is the length of the radio wave, K is wave number of a radio wave; r is radius of curvature; θ is angle of incidence. In this approximation, the normalized backscattering cross section is described by the expression

$$\sigma = \pi \sec^4 \theta |F_0|^2 P\left(\xi_\uparrow, \xi_\perp\right)\Big|_{\xi_\uparrow = tg\theta, \xi_\perp = 0}, \quad (4)$$

where F_0 is the Fresnel coefficient; P is two-dimensional probability density function of slopes; $\xi_\uparrow$ and $\xi_\perp$ are slopes of the sea surface in the sounding plane and in the orthogonal plane. Since the radar signal depends on the probability density function of sea surface slopes, this dependence allows you to restore statistical moments of slopes.

The "much more" condition in (3) is not strictly defined. We will assume that expression (3) is satisfied if $\lambda > n\Lambda$ and n is in the range from 3 to 5. For in situ measurements, there is also ambiguity in determining the value of λ_0 (Zapevalov 2020).To take this ambiguity into account in further analysis, we will assume that the value λ_0 is between two boundary values λ_{01} and λ_{02}. Boundary values λ_{01} and λ_{02} must be evaluated separately for each type of measuring equipment.

The regression equations that will be used for further analysis are presented in Table 1. The first three lines correspond to remote sensing, and the rest correspond to in situ measurements.

Table 1 includes the results of slope studies performed using string wave gauge (Kalinin and LeikinI 1988). In this paper, changes in the variance of slope components with changes in wind speed were not analyzed, but the dependence $\sigma_u^2 + \sigma_c^2 =$

Table 1 Regression equations describing the dependence of the variance of the slope components on the wind speed U

#	$\sigma_u^2 = \sigma_u^2(U)$	$\sigma_c^2 = \sigma_c^2(U)$	λ_{01}, m λ_{02}, m	Equipment, [source]
1.	$0.00316\,U + 0.000$	$0.00192\,U + 0.003$	0.001	Aerial camera, Cox and Munk (1954)
2.	$0.00316\,U + 0.001$	$0.00185\,U + 0.003$	0.001	Optical scanner, Bréon and Henriot (2006)
3.	$0.00079\,U + 0.0092$	$0.00053\,U + 0.0097$	0.065 0.109	Radar, Chen et al. (2015)
4.	$0.00231\,U + 0.0015$	$0.00166\,U + 0.0008$	0.004 0.006	Laser slopemeter, Hughes et al. (1977)
5.	$0.00205\,U + 0.0041$	$0.00162\,U + 0.00136$	0.004 0.006	Laser slopemeter, Khristophorov et al. (1992)
6.	$0.00146\,U - 0.0007$	$0.00064\,U - 0.0003$	0.4 1.6	String wave gauge, Kalinin and LeikinI (1988)
7.	$0.00060\,U - 0.00049$	$0.000287\,U + 0.0005$	3.7 6.2	Wave buoy, Longuett-Higgins et al. (1963)

0.0021 U − 0.001 is given and it is noted that the average ratio is σ_c^2/σ_u^2 =0.44. This allowed us to evaluate $\sigma_u^2 = \sigma_u^2(U)$ and $\sigma_c^2 = \sigma_c^2(U)$. The distance between the string wave gauge in different experiments was 20 cmor 40 cm, so the λ_{01} and λ_{02} were chosen corresponding to the minimum and maximum distance.

3 The Dispersion Slope of the Sea Surface

The regression equations $\sigma^2 = \sigma^2(U)$ given in the table are shown in Fig. 1. The regression dependencies obtained by equipment operating on the same physical mechanisms are close. Regressions calculated from optical images of the sea surface are almost identical. Regressions calculated from laser inclinometer measurements are also similar, despite the fact that in one case the measurements were made from a slow-moving vessel (Hughes et al. 1977) and in the other the measurements were made from a stationary base (Khristophorov et al. 1992). There is a well-pronounced tendency, the lower the value the λ_{01} and λ_{02}, the greater the variance of the slope. The regression equations of the upwind and crosswind components of the slope obtained using the radar are somewhat different.

It should be noted that linear regression equations constructed for a wide range of wind speeds are poor at describing slope changes in light winds. This

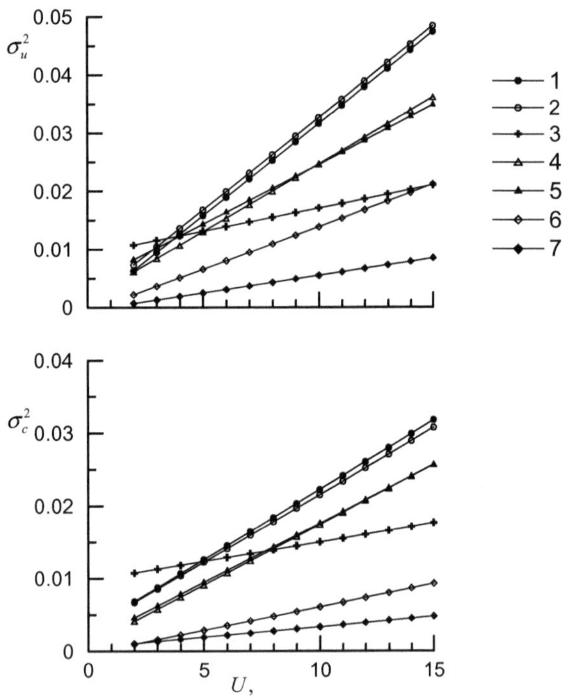

Fig. 1 Regression equations $\sigma^2 = \sigma^2(U)$. Line numbering corresponds to Table 1

is due to the fact that when the wind increases in the range of $U < 4–5$ m/s, the slope dispersion increases faster than at high wind speeds (Zapevalov 2002), i.e. the dependence $\sigma^2 = \sigma^2(U)$ is not linear. This relationship can be considered linear if the wind speed exceeds 8 m/s. In this case, the constant term in the regression equation can be ignored.

When describing the dependence of slope dispersion on length λ_0, it is customary to use its representation as part of the total variance of slopes created by waves of all scales (Zapevalov 2020; Wilheit 1979; Danilytchev et al. 2009)

$$\sigma^2(\lambda_0) = \chi(\lambda_0)\,\sigma^2_{All}. \tag{5}$$

The parameter is usually calculated using passive sensing data in the optical range (Cox and Munk 1954). Since the wavelength of the light wave is much less than the length of the shortest surface wave, all waves present on the surface create a specular reflection. The upper limit of the range of the wave spectrum on the sea surface is determined by the process of viscous dissipation of wave energy. A viscous cross section in the spectrum occurs at the $k_{dis} \approx 6283$ rad/m wave number, which corresponds to the m wave length $\lambda_{dis} \approx 0.001$ m [21]. In Table 1 for regressions based on passive sensing data in the optical range (# 1, 2), it is indicated $\lambda_0 = \lambda_{dis}$.

The parameters calculated under the assumption that the constant term in regressions can be ignored are shown in Fig. 2. The numbering corresponds to Table 1.

Regressions $\chi = \chi(\lambda_0)$ calculated for the upper and lower bounds of the parameter λ_0 have the form (Zapevalov et al. 2020)

$$\chi_u(\lambda_{01}) = 0.245\,\lambda_{01}^{-0.202}, \tag{6}$$

$$\chi_u(\lambda_{02}) = 0.292\,\lambda_{02}^{-0.177}, \tag{7}$$

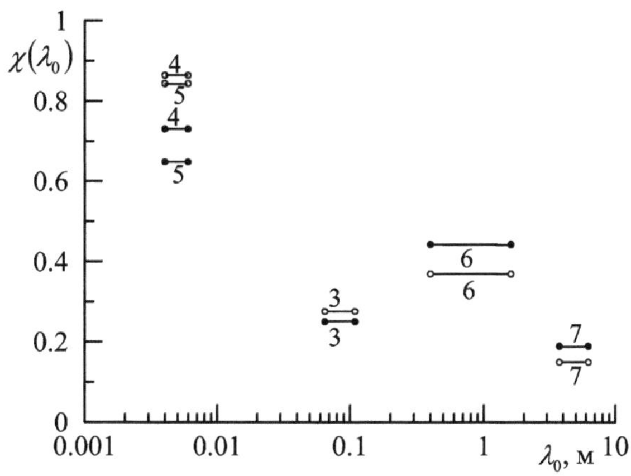

Fig. 2 Dependencies $\chi = \chi(\lambda_0)$. The black circle is χ_u, the white circle is χ_c Line numbering corresponds to Table 1

$$\chi_c(\lambda_{01}) = 0.219\,\lambda_{01}^{-0.220}, \tag{8}$$

$$\chi_c(\lambda_{02}) = 0.266\,\lambda_{02}^{-0.193}. \tag{9}$$

Since, as indicated above min $\lambda_0 = \lambda_{dis}$, when constructing regressions (6)-(9), the conditions were set: for $\lambda_{01} = 0.001$ m and $\lambda_{02} = 0.001$ m, the values χ_u and χ_c are equal to one.

4 Anisotropy

The anisotropy of sea surface slopes can be calculated using the same assumptions that were made when analyzing the proportion of dispersion of slope $\chi(\lambda_0)$ created by waves of a given scale

$$\gamma(\lambda_0) = \sigma_c^2(\lambda_0)/\sigma_u^2(\lambda_0). \tag{10}$$

The results of these calculations are shown in Fig. 3.

Used the data described in table. 1. The graph also shows the results of work (Pelevin and Burtzev 1975), in which slopes were measured by optical method at night under artificial light (#8). As for slope estimates based on the registration of reflected sunlight, it is assumed that under artificial lighting $\lambda_{01} = \lambda_{02} = 0.001$ m.

The general trend of changes in the coefficient γ is described by regression equations (Zapevalov et al. 2020)

$$\gamma(\lambda_{01}) = 0.508\,\lambda_{01}^{-0.0729}, \tag{11}$$

$$\gamma(\lambda_{02}) = 0.540\,\lambda_{02}^{-0.0671}. \tag{12}$$

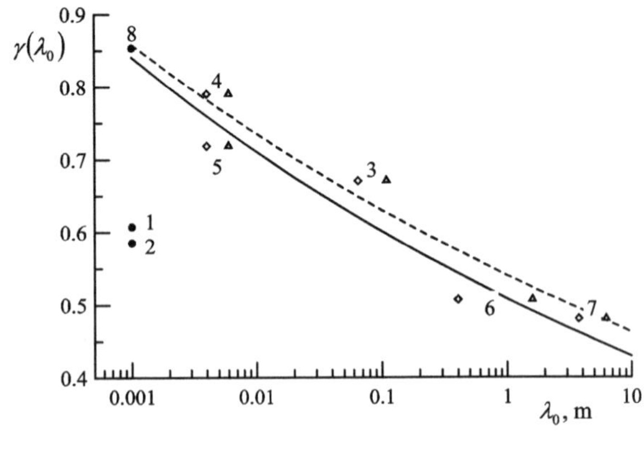

Fig. 3 Dependencies $\chi = \chi(\lambda_0)$. The black circle is χ_u, the white circle is χ_c Rhombus-triangle pairs, as well as solid and dashed lines correspond to $\lambda_{01} - \lambda_{02}$. Line numbering corresponds to Table 1

Regression Eqs. (11) and (12) are constructed without taking into account parameter estimates numbered 1 and 2.

Unexpected was the fact that the coefficient estimates γ obtained in Cox and Munk (1954); Bréon and Henriot 2006) do not correspond to the trend of its changes when changing λ_0. The method of determining the statistical characteristics of slope from aerial photographs of the sea surface or from satellite optical scanners is indirect. Its accuracy, in particular, depends on a priori representations of the slope probability density function, and on the range of slope values in which the approximation of this function is constructed (Zapevalov 2018). This may be one of the factors leading to deviation of the coefficient γ shown in Fig. 3 obtained by optical sensing from the regression curves (11) and (12).

5 Conclusion

The range of surface wavelengths that create a specular reflection of radio waves from the sea surface depends on the length of the sounding radio wave. This type of interaction of radio waves with the sea surface requires detailed information about the integral characteristics of the slope created by waves of a given range.

The analysis of changes in the variance of the components of the slope created by waves in the range with a changing boundary: from the longest waves to the wavelengths λ_0 is carried out. It is shown that, in the first approximation, the variance changes can be approximated by an exponential function with the base of the power λ_0 (Eqs. (6)–(9)). The parameters of this function are defined.

The coefficient of anisotropy of sea surface slopes (the ratio of the dispersion of the crosswind and upwind components of slope) increases with decreasing wavelength λ_0. It is shown that the estimation of the anisotropy coefficient obtained on the basis of the Cox-Munk model does not agree with the estimates of this coefficient calculated from in situ measurements and from the data of sounding the sea surface in the radio range.

This work was carried out as part of a state assignment on the topic No. 0827–2018-0003 "Fundamental studies of oceanological processes that determine the state and evolution of the marine environment under the influence of natural and anthropogenic factors, based on based methods of observation and modeling".

References

Apel JR (1994) An improved model of the ocean surface wave vector spectrum and its effects on radar backscatter. J Geophys Res 99(C8):16269–16291

Brekhovskikh LM (1952) The diffraction of waves by a rough surface. Zh Eksper i Tear Fiz 23:275–289

Bréon FM, Henriot N (2006) Spaceborne observations of ocean glint reflectance and modeling of wave slope distributions. J Geoph Res: Oceans 111(C06005) https://doi.org/10.1029/2005JC003343

Chen P, Yin Q, Huang P (2015) Effect of non-Gaussian properties of the sea surface on the low-incidence radar backscatter and its inversion in terms of wave spectra by an ocean wave. Chin J Oceanol Limnol 33(5):1142–1156

Cox C, Munk W (1954) Measurements of the roughness of the sea surface from photographs of the sun glitter. J Optical Soc Am 44(11):838–850

Danilychev MV, Kutuza BG, Nikolayev AG (2009) The use of the Kirchhoff method for practical calculations in microwave radiometry of rough sea surface. J Commun Technol Electron 54(8):915–926

Danilytchev MV, Kutuza BG, Nikolaev AG (2009) The application of sea wave slope distribution empirical dependences in estimation of interaction between microwave radiation and rough sea surface. IEEE Trans Geosci Remote Sens 47(2):652–661

Freilich MH, Dunbar RS (1999) The accuracy of the NSCAT 1 vector winds: comparisons with national data buoy center buoys. J Geophys Res 104(C5):11231–11246

Hughes BA, Grant HL, Chappell RWA (1977) A fast response surface-wave slope meter and measured wind-wave components. Deep-Sea Res 24(12):1211–1223

Kalinin SA, LeikinI (1988) A measurement of the slopes of wind waves in the Caspian Sea. Izvestiya of the Academy of Sciences of the USSR. Atmos Oceanic Phys 24(11):1210–1217

Khristophorov GN, Zapevalov AS, Babiy MV (1992) Statistics of sea-surface slope for different wind speeds. Okeanologiya 32(3):452–459

Longuett-Higgins MS, Cartwrighte DE, Smith ND (1963) Observation of the directional spectrum of sea waves using the motions of the floating buoy. Proceedings conference oceanwave spectra, EnglewoodCliffs. N. Y.: Prentice Hall, pp 111–132

Pelevin VN, Burtzev JG (1975) Measurements of elementary site slope. Optical investigations in ocean and atmosphere. Moscow: Institute Oceanology of the Academy of Sciences USSR. pp 202–218 [In Russian]

Valenzuela G (1978) Theories for the interaction of electromagnetic and ocean waves—a review. Bound Layer Meteorol 13(1–4):61–85

Wilheit TT (1979) A model for the microwave emissivity of the ocean's surface as a function of wind speed. IEEE Trans Geosci Electron GE-17(4)

Zapevalov AS (2002) Statistical characteristics of the moduli of slopes of the sea surface. Phys Oceanogr 12(1):24–31

Zapevalov AS (2018) Determination of the statistical moments of sea-surface slopes by optical scanners. Atmos Oceanic Opt 31(1):91–95

Zapevalov AS (2020) Distribution of variance of sea surface slopes by spatial wave range. Sovremennye Problemy Distantsionnogo Zondirovaniya Zemli Iz Kosmosa 17(1):211–219 [in Russian]

Zapevalov AS, Knyaz'kov AS, Shumeyko IP (2020) Describtion of sea surface slopes in applications related to radio wave reflection. Zhurnal radioelektroniki [electronic journal] (4). http://jre.cplire.ru/jre/apr20/8/text.pdf. https://doi.org/10.30898/1684-1719.2020.4.8. [In Russian]

Kinematics of the Polar Area of Lomonosov Ridge Bottom in Arctic

A. A. Schreider, A. L. Brehovskih, A. E. Sazhneva, M. S. Kluev, I. Ya. Rakitin, J. Galindo-Zaldivar, E. I. Evsenko, and O. V. Greenberg

Abstract The Arctic Ocean is crossed by the Lomonosov Ridge which plays an important role in all the evolution of region. In the range of 88° – 89° N (subpolar) where the position of the ridge top is located at depth of 1.5 km, the Intra Basin with depths over 2.6 km is settled. History of the Intra Basin occurrence is widely disputed. The lack of common view to the problem results in special attention to comprehensive interpretation of the entire geological and geophysical data set. Thus kinematic model of the Lomonosov Ridge subpolar segment evolution has been developed. Data analysis and of kinematic characteristics calculation led to first creation of a model of the Lomonosov Ridge subpolar region evolution on basis of quantitative sources.

Keywords Lomonosov Ridge · Intra Basin · Aeromagnetic and Seismic Profiles · Gravity Anomalies · Euler Poles · Reconstruction of Paleobathymetry

1 Introduction

In the subpolar region of the Lomonosov Ridge in the range between 88 and 89°N the Intra Basin (Cochran et al. 2006; Bjork et al. 2007) is located. Depths of which exceeds 2.6 km and has flat bottom and dimensions of 100 × 30 km. The position of the Lomonosov ridge crest is at depth of about 1.5 km. The origin of the Intra Basin and its role in the evolution of the polar region of the Arctic Ocean has been discussed in many studies (Cochran et al. 2006; Bjork et al. 2007; Backman et al. 2006; Jakobsson, et al. 2008; Lebedeva-Ivanova 2010; Truhin et al. 2004). In these studies, the origin of the Intra Basin is associated with the deformation of the

A. A. Schreider · A. L. Brehovskih · A. E. Sazhneva (✉) · M. S. Kluev · I. Ya. Rakitin · E. I. Evsenko · O. V. Greenberg
Shirshova Institute of Oceanology, RAS, Nakhimovsky Prospect, 36, 117997 Moscow, Russia
e-mail: aschr@ocean.ru

J. Galindo-Zaldivar
University of Granada, Via del Ospicio, 18010 Granada, CP, Spain

T. Chaplina (ed.), *Processes in GeoMedia—Volume IV*, Springer Geology,
https://doi.org/10.1007/978-3-030-76328-2_29

Lomonosov Ridge that occurred in the pre-Cenozoic era as a result of the gravitational slippage of one wing of the ridge along the other, stretching and splitting of the ridge as a piece of continental crust from the Barents Sea shelf, etc. Available data on sedimentary structure obtained as a result of analysis of seismic sounding, of magnetic and gravitational fields, of bottom relief by the international database IBCAO (IBCAO 2020) can be used to restore the peculiarities of evolution of the polar region of the Arctic Ocean (at the same time it is flagged that bathymetry in the IBCAO project framework is a subject to prognosis and continuous refinement), which is the goal of the presented study.

2 Research Methodology

A computer method was proposed in Bullard et al. (1965) for the best isobaths combination on the example of the Atlantic Ocean. The alignment was done by trial and error method, by minimizing the angular misalignment measured along the Euler latitudes. The technique illustrates the principle of the best match, which helps to restore the primary continuity of any contours, including isochronous, isobaths, isohypses, etc. Here this technique (Bullard et al. 1965) is applied for the first time for overlapping slope zones of the Intra Basin in the subpolar region of the Lomonosov Ridge. Calculations of the Euler poles and angles of rotation are performed using original computer programs of the Laboratory of Geophysics and Tectonics of the World Ocean Floor of the P.P. Shirshov Institute of Oceanology of RAS.

Complex geological and geophysical interpretation of data from international banks of digital and analogue information IBCAO (IBCAO 2020) on bathymetry, magnetometry, distribution of gravity, seismic data on sedimentary rock power and structure of the crust is the basis for creating a new geodynamic model of the Lomonosov Ridge subpolar region evolution. As a result, for the first time the evolution of the Intra Basin bottom and adjacent areas of the Lomonosov Ridge has been restored quantitatively.

The trough is stretched along the axis of the ridge and is surrounded by elevations with peaks at depth of about 1.5 km (Fig. 1). The slopes of the Intra Basin from the side of the Amerasian and Eurasian basins are rather steep and have a slope of up to 10°–20°. On profiles J-K (Cavalla7) and G-H (HB 9801), sublatitudinally crossing the Lomonosov Ridge, in the area of the Intra Basin (Figs. 1, 2) according to Cochran et al. (2006) there is a minimum of gravity field in the Fai reduction.

Its relative amplitude does not exceed 80 mGal. Absolute values above the depression are – 50 mGal, and above the adjacent areas with the continental crust of the Lomonosov Ridge from the side of Makarov and Amundsen Basins the values are 30–50 mGal (Fig. 2).

There are 8 aeromagnetic profiles known in the area of the Intra Basin (Fig. 3) (Brozena et al. 2003). The decrease of the anomalous magnetic field with amplitude about hundred nT is associated with the central part and sides of the trough. It has rather complex configuration and, in our opinion, does not possess any signs

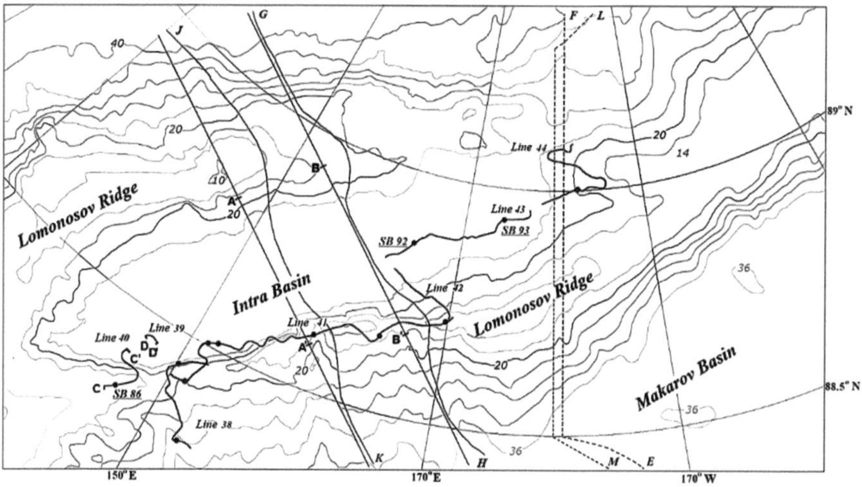

Fig. 1 Bathymetry of the Intra Basin (isobaths in hundreds of meters). In accordance with data (Artjushkov and Poselov 2009; Bullard et al. 1965; IBCAO 2020 with changes) the position of the main profiles of seismic (thick lines 38 – 44) and gravimetric F-E, L-M (NP-28) (dashed lines) measurements are shown, as well as profiles of anomalous values in Fai reduction along the profiles of G-H, J-K. The points of hydroacoustic probing SB 86, SB 92 and SB 93 by sonobuoy are marked with dashes. Letter designations with dashes correspond to end-points of profiles shown in Figs. 2, 4 and 5

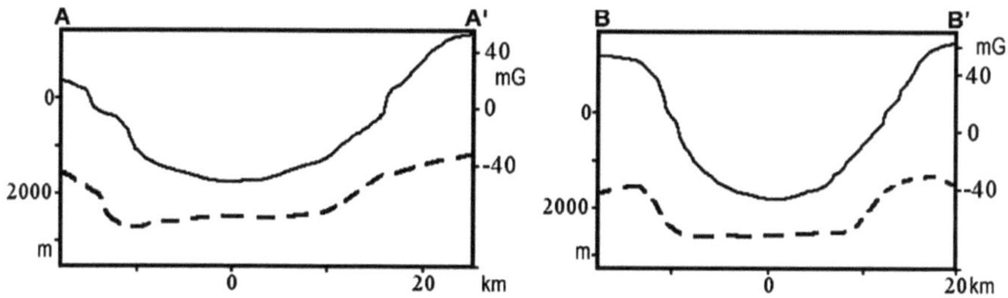

Fig. 2 Bottom relief (dashed line) and anomalous values of gravity field in milliGals (solid line) at sections A-A1 and B-B1 of I-K and L-Z profiles, shown in Fig. 1

of a linear magnetic anomaly born in the process of bottom spreading. Estimation modelling suggests the sources of the anomalous field to be related to tectonic heterogeneities in the crystal foundation of the sides and bottom of the trough.

Six seismic profiles (lines 39–44) of the HOTRAX project are acquired and processed together with data from the ACEX project (Arctic Coring Expedition, IODP, Leg 302) (Lebedeva-Ivanova 2010). In this case line 40 (Fig. 1) is the closest to the extraction points of geological samples of the ACEX drilling project and is characterized by high quality of seismic recording. Comparison of seismic signals and drilling geological data shows (Fig. 4) that among the fixed seismic layers (units 1–4) the most near-surface of them has the power of about 0.15 km with the velocity of

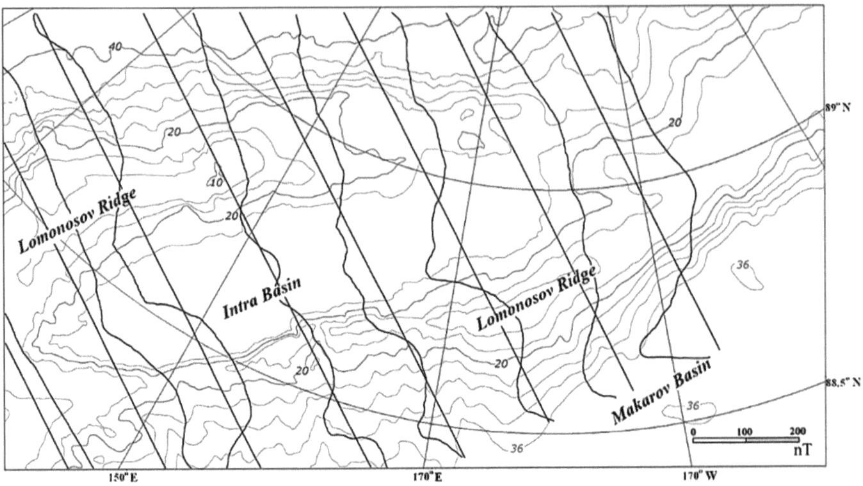

Fig. 3 Profiles of the anomalous magnetic field along the aeromagnetic observation lines according to data (Brozena et al. 2003; Schreider et al. 2020), isobaths in hundreds of meters. The scale in nanotesla (nT) is given

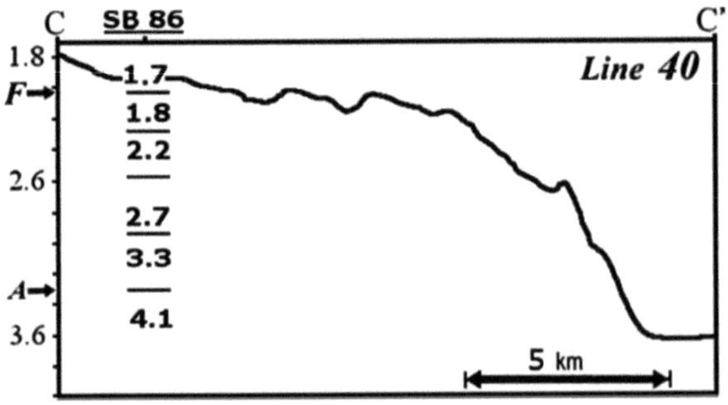

Fig. 4 Longitudinal seismic wave velocities in the region of multichannel seismic profiling point SB 86 (modified by Lebedeva-Ivanova (2010)). Position of line 40 and points of profiling are shown in Fig. 1. Vertical axis represents time of double run of seismic signal in seconds

longitudinal seismic waves up to 1.7 km/s. At the same time, the underlying reflector "D" (Fig. 4) correlates with hiatus, which reflects a break in sediment accumulation, dated to the interval of 16–44 mln years (Lebedeva-Ivanova 2010).

Seismically more transparent layers with velocities of 1.8 km/s underlie further. This layers are most likely composed of muddy rocks observed when sampling columns according to ACEX drilling project. A pack of Cenozoic sedimentary deposits of more than 0.7 km in thickness with velocities close to 2.2 km/s is located on boards and inside Intra Basin. The reflector 'A' in Fig. 4 corresponds to the boundary between Cenozoic and Mesozoic sediments (Lebedeva-Ivanova 2010).

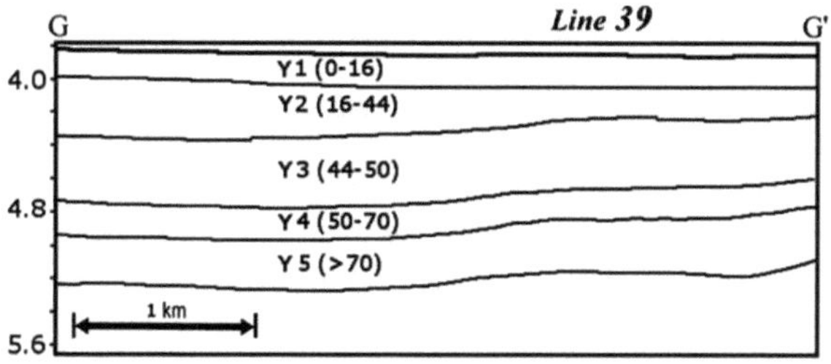

Fig. 5 Sedimentary layers Y1-Y5 (numbers in parentheses indicate millions of years) and estimated geochronological interpretation along line 39 between points G-G' (Fig. 1) at the Eurasian slope of the Intra Basin (modified by Lebedeva-Ivanova (2010); Schreider et al. 2020))

Seismic records in the peripheral part of line 39 illustrate the most complete cross-section of the Cenozoic sediments (Fig. 5). The slope of sediment deposits recorded at line 39 for Middle and Late – Cenozoic deposits allows to assume (Lebedeva-Ivanova 2010), that the sedimentaion was mostly filled from the North-West on the West, in the Siberian part of the trough.

Data interpretation along line 39 (Lebedeva-Ivanova 2010) serves for expansion of sedimentary layer structure vision in the subpole region of the Lomonosov Ridge known along line 40 and allows detecting a number of inter-layers (units 1–5). Unit 1 is represented by silt with thickness up to 0.5 km and age less than 16 mln years. Below its thickened silts are of lower depth (unit 2), and the underlying reflector correlates with hiatus reflecting a break in sedimentation process. Estimation calculations show that unit 3 is composed of highly compacted silt and loams with thickness of about one kilometer and age 44–50 mln years. Unit 4 is composed of clay (?) material with the age of 50–70 mln years. Unit 5 is represented by consolidated (?) sedimentary rocks with thickness of more than one kilometer and age of more than 70 mln years. The pattern of seismic recording suggests the presence of even more ancient sediments, underlying unit 5. Earlier sediments of varying thickness (from 0.5 to 1.5 km) with velocities of 2.6–3.9 km/s are characterized by moderate deformations.

Seismic lines on the ridge heights show an acoustic foundation at a depth of about 3 km, and the position of reflecting horizons in the trough, usually parallel to the bottom surface, shows depth of 4.5 km. This fact and the presence of layers (Lebedeva-Ivanova 2010) with higher velocities (about 4–5 km/sec) suggest, as mentioned above, the presence of well consolidated even older sediments, with possible ages of up to 110 mln years or even older [5, etc.]. Our estimates and data from Cochran et al. (2006); Langinen et al. 2009; Brozena et al. 2003; Langinen et al. 2006; Jokat et al. 1992) do not allow to identify with certainty the material underlying the trough sediments as fragments of the oceanic crust.

3 Paleogeodynamic Calculation Results

The analysis of the geomorphology of the Intra Basin bottom allows to highlight a number of bathymetric features that can be interpreted within the framework of the lithospheric plate tectonics concept. The most important are the similarity of the isobaths patterns on opposite slopes of the trough (Fig. 6a). The maximum similarity has the isobaths of 1.8 km for about forty kilometers (Fig. 6b).

The junction of conjugate isobaths is registered for more than 40 km and can be described by the Euler Pole of finite rotation at a point with coordinates 82.43°N, 166.14°W. The angle of rotation was 14.15° ± 3.2°. The point alignment error was 1.2 ± 0.7 km (9 values). As a result, for Intra Basin the position of axis of split zone is restored and mutual position of sides of the depression is reconstructed (Fig. 7a). Also the profile of the bottom before the split along D–D[1] line is obtained.

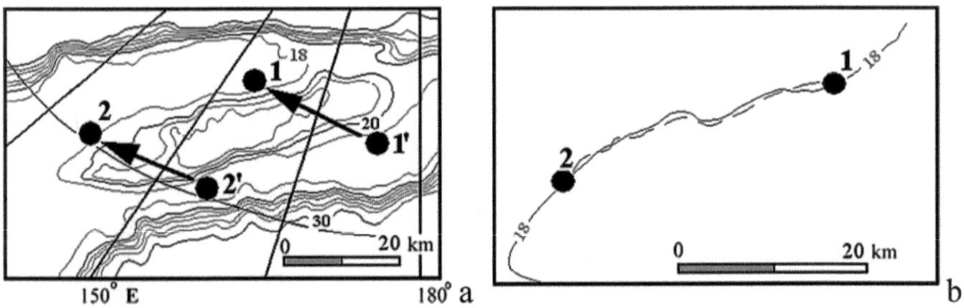

Fig. 6 Intra Basin bottom on the Lomonosov Ridge: **a** bathymetry (hundreds of meters deep) of according to data (IBCAO 2020; Schreider et al. 2020) and the position of the end-points of conjugate isobaths 1.8 km 1–2 on the Southern and 1^1–2^1 on the Northern slopes of the trough, as well as the direction (arrows) of the closure points, **b** junction of conjugate isobaths 1.8 km Southern (solid line) and Northern (dashed) slopes of the trough, location of points is shown in Fig. 6a

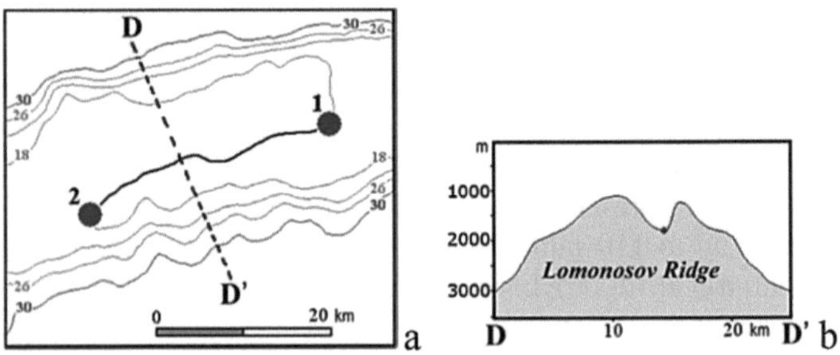

Fig. 7 Slopes of the Intra Basin: **a** paleogeodynamic reconstruction of oncoming slope closure, location of points is shown in Fig. 6a, black bold line shows the restored position of the rift axis, isobaths in hundreds of meters; **b** cut of the primary relief of the bottom before the breakup of fragments of the Lomonosov Ridge along the D–D' line, which is shown in Fig. 7a, the point indicates the exit of the rift axis to the bottom surface

As the crust of the Lomonosov Ridge is continental and, according to the results of the above mentioned geological and geophysical studies, there are no obvious signs of oceanic crust fragments under the Intra Basin, the crust underlying the depression should also have continental origin. Thus, the process of destruction has not led to the splitting of the continental lithosphere in the area of the Intra Basin and the formation of the Oceanic crust. The above calculations also allow restoring the peculiarities of mutual position of the primary relief of the trough bottom that existed before the split (Fig. 7b) and marking the exit of the split axis to the bottom. These results seem to be very important in restoration of parameters of the stress state of the crust at the Lomonosov Ridge breakup from the periphery of the Eurasian continent.

4 Conclusion

Thus, the computer technique for the best isobath alignment (Bullard et al. 1965; Shreider et al. 2016) is first applied to the case of isobath alignment in the Intra Basin at the Lomonosov Ridge subpolar region of the Arctic Ocean. Numerous studies of the connectivity of different parts of different and homonymous isobaths have shown that the most suitable for the purposes of paleogeodynamic analysis are parts of isobaths in the range of 1.6–2.2 km, which correspond to the steepest slope section with the lowest sediments thickness. Therefore isobate areas of 1.8 km are identified on the counter slopes of the Intra Basin to show good connection with each other. For these sections, Euler's finite poles are calculated and the rotation angles are determined, the axis of the split zone is restored in the crust of the Intra Basin, and the features of the paleo-relief of the bottom are presented. No clear evidence of oceanic crust fragments in the trough is found.

The reconstructions made possible to restore paleobathymetry in the subpolar area of the Lomonosov Ridge and to estimate the age of the bottom split in the Intra Basin as possibly reaching 100 mln years or even more. The conclusions obtained do not contradict the results (Artjushkov and Poselov 2009). The results of the studies seem undoubtedly important in the face of the discussion of justifying issues for the position of the Russian continental shelf in the Arctic Ocean outer limit.

Acknowledgements This work was performed as part of the State assignment (Project No. 0149-2019-0005). The methodological issues of combining conjugate isobaths on opposite slopes were studied under RFBR Project № 20-05-00089.

References

Artjushkov EV, Poselov VA (2009) Kontinental'naja kora v glubokovodnyh vpadinah na severo-vostoke Rossijskogo sektora Arktiki. Geologija poljarnyh oblastej Zemli. Mat. XLII Tektonich. soveshh. 1:24–27. [In Russian]

Backman J, Moran A. et al (2006) Proceeding IODP. 302. Edinburgh (Integrated ocean drilling program management international, Inc.) https://doi.org/10.2204/iodp.proc.302.2006

Bjork G, Jakobsson M, Rudels B. et al (2007) Bathymetry and deep-water exchange across the central Lomonosov ridge at 88–89°N. Deep-Sea Res 54:1197–1208

Brozena JM, Childers VA, Lawver LA, Gahagan LM, Forsberg R, Faleide JI, Eldholm O (2003) New aerogeophysical study of the Eurasia basin and Lomonosov ridge: implications for basin development. Geology 31(9):825–828

Bullard E, Everett J, Smith A (1965) The fit of continents around Atlantic. Symphosium on continental drift. Phil Trans Roy Soc London 258A:41–51

Cochran J, Edwards M, Coakley B (2006) Morphology and structure of the Lomonosov ridge, Arctic Ocean. Geochem Geophys Geosys 7(Q05019):26. https://doi.org/10.1029/2005GC001114

Jakobsson M, Macnab R. et al (2008) An improved bathymetric portrayal of the Arctic ocean: implications for ocean modeling and geological, geophysical and oceanographic analyses. 35:L07602. https://doi.org/10.1029/2008GL033520

Jokat W, Uenzelmann-Neben G, Kristoffersen Y, Rasmussen T (1992) ARCTIC'91: Lomonosov ridge—a double sided continental margin. Geology 20:887–890

Langinen AE, Gee DG, Lebedeva-Ivanova NN, Zamansky YY (2006) Velocity structure and correlation of the sedimentary cover on the Lomonosov ridge and in the Amerasian basin, Arctic Ocean. In: Scott RA, Thurston DK (eds) Proceeding fourth international conference on Arctic margins, OCS Study MMS. 179–188

Langinen A, Lebedeva-Ivanova N, Gee D, Zamansky Y (2009) Correlations between the Lomonosov ridge, Marvin Spur and adjacent basins of the Arctic ocean based on seismic data. Tectonophysics. 472:309–322

Lebedeva-Ivanova N (2010) Geophysical studies bearing on the origin of the Arctic basin. Uppsala University Press, Uppsala, p 79

Schreider AA, Brehovskih AL, Sazhneva AE, Kluev MS, Rakitin IY, Galindo-Zaldivar J, Evsenko EI, Greenberg OV (2020) Kinematics of the bottom of the appearance area of the lomonosov ridge in the arctic. Processes in GeoMedia. 3(25):823–828. [In Russian]

Shreider AlA. et al (2016) Bottom kinematics near the Svalbard region of the Eurasian basin. Okeanol. 56(5):791–803. [In Russian]

Truhin VI, Pokazeev KV, Kunicyn VE, Schreider AA (2004) Osnovy ekologicheskoi geofiziki. St.-Petersburg, Lan, p 384 [In Russian]

www.ngdc.noaa.gov/mgg/bathymetry/arctic/, (IBCAO), 2020

Variations in the Electric Field Parameters During Magnetic Storms in 2018

Svetlana Riabova

Abstract The presented research provides an analysis of electric parameter anomalies during magnetic storms in 2018 selected from weather data. As the initial data, we used the time series of instrumental observations of vertical components of the electric field strength and atmospheric current, carried out at the Center for Geophysical Monitoring in Moscow of Sadovsky Institute of Geosphere Dynamics of Russian Academy of Sciences and Mikhnevo Geophysical Observatory of Sadovsky Institute of Geosphere of Russian Academy of Sciences. The changes in the parameters of the electric field during the one of strongest magnetic storm (August 25–26, 2018) in solar cycle No. 24 are discussed in more detail.

Keywords Electric field strength · Atmospheric current · Magnetic storm · Instrumental observations

1 Introduction

Processes in the upper atmosphere and ionosphere of the Earth are mainly due to the interaction of the Sun and the Earth (Basavaiah 2011). Strong disturbances in the upper layers are observed during events associated with space weather and called magnetic storms (Carlowicz and Lopez 2002). A magnetic storm is a disturbance of the Earth's magnetosphere caused by the effective effect of the solar wind on the space environment and associated with an intense energy release in the magnetosphere (Lazutin 2012).

Plasma streams from coronal mass ejections from the Sun and high-velocity streams of the solar wind following the shock wave of flare events exert sudden dynamic pressure on the Earth's magnetosphere (Buonsanto 1999). In response to changes in the parameters of the interplanetary medium, magnetospheric currents change (their shape, spatial position and intensity change). In addition, new current systems are formed (Curto et al. 2007). A magnetic storm is characterized by the

S. Riabova (✉)
Sadovsky Institute of Geospheres Dynamics of Russian Academy of Science, Leninsky Prospekt, 38, building 1, Moscow, Russia

T. Chaplina (ed.), *Processes in GeoMedia—Volume IV*, Springer Geology,
https://doi.org/10.1007/978-3-030-76328-2_30

formation of a powerful ring current in the inner magnetosphere, at a distance of ~3–7 Earth radii from the center of the Earth, which leads to a planetary depression of the magnetic field (Daglis et al. 1999). All changes in current systems are recorded by magnetometers (Gonzalez et al. 1994).

Recently, in the study of geomagnetic disturbances, much attention has been paid to the study of the effects of solar activity and associated geomagnetic storms in atmospheric electricity, mainly in high and middle latitudes (Sheftel et al. 1992; Spivak and Riabova 2019). In particular, it was shown that during the maximum of the Forbush effect during strong geomagnetic storms, positive disturbances of the electric field are observed, which exceed the background ones by ~2%, followed by a gradual restoration of the background electric field variations (Marcz 1997). This behavior of the electric field is explained by the effect on the atmosphere of the hard muon component of the GCR, which reaches sea level and determines their contribution to the atmospheric conductivity at the place of registration of the electric field (Aplin and Harrison 2014). The decrease in the gradient of the electric field potential noted in Kleimenova et al. (2008) was explained by the authors by an increase in the conductivity of the upper atmosphere caused by the precipitation of energetic electrons into subauroral latitudes. Interesting results were obtained in articles (Smirnov 2014a, b), where it was shown that during magnetic storms an anomalous increase in air temperature and humidity was observed, which led to the formation of clouds of various shapes. The author suggested that meteorological processes are sources of additional energy inflow into the lower atmosphere, which could be reflected in changes in electrical parameters.

In this work, the registration data for the parameters of the electric field at the Mikhnevo Geophysical Observatory and at the Center for Geophysical Monitoring in Moscow of Sadovsky Institute of Geosphere Dynamics of Russian Academy of Sciences during magnetic storms in 2018 are analyzed.

2 Initial Data and Data Processing Methods

In carrying out these studies, the data of registration of the Earth's magnetic field at the Mikhnevo Geophysical Observatory of Sadovsky Institute of Geosphere Dynamics of Russian Academy of Sciences (IDG RAS), data of registration of the vertical component of the atmospheric electric current at the Mikhnevo Geophysical Observatory of IDG RAS and the registration data of the vertical component of the electric field strength at the Mikhnevo Geophysical Observatory and at the Center for Geophysical Monitoring in Moscow of IDG RAS. The geographic coordinates of the Mikhnevo Observatory is 54.94°N; 37.73°E; the geographic coordinates of Center for Geophysical Monitoring of Moscow is 55.71°N; 37.57°E.

The list of instruments used to register the parameters under consideration at both observation points is given in Table 1. Time series of variations in the parameters of the electric field and three components of the magnetic field were generated with 5 s discretization.

Table 1 List of means of electromagnetic monitoring at the Earth's surface layer at the Mikhnevo Geophysical Observatory of IDG RAS and at the Center for Geophysical Monitoring in Moscow of IDG RAS

Point	Parameter		Mean
Mikhnevo Geophysical Observatory	Magnetic monitoring	Three components of the magnetic field	LEMI-018 fluxgate magnetometer
	Electrical monitoring	Vertical component of the electric field strength	INEP electrostatic fluxmeters
		Vertical atmospheric current	Compensation current recorder
Center for Geophysical Monitoring in Moscow	Electrical monitoring	Vertical component of the electric field strength	INEP electrostatic fluxmeters

The data were processed and analyzed during the period of the strongest magnetic storms in 2018. The list of events is given in Table 2. Table 2 also shows the values of the Kp-index (GFZ, Potsdam, Germany) and Ap-index (average daily value of geomagnetic activity).

3 Results

In this work, when carrying out research, due to the high variability from day to day of variations in the vertical component of the electric field strength and the vertical atmospheric current, not the relative values of the electric field parameters (deviation from the daily average) are considered, but the absolute ones.

The variations in the parameters of the electric field are strongly influenced by disturbances of different nature (Spivak et al. 2019); therefore, only the sections of the records obtained on the days of the so-called "fair weather" (Adushkin et al. 2018) were selected. As you can see from the Table 3, some of the events were accompanied by precipitation in the form of snow or rain, and some of the events were observed during the cloudy period; therefore, when analyzing the variations in electrical parameters, such events were rejected.

The results of processing and analysis of geomagnetic storms selected from the weather data in 2018 demonstrated a complex picture of the development of variations in electrical parameters, which manifest themselves in the form of alternating sign variations or in the form of positive or negative bay-like disturbances. This behavior can be explained both by different sources of the storms themselves, and by specific geophysical conditions during a particular storm. This requires further research, in particular, with the use of data for a different period of time.

As an example, consider the anomalies in the Earth's electric field in the surface layer of the atmosphere caused by the magnetic storm G3 on August 25–26, 2018

Table 2 List of geomagnetic storms for 2018 year considered in the course of these studies; Kp values (GFZ, Potsdam, Germany); Ap-index is the average daily geomagnetic activity; times are in UT

No	Date	Ap	0–3	3–6	6–9	9–12	12–15	15–18	18–21	21–0	Kp_{max}
1	2018/08/26	67	5	7−	7+	5−	5−	6	5	3	7+
2	2018/04/20	43	3+	4	6	6−	4+	5	5+	3+	6
3	2018/09/11	35	4−	5	6	6−	3	3	4−	2+	6
4	2018/11/05	32	4+	6−	5	5−	3+	3+	4−	2+	6−
5	2018/09/22	27	5−	4+	4−	4	4+	3−	4	4−	5−
6	2018/05/06	26	6−	4	3+	3−	3	2+	4+	4+	6−
7	2018/08/27	25	1+	2+	3−	4−	4−	6−	5	3−	6−
8	2018/10/07	24	0+	1−	1	2+	4−	5+	5+	5−	5+
9	2018/06/01	23	3	4−	5−	4+	5−	4−	2+	2+	5−
10	2018/05/05	23	1−	1−	0+	3−	4	5−	4+	6−	6−
11	2018/12/28	22	4	4	4+	3	4−	3	3+	3+	4+
12	2018/03/18	22	3−	2+	1−	1	3+	4−	4+	6	6
13	2018/09/10	20	0+	0+	1−	2+	3−	5	4+	5	5
14	2018/06/26	20	5−	4−	3−	4−	4−	3	3+	2	5−
15	2018/10/08	19	4	4+	4−	2	4+	4−	2	2−	4+
16	2018/06/23	18	2+	3+	2−	4	4	3	3+	4−	4
17	2018/03/16	18	3	4+	3−	2+	3	4	3	4−	4+
18	2018/02/27	18	4	5+	2−	2	4+	2+	1+	1	5+
19	2018/09/13	17	4	4	3	3	3	3+	2−	3	4
20	2018/07/24	17	4−	3	3+	3−	2	3	3+	4	4
21	2018/07/05	17	2+	1	0+	2	4−	4+	5−	4−	5−
22	2018/04/10	17	2	4	4	3−	3+	3	2+	4−	4
23	2018/11/04	16	1	1−	1−	1−	1	3−	5−	5+	5+
24	2018/10/09	16	3	3−	3	4−	3	4−	3−	3	4−
25	2018/06/18	16	4	3+	4−	3+	3	2+	2+	2	4
26	2018/06/02	16	3−	4	3−	2	3−	2+	4	4−	4
27	2018/05/07	16	4−	4	3	2−	3	3+	2+	3+	4
28	2018/03/25	16	3+	4−	3	3+	2	3−	2+	4	4
29	2018/02/19	16	4	4	2+	2−	3−	2	4	3−	4
30	2018/10/10	15	3+	1	2−	2	3+	3	5−	3−	5−
31	2018/03/23	15	4	3	1+	3	2	2−	4−	4−	4
32	2018/03/15	15	4	2+	2−	3−	2−	3+	3+	4−	4
33	2018/02/23	15	2−	4−	3	4+	1+	2	3+	3−	4+
34	2018/09/14	14	4+	4	2+	2	2	2−	3−	3	4+
35	2018/08/20	14	4−	3+	2−	2	3−	3	4	2+	4

(continued)

Table 2 (continued)

No	Date	Ap	0–3	3–6	6–9	9–12	12–15	15–18	18–21	21–0	Kp_{max}
36	2018/05/11	14	2	3+	4−	2−	3	2	4−	3+	4−
37	2018/05/09	14	3+	3−	2+	2−	2+	2+	4	4−	4
38	2018/05/08	14	3−	3	2+	3−	3−	4−	4−	2	4−
39	2018/03/19	14	4+	4−	3	1	2−	2	2+	3	4+
40	2018/03/17	14	4+	3	2+	2−	2+	2−	2+	4−	4+
41	2018/11/10	13	2	2−	2	4+	2	1+	4−	3+	4+
42	2018/10/13	13	1	2+	1+	0+	2−	4+	4	3+	4+
43	2018/08/15	13	2	1	2−	3−	4	3	3	3+	4
44	2018/02/18	13	4	2	2+	2	3	4−	2+	2+	4
45	2018/01/14	13	5−	4	2	1+	2−	2	1	2	5−
46	2018/08/25	12	1−	1	2	2+	3−	2−	4−	4+	4+
47	2018/08/17	12	3	4	2+	2	1	0+	3+	3+	4
48	2018/06/25	12	1	1−	1−	2−	3−	2	3	5	5
49	2018/05/31	12	1+	1+	1−	1+	2−	4−	4+	3	4+
50	2018/04/21	12	3	3−	3−	2	2+	3+	3	1+	3+

($Kp_{max} = 7$ and $DST_{min} = -174$ nT). This storm was caused by an extremely slow solar coronal mass ejection (CME) (Vanlommel 2018). Low velocity of CME could not be automatically identified. The most likely source of CME is the filament ejection observed on August 20, 2018 at 8:00 UT on the heliographic coordinates $\theta_{Sun} = 16°$, $\varphi_{Sun} = 14°$ on the solar surface (Abunin et al. 2020).

The results of observations of the parameters of the interplanetary medium and the geomagnetic field during the magnetic storm on August 25–26, 2018 are shown in Fig. 1. Data have taken from the Space Physics Data Center of NASA website (https://omniweb.gsfc.nasa.gov/html/ow_data.html).

The CME shock wave approached the Earth at 2:45 UT on August 25, 2018, but did not cause a sudden geomagnetic impulse (Blagoveshchensky and Sergeeva 2019). Analysis of the behavior of the SYM-H index showed that from 3:51 UT its values began to increase to 9 nT at 4:01 UT (Lissa et al. 2020).

The interplanetary shock wave was recorded by three spacecraft at ~5:37 UT, ~5:42 UT, and ~5:43 UT on August 24, 2018 and was characterized by a slight change in the solar wind density. The calculated velocity of the shock wave is ~300–350 km/s (Piersanti et al. 2020). The shock wave was followed by the arrival of material from a coronal mass ejection at ~7:45 UT, which compressed the magnetosphere and marked the beginning of the initial phase of the storm (Astafyeva et al. 2020). A significant magnetic cloud was observed in the Earth's orbit from ~12:15 UT on August 25 (sudden onset of a magnetic storm) to ~10:00 UT on August 26 (Lissa et al. 2020). According to the registration data with ground-based magnetometers, the onset of a magnetic storm in the form of a sudden commencement was recorded

Table 3 List of geomagnetic storms for 2018 year, indicating the meteorological conditions during the observation period; the events for which the analysis of electric field anomalies is performed are highlighted in bold

No	Date	Weather	No	Date	Weather
1	**2018/08/26**	**Mostly clear**	26	2018/06/02	Mostly cloudy
2	2018/04/20	Mostly cloudy	27	2018/05/07	Mostly cloudy
3	2018/09/11	Cloudy, rain	28	2018/03/25	Mostly cloudy
4	2018/11/05	Cloudy	29	2018/02/19	Light snow
5	**2018/09/22**	**Clear**	30	2018/10/10	Mostly cloudy
6	2018/05/06	Rain with thunderstorms	31	2018/03/23	Mostly clear, snow at times
7	**2018/08/27**	**Clear**	32	2018/03/15	Cloudy, snow at times
8	**2018/10/07**	**Clear**	**33**	**2018/02/23**	**Clear**
9	2018/06/01	Mostly cloudy	34	2018/09/14	Partly cloudy
10	2018/05/05	Cloudy, rain at times	35	2018/08/20	Partly cloudy
11	2018/12/28	Cloudy, snow at times	36	2018/05/11	Mostly clear
12	**2018/03/18**	**Clear**	**37**	**2018/05/09**	**Clear**
13	2018/09/10	Cloudy, rain	**38**	**2018/05/08**	**Clear**
14	2018/06/26	Mostly cloudy	**39**	**2018/03/19**	**Clear**
15	2018/10/08	Rain	40	2018/03/17	Cloudy
16	**2018/06/23**	**Mostly clear**	41	2018/11/10	Cloudy
17	2018/03/16	Cloudy	42	2018/10/13	Partly cloudy
18	2018/02/27	Clear	43	2018/08/15	Mostly clear, light rain at times
19	2018/09/13	Partly cloudy	44	2018/02/18	Cloudy, snow at times
20	2018/07/24	Cloudy, rain	45	2018/01/14	Cloudy, snow at times
21	2018/07/05	Partly cloudy, rain	**46**	**2018/08/25**	**Clear**
22	**2018/04/10**	**Mostly clear**	**47**	**2018/08/17**	**Clear**
23	2018/11/04	Cloudy	48	2018/06/25	Mostly clear, rain at times
24	2018/10/09	Mostly cloudy	49	2018/05/31	Mostly cloudy
25	**2018/06/18**	**Clear**	50	2018/04/21	Partly cloudy

at ~8:00 UT. The dynamic pressure jump was 6 nPa, and the solar wind density increased to 20 cm^{-3} (Kleimenova et al. 2019).

As seen from Fig. 1, while the plasma temperature decreased from $\sim 9 \times 10^4$ K to $\sim 1.5 \times 10^4$ K; the strength of the total magnetic field increased to 16 nT and was maintained for ~12 h; the magnetic field rotated smoothly, leading to a pronounced and prolonged orientation to the south (beginning at ~15:00 UT on August 25, the beginning of the main phase of the storm) for about 22 h; the solar wind velocity ranged from ~450 km/s to ~370 km/s. It should be noted that no strong geomagnetic variations are noted at the initial stage (in particular, at the Mikhnevo observatory,

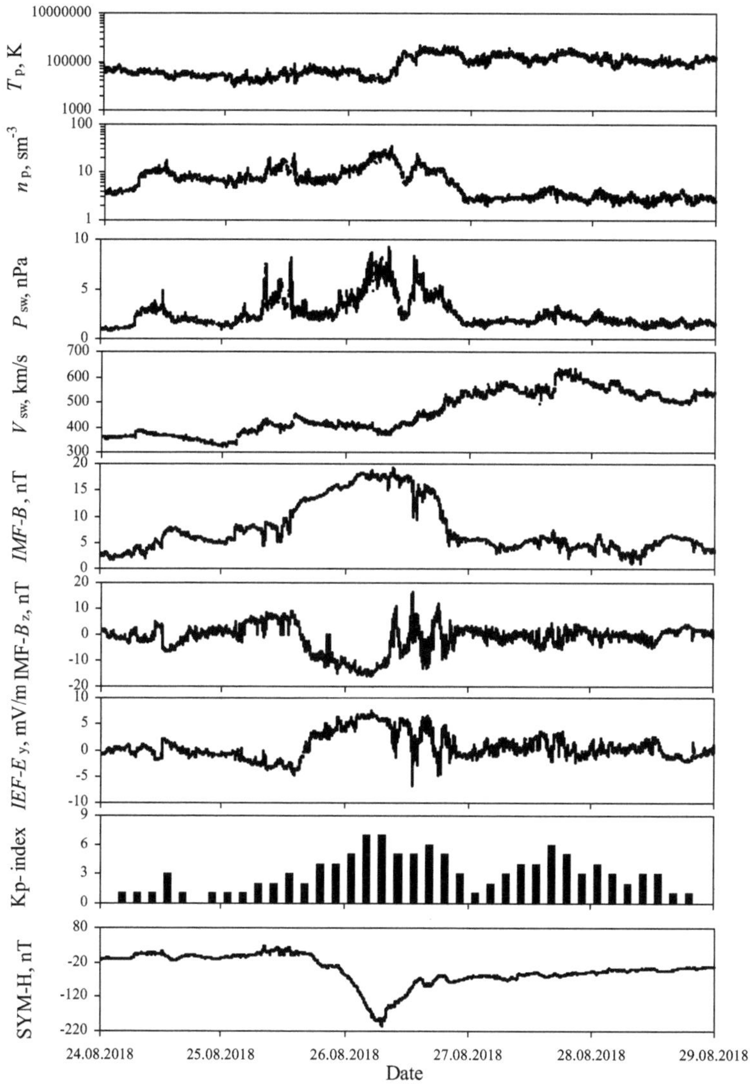

Fig. 1 Time variations of the solar wind parameters observed by spacecrafts: proton temperature T_p; proton density n_p; dynamic pressure P_{SW}; velocity V_{SW}; strength of the total magnetic field IMF-B, the strength of z-component of the IMF IMF-B_z; the intensity of y-component of the IEF IEF-E_y; as well as variations in the Kp and SYM-H indices for the period from August 24 to August 28, 2018

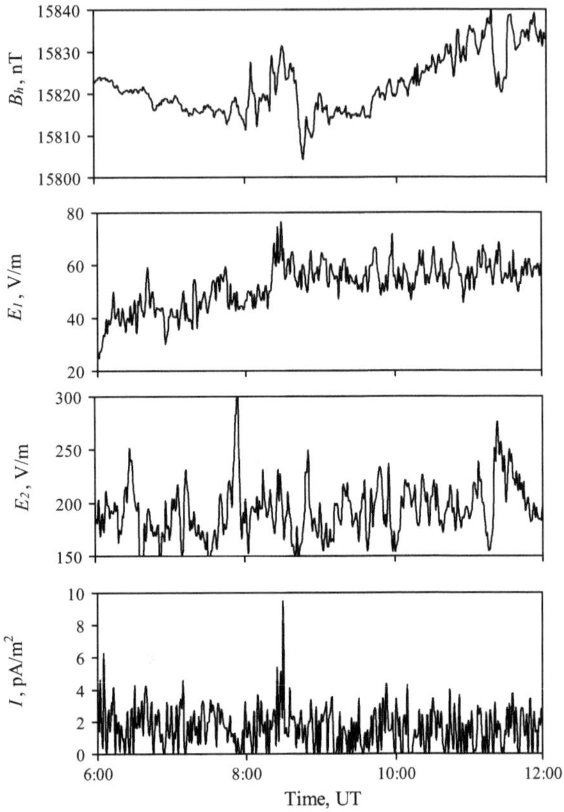

Fig. 2 Variations in the horizontal component of the geomagnetic field B_h, vertical components of the electric field strength E_1 and atmospheric current I at the Mikhnevo Observatory, variations in the vertical component of the electric field strength E_2 recorded at the Center for Geophysical Monitoring in Moscow during the initial phase of a magnetic storm 25–26 August 2018

Fig. 2). This behavior can be explained by the northern direction of the interplanetary magnetic field (Younas et al. 2020).

The main phase of the magnetic storm began with a southward turn of the IMF at about 15:00 UT and its strengthening to −17 nT, which led to the development of substorms on the night side of the Earth (Astafyeva et al. 2020). IMF remained mostly negative until ~10:00 UT on August 26, when it returned to zero for a short time. Such a long negative B_z indicates that the solar wind flux near the Earth was oriented almost perpendicular to the ecliptic plane (Abunin et al. 2020).

From ~17:30 UT, the SYM-H index began to gradually decrease, from ~20:00 to ~22:30 UT it remained at the "plateau" level of about −28––30 nT. From ~23:00 UT SYM-H began to fall faster and reached a minimum of −206 nT at 7:00 UT on August 26, 2018 (Krauss et al. August 2018). From about 7:00 UT, the SYM-H index began to rise slowly, but remained below zero until August 31, 2018. In terms of minimum SYM-H excursion, this storm was the third strongest in the 24th solar cycle.

The interplanetary electric field was directed to the east during the main phase of the storm and was characterized by an amplitude of 5 mV/m. On August 26, 2018 a corotating interaction region (CIR) was formed; the velocity and temperature of the solar wind plasma increased from ~370 km/s and ~4 × 10^4 K at ~10:00 UT to ~550 km/s and ~30 × 10^4 K at ~12:20 UT, respectively, the density decreased from ~11 cm^{-3} to ~3 cm^{-3}, because a high-velocity flow from a coronal hole of negative polarity (CH HSS) approached the Earth (Piersanti et al. 2020).

On August 26, 2018, at 7:20 UT, the recovery phase began and lasted for several days. During the recovery phase, the solar wind velocity increased to a maximum of 633 km/s at about 17:00 UT on August 27, 2018, and then it was maintained for several days at an average level of ~520 km/s. IMF-B_z demonstrates fluctuations in the north and south directions and has three positive peaks (more than 10 nT) from 9:00 to 21:00 UT on August 26, 2018. It should be noted here that during the magnetic storm recovery phase on August 25–26, 2018, synchronous negative magnetic bays were observed both in the auroral night sector and at the polar latitudes of the day sector (Kleimenova et al. 2019).

Recent studies (Piersanti et al. 2020) showed that on August 26, 2018 the western electric jet was intense, and around midnight the effect was also noticeable near the Earth's surface due to the substorm electric jet current.

Analysis of the temporal dynamics of the AU and AL indices characterizing the maximum magnetic effects of the eastern and western auroral electrojets, respectively, showed that they began to grow from 6:00 UT, and synchronously fluctuated from −500 to 500 nT until ~23:00 UT on August 25, 2018. Starting from ~3:30 UT on August 26, the AU index decreased to ~50 nT, while the AL index increased in intensity and its values varied within 1500–2000 nT. The maximum deviation of the AL index values reached 7.5–8.5 nT. Electrojet-related disturbances were observed at geomagnetic latitudes from 50° to 75° on the night side. The zone of auroral convection expanded towards the equator to 50° geomagnetic latitude during a geomagnetic storm, which is associated with a daytime reconnection, which forms new open field lines after the IMF turned south at the end of August 25, 2018 (Piersanti et al. 2020).

The intensity of the ring current increases during the main phase of the geomagnetic storm due to the injection of energetic particles from the tail of the magnetosphere in the equatorial plane, and gradually decays during the recovery phase (Gnushkina et al. 2018; de Michelis et al. 1997). The time evolution of the ring current, through the time evolution of associated magnetic disturbances in the surface layer of the atmosphere, is clearly visible in the magnetic data at Mikhnevo Observatory (middle latitudes). During the main phase of the storm (August 26, 2018), an increase in the ring current flowing in the westerly direction caused a strong depression of the horizontal component of the magnetic field.

The registration data of the magnetic and electric field during the initial and main phases of the magnetic storm are shown in Figs. 2 and 3. As seen from Fig. 2, during the SSC, which manifested itself as an increase (although not very brightly) of the vertical component of the magnetic field by ~22 nT at ~8:10 UT, positive changes are observed in the vertical components of the electric field strength and atmospheric current. Analysis of the data shown in Fig. 3, demonstrates the manifestation of the arrival of a high-velocity solar wind to the Earth (~12:00 UT on August 26, 2018),

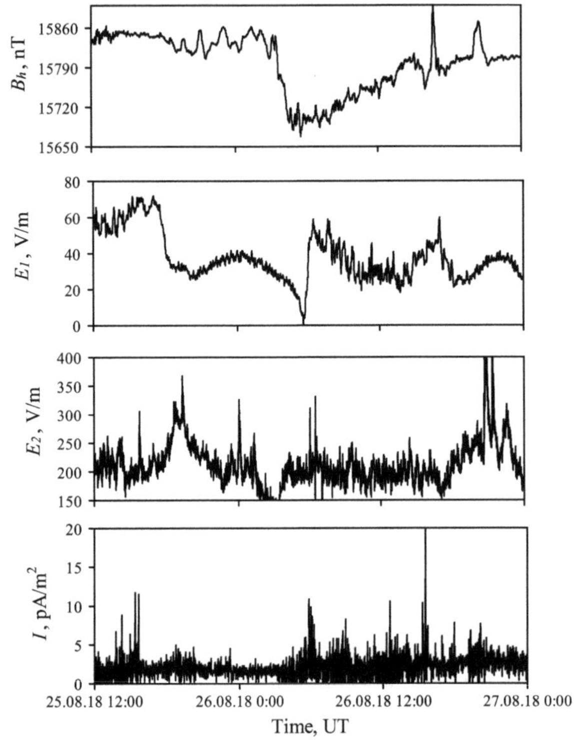

Fig. 3 Variations in the horizontal component of the geomagnetic field B_h, vertical components of the electric field strength E_1 and atmospheric current I at the Mikhnevo Observatory, variations in the vertical component of the electric field strength E_2 recorded at the Center for Geophysical Monitoring in Moscow during the main phase of the magnetic storm 25–26 August, 2018

intensification of magnetotail currents, and the development of a ring current (~5:00–7:00 UT on August 26, 2018) in variations of electrical parameters. In addition, electrical anomalies are recorded, probably associated with a change in the IMF orientation (~15:00 UT on August 25, 2018).

4 Conclusion

Thus, when performing these studies, the records of variations in electrical parameters were analyzed during the period of 14 strong geomagnetic disturbances (magnetic storms).

In more detail, the article discusses changes in the parameters of the electric field during the strongest magnetic storm (August 25–26, 2018) for the period under consideration. Using data on the state of the interplanetary medium and the geomagnetic environment, anomalies in the variations of the vertical components of the electric field strength and atmospheric current in the initial phase and the main phase of the magnetic storm were identified.

In general, the processing and analysis of instrumental observation data have shown a complex picture of electrical anomalies during magnetic storms, the explanation of which requires further research, including the accumulation of statistical material.

References

Abunin AA, Abunina MA, Belov AV, Chertok IM (2020) Peculiar solar sources and geospace disturbances on 20–26 August 2018. Solar Phys 295(7). https://doi.org/10.1007/s11207-019-1574-8

Adushkin VV, Soloviev SP, Spivak AA (2018) Electric fields of technogenic and natural processes. GEOS, Moscow [In Russian]

Aplin KL, Harrison RG (2014) Atmospheric electric fields during the Carrington flare. Astron Geophys 55(5):5.32–5.37. https://doi.org/10.1093/astrogeo/atu218

Astafyeva E, Bagiya MS, Förster M, Nishitani N (2020) Unprecedented hemispheric asymmetries during a surprise ionospheric storm: a game of drivers. J Geophys Res: Space Phys 125(3). https://doi.org/10.1029/2019JA027261

Basavaiah N (2011) Geomagnetism: solid earth and upper atmosphere perspectives. Springer Science and Business Media, Berlin

Blagoveshchensky DV, Sergeeva MA (2019) Ionospheric parameters in the European sector during the magnetic storm of August 25–26, 2018. Adv Space Res 65(1):11–18. https://doi.org/10.1016/j.asr.2019.07.044

Buonsanto M (1999) Ionospheric storms: a review. Space Sci Rev 88(3–4):563–601. https://doi.org/10.1023/A:1005107532631

Carlowicz MJ, Lopez RE (2002) Storms from the Sun: the emerging science of space weather. Joseph Henry Press, Washington, D.C

Curto JJ, Araki T, Alberca LF (2007) Evolution of the concept of sudden storm commencements and their operative identification. Earth, Planet Space 59(i–xii).https://doi.org/10.1186/BF03352059

Daglis IA, Thorne RM, Baumjohann W, Orsini S (1999) The terrestrial ring current: Origin, formation and decay. Rev Geophys 37(4):407–438. https://doi.org/10.1029/1999RG900009

De Michelis P, Daglis I, Consolini G (1997) Average terrestrial ring current derived from AMPTE/CCE-CHEM measurements. J Geophys Res Space Phys 102(A7):14103–21411. https://doi.org/10.1029/96JA03743

Gnushkina NY, Liemohn MW, Dubyagin S (2018) Current systems in the Earth's magnetosphere. Rev Geophys 56(2):309–332. https://doi.org/10.1002/2017RG000590

Gonzalez WD, Joselyn JA, Kamide Y, Kroehl HW, Rostoker G, Tsurutani BT, Vasyliunas VM (1994) What is a geomagnetic storm? J Geophys Res Space Phys 99(A4):5771–5792. https://doi.org/10.1029/93JA02867

Kleimenova NG, Kozyreva OV, Mikhnovski S, Kubicki M (2008) Effect of magnetic storms in variations in the atmospheric electric field at midlatitudes. Geomag Aeron 48(5):622–630. https://doi.org/10.1134/S0016793208050071

Kleimenova NG, Gromova LI, Gromov SV, Malysheva LM (2019) The magnetic storm of August 25–26, 2018: dayside high latitude geomagnetic variations and pulsations. Geomag Aeron 59(6):706–713. https://doi.org/10.1134/S0016793219060070

Krauss S, Behzadpour S, Temmer M, Lhotka C (2020) Exploring thermospheric variations triggered by severe geomagnetic storm on 26 August 2018 using GRACE follow-on data. J Geophys Res: Space Phys 125(5). https://doi.org/10.1029/2019JA027731

Lazutin LL (2012) World and polar magnetic storms. SINP MSU, Moscow

Lissa D, Srinivasu VKD, Prasad DSVVD, Niranjan K (2020) Ionospheric response to the 26 August 2018 geomagnetic storm using GPS-TEC observations along 80°E and 120°E longitudes in the Asian sector. Advan Space Res 66:1427–1440. https://doi.org/10.1016/j.asr.2020.05.025

Marcz F (1997) Short-term changes in atmospheric electricity associated with Forbush decreases. J Atmos Solar Terr Phys 59(9):975–982. https://doi.org/10.1016/S1364-6826(96)00076-4

Piersanti M, De Michelis P, Del Moro D, Tozzi R, Pezzopane M, Consolini G, Marcucci MF, Laurenza M, Di Matteo S, Pignalberi A, Quattrociocchi V, Diego P (2020) From the Sun to the Earth: August 25, 2018 geomagnetic storm effects. Ann Geophys 38:703–724. https://doi.org/10.5194/angeo-38-703-2020

Sheftel VM, Bandilet OM, Chernyshev AK (1992) Effects of planetary magnetic storms in atmospheric electricity near the Earth's surface. Geomag Aeron 32(1):186–188 [In Russian]

Smirnov S (2014a) Reaction of electric and meteorological states of the near-ground atmosphere during a geomagnetic storm on 5 April 2010. Earth, Planets Space 66(154). https://doi.org/10.1186/s40623-014-0154-2

Smirnov SE (2014b) Response of the electric state of the surface atmosphere to the geomagnetic storm of April 5, 2010. Dokl Earth Sci 456(1):622–626. https://doi.org/10.1134/S1028334X14050377

Spivak AA, Riabova SA (2019) Variations in electrical characteristics of the near-ground atmosphere during periods of magnetic storms. Dynamic processes in geospheres, Grafiteks, Moscow, pp 142–150. [In Russian]

Spivak AA, Riabova SA, Kharlamov VA (2019) The electric field in the surface atmosphere of the megapolis of Moscow. Geomag Aeron 59(4):467–478. https://doi.org/10.1134/S0016793219040169

Vanlommel P (2018) Solar-terrestrial center of excellence (STCE) Newsletter, 20–26 Aug 2018, www.stce.be/newsletter/pdf/2018/STCEnews20180831

Younas W, Amory-Mazaudier C, Khan M, Fleury R (2020) Ionospheric and magnetic signatures of a space weather event on 25–29 August 2018: CME and HSSWs. J Geophys Res: Space Phys 125(8). https://doi.org/10.1029/2020JA027981

Hydrological Parameters Measuring and Gas Fluxes Quantification of Shallow Gas Seepage at Cape Fiolent

A. A. Budnikov, T. V. Malakhova, I. N. Ivanova, and A. I. Murashova

Abstract Complex measurements of hydrological parameters and gas fluxes at the new methane seep recently discovered at Cape Fiolent (Crimean coast) were carried out. A lower values of water temperature T and concentration of dissolved oxygen O_2 outer of the seepage area in comparison with the values above the gas point were registered. Episodes of an abrupt T and O_2 increase with a subsequent decrease were noted. The estimation of the bubble gas fluxes was made by two different methods: video registration and trap method. The estimated flux was about 60 and 22.5 L/day out of entire seepage area respectively. High CH_4 concentrations in sea-surface water above seep points were measured.

Keywords Shallow methane seep · Gas fluxes · Gas trap · Hydrological parameters · Cape Fiolent · Black sea

1 Introduction

Over the last decade, a number of bubble gas emissions (cold seeps) have been discovered in coastal shallow-water areas of the Crimean peninsula (Malakhova et al. 2015, 2020; Pimenov et al. 2013, 2018; Tarnovetskii et al. 2018). It was found that methane predominates in the gas composition of the emitted bubbles. This methane can be of biogenic or thermocatalytic origin. Coastal seeps observed at depths from 1.5 to 30 m and deeper, have different intensity and frequency, as well as differ in the gas nature. Generally the process of bubble gas emission is accompanied by the fluid gas flow from the seabed in a certain area around the bubbles seepage point,

A. A. Budnikov (✉) · I. N. Ivanova
Faculty of Physics, M.V. Lomonosov Moscow State University, Moscow, Russia
e-mail: aa.budnikov@physics.msu.ru

T. V. Malakhova · A. I. Murashova
A.O. Kovalevsky Institute of Biology of the Southern Seas of RAS, Moscow, Russia
e-mail: t.malakhova@imbr-ras.ru

A. I. Murashova
e-mail: alenamyra@mail.ru

T. Chaplina (ed.), *Processes in GeoMedia—Volume IV*, Springer Geology,
https://doi.org/10.1007/978-3-030-76328-2_31

and the magnitude of this flow can be comparable with the magnitude of the bubble flow itself (Malakhova et al. 2020; Torres et al. 2002). Monitoring of bubble gas emission sites allows to clarify estimates of methane flows from the coastal area and determine the patterns of shallow seep behavior.

2 Study Site and Research Methods

In August 2019, in the coastal area of the southwestern Crimea at Cape Fiolent a new sites of bubble gas emissions were discovered ("Main Site" 44.522474 N, 33.466825 E, and "Back Site" 44.519912 N, 33.468017 E Fig. 1a–c). Seepages were further observed during spring and summer seasons 2020. The seafloor of gas emission area was sandy clearing with boulders of various sizes and rock outcrops. The seeps site is located at a depth of 2.5 m with an area of about 20 m^2, adjacent to the rock, which has a slight slope from the rock into the bay. The carbon isotopic composition of methane $\delta^{13}C\text{–}CH_4$ of bubble gas was −60.3 ‰ VPDB, which indicates the biogenic origin of methane (Malakhova et al., inprint). Presumably, seeps could be associated with discharge of groundwater sources in the area (Malakhova et al., inprint).

On August 13, 2020, complex measurements of hydrological parameters were carried out at "Main Site" from 10:00 to 17:30 local time. The Recording Current

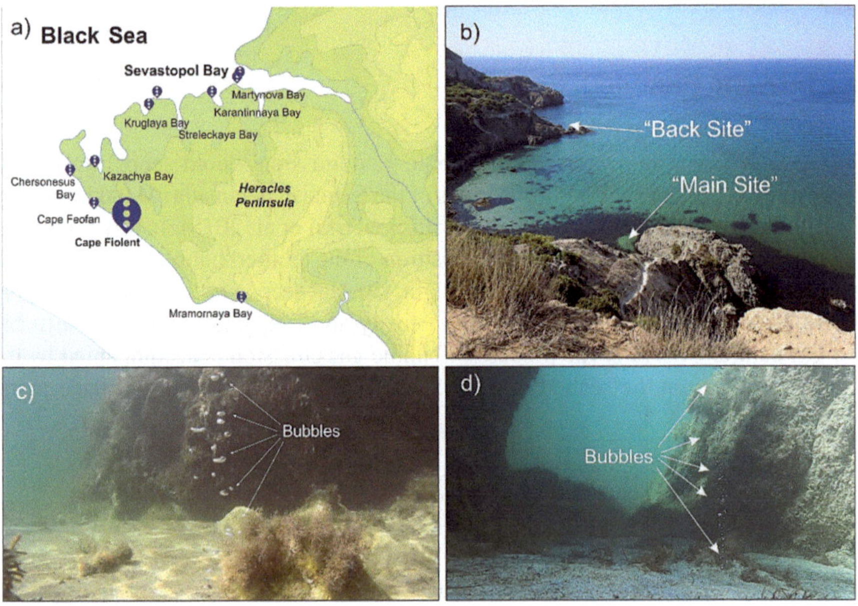

Fig. 1 A map of methane seeps sites of the Crimean coast, a big marker indicates the study area (**a**); overview photo of study area (**b**); underwater photos of bubble gas emissions at "Main Site" (**c**) and "Back Site" (**d**) at Cape Fiolent

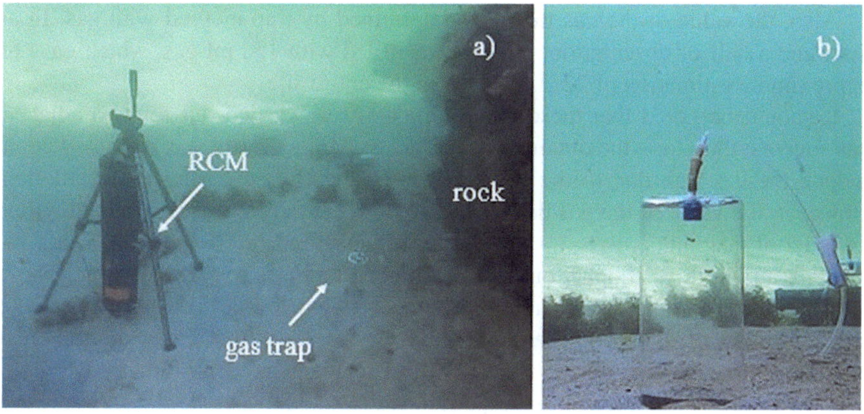

Fig. 2 Underwater photo of the seepage area at Cape Fiolent: **a** Overview photo of RCM9 LW AANDERAA INSTRUMENTS (Norway); **b** a gas trap installed above the seep

Meter RCM9 LW AANDERAA INSTRUMENTS (Norway) was used to measure the current velocity modulus V, temperature T, water conductivity C, dissolved oxygen content O_2, and turbidity Tu. The measurement accuracy was 0.5 cm/s, 0.02 °C, 0.02 mS/cm, 0.25 mg/L, and 0.4 NTU (in international turbidity units), respectively. During measurements, the sensors of the device were located directly at the one of the bubble gas points at the 0.5 m depth above the seabed (Fig. 2a).

The outgassing process was recorded using a GoPro4 Action Camera. Also the gas traps which were plastic 1600-ml cylindrical containers with an open lower base were installed to estimate the gas flux (Fig. 2b).

To determine CH_4 concentrations precision sampling of pore water and surface water was performed by a diver with 60-mL syringe. The background samples of bottom sediments were taken from the regions close to the sites of gas emission but without the signs of bubble gas. The CH_4 content was measured by phase-equilibrium degassing (Bolshakov and Egorov 1987) on an HP 5890 chromatograph with a packed column and a flame ionization detector.

3 Results and Discussions

About 10 bubble gas exit points were observed at the site during the measurements. The bubbles were emitted in portions with duration from 1 to 5 s, pauses between portions were from 2 s to 1 min. The average bubbles size was approximately 2 mm. Video recordings of bubble discharge were used to estimate the gas flux value from one of the emission points and approximate this value for the entire gas-generating area. The estimated bubble flux was about 60 L/day.

Also, the value of the gas flux was determined by trap method with gas. In an hour and a half of observation, the trap was filled with 150 ml gas, which gave an approximate estimation of 22.5 L/day from the entire site.

It should be noted that these measurements were carried out for two adjacent gas sources; therefore, the obtained values differ. Moreover, the gas trap itself could reduce the gas flux due to the deformation of the seabed during its installation. However, the value of the gas flow obtained in two different ways coincided within an order of magnitude.

Such a high fluxes impact on the CH_4 concentration in sea-surface water above seep points resulting in oversaturation concentration levels (Table 1). Even at a shallow depth of the studied areas, bubble gas enriches the water column. It is also likely that fluid discharge contributes to methane concentration in the water column as shown for Laspi Bay seepages area (Malakhova et al. 2020). Porewater CH_4 concentration at seep points at both sites was 2 orders of magnitude higher than background porewater concentration (Table 1), but still did not reach saturation values. This indicates that the bubbles are not formed in the upper layer of the tested sediments, but probably rise from deeper layers, trapping at the water-bottom sediments interface and partially dissolving in pore water.

The data obtained using the multiparameter platform RCM9 is shown in Figs. 3 and 4. The change in temperature T and concentration of dissolved oxygen O_2 in near-bottom water above methane seep point during continuous 6-h recording is shown (Fig. 3). At the beginning and at the end of the measurements the background

Table 1 CH_4 concentration (nM/L) in porewater at seep point, sea surface water above seep point and background station porewater at "Main Site" and "Back Site" seep areas at Cape Fiolent

	Porewater at seep point, nM/L	Sea surface water above seep point, nM/L	Background station porewater, nM/L
"Main Site"	448×10^3	353	3115
"Back Site"	447×10^3	262	3533

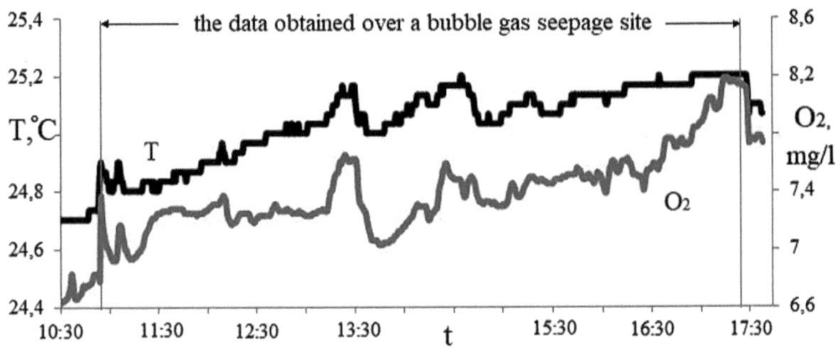

Fig. 3 Temperature T (black graph) and dissolved oxygen concentration O_2 (gray graph) 6-h variation during continuous in situ monitoring. The highlighted area shows the time interval when measurements were taken directly above the bubble gas emission point

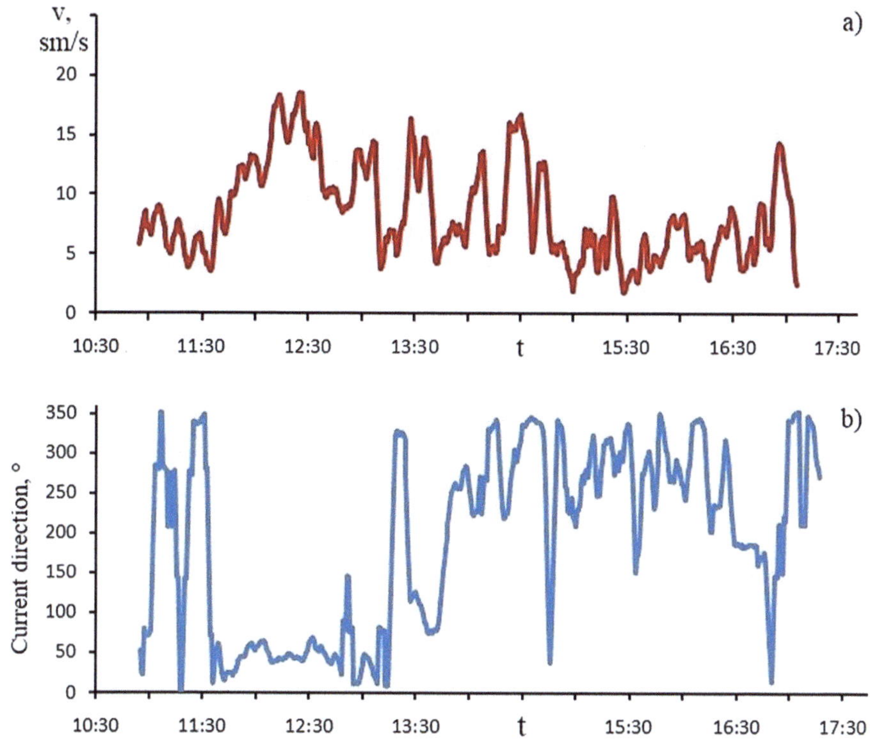

Fig. 4 Horizontal current velocity (**a**) and direction of current (**b**) at the measurement point

station was worked on. For 20 min RCM9 was installed at the edge of the sandy area where gas bubbles weren't observed. The depth at the background station was 0.3 m deeper than at the main station.

In general, throughout the entire record above the seep, an increase in T and O_2 values was observed from 24.8 to 25.5 °C and 7.1 to 8.2 mg/L, respectively. Two episodes (at 13.30 and 14.30) of an abrupt T and O_2 increase with a subsequent decrease were noted. The correlation coefficient for T and O_2 was 0.76, which indicates a strong connection between the processes that affect these parameters. The salinity value calculated from the measured electrical conductivity almost did not change and amounted to 18.4 ‰. The turbidity values were below the sensor's threshold.

During measurements at the background site, lower T and O_2 values were observed compared to the values above the bubble gas emission point. This may indicate a difference in hydrodynamic conditions at the point of gas release and at the background station. Probably, the nature of the change in water temperature and oxygen content in it is determined by strong mixing of the vertical layer above the site. In addition, the measured parameters can be influenced by horizontal currents (Fig. 4).

4 Conclusion

At the new sites of bubble methane emissions, hydrologic parameters were measured and gas flux were quantified. Measurements have shown that the values of hydrologic parameters in the immediate vicinity to the bubble fluxes differ from the background values. The measured parameters can be influenced by the current speed and direction. The measurements also showed an increased concentration of methane in porewater at seepages. Bubble gas fluxes may differ at different bubble gas emission point. For a more comprehensive gas flux studies, several methods should be used simultaneously.

Acknowledgements This work was carried out with financial support from the topic of the state task IBSS RAS "Molismological and biogeochemical principles of the homeostasis of marine ecosystems", state number registration AAAA-A18-118020890090-2.

References

Bolshakov AM, Egorov AV (1987) On the application of phase equilibrium degasing method for gasometric investigation. Okeanologiya 27(5):861–862 (in Russian)

Malakhova TV, Egorov VN, Malakhova LV, Artemov YuG, Pimenov NV, Biogeochemical characteristics of shallow methane seepages of the Crimean coastal areas in comparison with deep-sea seeps of the Black Sea. Mar Biol J 5(4) (in Print)

Malakhova TV, Egorov VN, Malakhova LV, Artemov YG, Evtushenko DB, Gulin SB, Kanapatskii TA, Pimenov NV (2015) Microbial processes and genesis of methane gas jets in the coastal areas of the Crimean Peninsula. Microbiology (moscow) 84(6):838–845

Malakhova TV, Budnikov AA, Ivanova IN, Murashova AI, Methane fluid flow from seafloor: data from Laspi Bay seepage area compared to other gas emission regions. Protsessy v geosredakh 3(25):822–830

Pimenov NV, Egorov VN, Kanapatskii TA, Malakhova TV, Artemov JuG, Sigalevich PA, Malakhova LV (2013) Sulfate reduction and microbial processes of the methane cycle in the sediments of the Sevastopol Bay. Microbiology (moscow) 82(5):614–624

Pimenov NV, Merkel AYu, Tarnovetskii IYu, Malakhova TV, Samylina OS, Kanapatskii TA, Tikhonova EN, Vlasova MA (2018) Structure of microbial mats in the Mramornaya Bay (Crimea) coastal areas. Microbiology (mikrobiologiya) 87(5):681–691

Tarnovetskii IYu, Merkel AYu, Kanapatskiy TA, Ivanova EA, Gulin MB, Toshchakov S, Pimenov NV (2018) Decoupling between sulfate reduction and the anaerobic oxidation of methane in the shallow methane seep of the Black Sea. FEMS Microbiol Lett 365(21)m article no. fny235. https://doi.org/10.1093/femsle/fny235

Torres ME, McManus J, Hammond DE et al (2002) Fluid and chemical fluxes in and out of sediments hosting methane hydrate deposits on Hydrate Ridge, OR, I: Hydrological provinces. Earth Planetary Sci Lett 201(3–4):525–540. https://doi.org/10.1016/S0012-821X(02)00733-1

On the Stability of the Interface Between Two Heavy Fluids in a Fast Oscillating Vessel

S. V. Nesterov and V. G. Baydulov

Abstract The solution of the problem of dynamic stabilization of the Rayleigh–Taylor instability of the interface between two heavy fluids with an inverse density distribution is considered within the framework of the linear model of an ideal fluid. A necessity condition for the stabilization of an unstable equilibrium is concluded for the frequency and amplitude of vertical oscillations of a vessel with a liquid, which is in agreement with the previously obtained experimental data.

Keywords Ideal liquid oscillations · Liquid/liquid interface · Rayleigh–Taylor instability

1 Introduction

Previously it was theoretically and experimentally shown that the interface between two heavy fluids in a vertically oscillating vessel loses stability, and parametric oscillations are excited (Fig. 1). A characteristic feature of the oscillations is approximately a double excess of the frequency of vertical oscillations over the frequency of a standing wave.

If vertical oscillations of the vessel occur with a high frequency, then it is possible to suppress oscillations of the interface between two heavy liquids, even in the case when a denser liquid is located above a less dense one. A similar effect of stabilization of an unstable equilibrium due to fast oscillations is well known in the mechanics of systems with a finite number of degrees of freedom and in the theory of elasticity (Kapitsa 1951).

The study of the problem of oscillations of the interface between two fluids began with determine the critical conditions for the appearance of oscillations of the free surface of a fluid. The results of a series of experiments on the standing waves excitation on the surface of a fluid poured onto a plate vibrating in a vertical direction

S. V. Nesterov · V. G. Baydulov (✉)
Ishlinsky Institute for Problems in Mechanics RAS, Moscow, Russia

V. G. Baydulov
Moscow, Russia

T. Chaplina (ed.), *Processes in GeoMedia—Volume IV*, Springer Geology,
https://doi.org/10.1007/978-3-030-76328-2_32

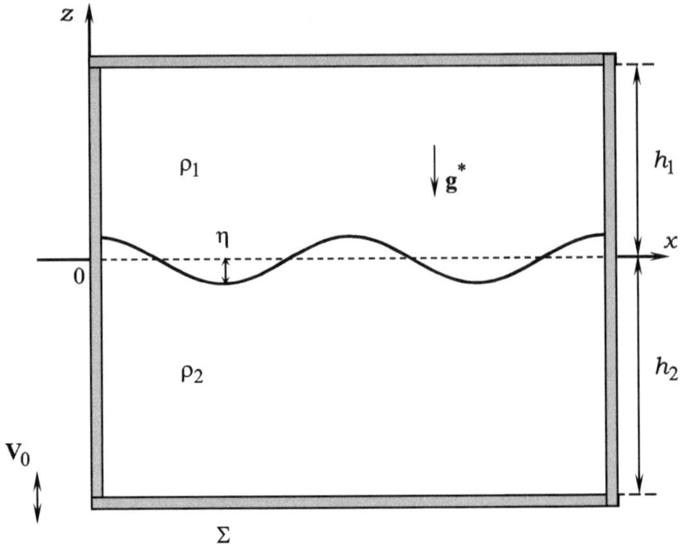

Fig. 1 Geometry of the problem

were presented in the first work devoted to parametric resonance (Faraday 1831). A theoretical explanation was given by Rayleigh (Rayleigh 1896; Mason 1965) in the framework of the ideal fluid model. It was shown that resonance occurs when the frequency of the excited wave is half the frequency of the plate vibrations. Later it was obtained (Benjamin and Ursell 1954) that the excitation of a wave on a free surface becomes possible even at an infinitely small amplitude of oscillations.

The presence of the experimentally observed threshold value of the amplitude of oscillations was associated with dissipative effects. The threshold value was determined (Sorokin 1957) within the framework of a viscous fluid model by using a phenomenological approach. As a result the problem was reduced to the Mathieu equation. An analytical solution to the problem of the appearance of parametric resonance was constructed for both cases: a free surface (Bezdenezhnykh et al. 1982; Lyubimov et al. 2003) and the interface between two heavy liquids (Sekerzh-Zenkovich and Kalinichenko 1979; Sekerzh-Zenkovich 1983).

A nonlinear theory of free surface vibrations was constructed (Ockendon and Ockendon 1973) in framework of extended ideal fluid model, including the phenomenological consideration of viscosity effects. This approach has been further developed (Chen and Vinals 1994; Edwards and Fauve 1994; Kumar and Tuckerman 1994; Cerda and Tirapegui 1998; Becchoffer et al. 1998; Wright et al. 1999). The stability of a flat interface was studied (Kelly 1965) for two infinitely deep ideal fluids. The solution of the problem was reduced to the Mathieu equation.

The possibility of suppressing the development of the Rayleigh–Taylor instability of the interface between two heavy fluids by high-frequency vertical oscillations of the vessel was shown experimentally (Benjamin and Ursell 1954; Wolf 1969, 1970), in the case when a denser fluid was located above a less dense one. The margins of stability regions were determined by numerical methods (Troyon and Gruber 1971) within the framework of a viscous fluid model. In this paper, the problem of determining the critical conditions for the appearance of the Rayleigh–Taylor instability in a two-layer fluid with an inverse density distribution is carried out within the framework of a linearized model of an ideal incompressible fluid.

2 Statement of the Problem

Consider a cylindrical vessel filled with two fluids with densities ρ_1 and ρ_2, which moves vertically according to the law $a \cos \omega t$. Let us denote by Φ_1 and Φ_2 the velocity potentials for the upper and lower fluids. The linearized boundary condition on the surface of the fluid has the form

$$\rho_1\left(\frac{\partial \Phi_1}{\partial t} + g^*\eta\right) - \rho_2\left(\frac{\partial \Phi_2}{\partial t} + g^*\eta\right)\bigg|_{z=0} = 0 \tag{1}$$

$$\frac{\partial \eta}{\partial t} = \frac{\partial \Phi_1}{\partial z} = \frac{\partial \Phi_2}{\partial z}\bigg|_{z=0} \tag{2}$$

The impermeability condition is valid on the solid walls of the vessel

$$\frac{\partial \Phi_1}{\partial z}\bigg|_{z=h_1} = \frac{\partial \Phi_2}{\partial z}\bigg|_{z=h_2} = 0; \quad \frac{\partial \Phi_{1,2}}{\partial z}\bigg|_{\Sigma} = 0 \tag{3}$$

The expression for the effective value of the acceleration of gravity g in the coordinate system associated with the oscillating vessel has the form:

$$g* = g - \omega^2 a \cos \omega t.$$

We denote by λ_m^2 the eigenvalues of the Laplace operator for the boundary value problem

$$\Delta\psi_m + \lambda_m^2\psi_m = 0, \quad \frac{\partial \psi_m}{\partial n}\bigg|_{\Sigma} = 0 \tag{4}$$

where ψ_m eigenfunctions of the region bounding the cross-section of the vessel.

The velocity potentials Φ_1 and Φ_2, as well as the elevation of the interface, are sought in the form

$$\begin{aligned} \Phi_1 &= \phi_1(t)\psi_m(x, z)\mathrm{ch}\lambda_m(h_1 - z), \\ \Phi_2 &= \phi_z(t)\psi_m(x, z)\mathrm{ch}\lambda_m(h_2 + z), \\ \eta &= h(t)\psi_m(x, z) \end{aligned} \tag{5}$$

Substituting functions (5) into boundary conditions (1) and (2), we obtain the equation for the function $h(t)$

$$\frac{d^2h}{dt^2} = -\frac{g^*(\rho_2 - \rho_1)\lambda_m \mathrm{th}\lambda_m h_1 \mathrm{th}\lambda_m h_2}{\rho_2 \mathrm{th}\lambda_m h_1 + \rho_1 \mathrm{th}\lambda_m h_2} h \tag{6}$$

Further we will consider the case $\lambda_m h_{1,2} \gg 1$. Then Eq. (6), taking into account the expression for $g*$, takes the form

$$\frac{d^2h}{dt^2} = -\frac{\lambda_m(\rho_2 - \rho_1)\left(g - \omega^2 a \cos\omega t\right)}{\rho_2 + \rho_1} h \tag{7}$$

We introduce new variable $\tau = \omega t$ and parameters $\lambda_m a = \varepsilon \ll 1$, $b^2 = g/(\varepsilon a \omega^2) = g/(\lambda_m a^2 \omega^2)$, in which the Eq. (7) is written as

$$\frac{d^2h}{d\tau^2} = -\delta\left[\varepsilon^2 b^2 - \varepsilon\cos\tau\right]h, \quad where \;\; \delta = \frac{\rho_2 - \rho_1}{\rho_2 + \rho_1}. \tag{8}$$

We also introduce new unknown functions Θ and Ψ such that

$$h = \Theta(\tau)(1 - \varepsilon\cos\tau), \qquad \frac{dh}{d\tau} = \varepsilon\Psi(\tau) + \varepsilon\delta\sin\tau\,\Theta(\tau) \tag{9}$$

for which, for calculating the derivative dh/dt of the first expression of the substitution (9) we find

$$\begin{aligned} \frac{d\Theta}{d\tau} &= \varepsilon\frac{\Psi + (\delta - 1)\Theta\sin\tau}{1 - \varepsilon\cos\tau} \\ &= \varepsilon\Psi + \varepsilon\Theta\sin\tau(\delta - 1) \\ &\quad + \varepsilon^2(\Psi\cos\tau + \Theta(\delta - 1)\sin\tau\cos\tau) + \bar{\bar{o}}(\varepsilon^2) \end{aligned} \tag{10}$$

Calculation of the second time derivative of the function h in the substitution (9) leads to the equation

$$\varepsilon\frac{d\Psi}{d\tau} + \varepsilon^2\delta\sin\tau\left(\Psi\cos\tau + (\delta - 1)\Theta\sin\tau + \bar{\bar{o}}(1)\right) = -\varepsilon^2\delta b^2\left(\Theta + \cos^2\tau\right). \tag{11}$$

Averaging Eqs. (10) and (11) over the explicitly included time variable with the preservation of terms of the first order of smallness in the parameter ε leads to the system of equations

$$\frac{d\Theta}{d\tau} = \varepsilon\Psi, \quad \frac{d\Psi}{d\tau} = -\varepsilon\delta\left(b^2 + \frac{1}{2}\delta\right)\Theta \tag{12}$$

which can be written as a second order equation

$$\frac{d^2\Theta}{d\tau^2} = -\varepsilon\delta\left(b^2 + \frac{1}{2}\delta\right)\Theta \tag{13}$$

Let $\rho_2 > \rho_1$ ($\delta > 0$), then the solution $\Theta = 0$ is stable for any value of the parameter b. Let us now consider the case $\delta < 0$ when a liquid with a lower density is located under a denser liquid. In a stationary vessel, in this case, the equilibrium of such a system is known to be unstable (the Rayleigh–Taylor instability).

However, in a rapidly oscillating vessel equilibrium with the inverse by the density distribution of the layers of liquid ($\rho_2 < \rho_1$) is stable, if the equality

$$b^2 < \frac{1}{2}|\delta|$$

Taking into account the previously introduced notation, we obtain for the dimensional quantities the condition under which the solution $\Theta = 0$ is stable in the case $\rho_2 < \rho_1$

$$\frac{1}{2}\frac{\rho_1 - \rho_2}{\rho_1 + \rho_2}\lambda_m a^2 \omega^2 > g \tag{14}$$

Condition (14) shows stabilization of an unstable equilibrium in the hydrodynamic system under consideration is possible with fast oscillations of the vessel. It should be noted that, linearized equations of hydrodynamics were considered under the deduction of condition (14). The solution to system (12) $\Theta = 0$, $\Psi = 0$ is the center; therefore, condition (14) is a necessary condition of dynamic stabilization of the equilibrium state.

Thus, the threshold condition for the suppression of the Rayleigh–Taylor instability was determined within the framework of the linear model of an ideal fluid without taking into account the effects of viscosity and the obtained criterion (14) is in a good agreement with the results of an experimental study (Wolf 1969, 1970), where the critical condition is given for the first modes of natural vibrations of the interface of fluids.

Acknowledgements The work was carried out within the framework of state tasks No. AAAA-A20-120011690138-6 and AAAA-A20-120011690132-4.

References

Becchoffer J, Ego V, Manneville S, Jonson B (1998) An experimental study of the onset of parametrically pumped surface waves in viscous fluids. J Fluid Mech 288:325

Benjamin TB, Ursell F (1954) The stability of the plane free surface of a liquid in vertical periodic motion. Proc Roy Soc A225(1163):505

Bezdenezhnykh NA, Briskman VA, Puzanov GV, Cherepanov AA, Shaidurov GF (1982) On the influence of high-frequency vibrations on the stability of the interface of liquids. In: Hydrodynamics and mass-heat transfer in zero gravity. Book, Moscow, p 34 (in Russian)
Cerda EA, Tirapegui EL (1998) Faraday's instability in viscous fluids. J Fluid Mech 368:195
Chen P, Vinals J (1994) Pattern selection in Faraday waves. Phys Rev Lett 14:2670
Edwards WS, Fauve S (1994) Patterns and quasipatterns in Faraday experiments. J Fluid Mech 278:123
Faraday M (1831) On peculiar class acoustical figures and on certain forms assumed by a group of particles upon elastic surface. Phil Trans Roy Soc London 121:209
Kapitsa PL (1951) Pendulum with vibrating weight. UFN 44(1):7 (in Russian)
Kelly RE (1965) The stability of an unsteady Kelvin-Helmholtz flow, part 3. J Fluid Mech 22:547
Kumar K, Tuckerman LS (1994) Parametric instability of the interface between two fluids. J Fluid Mech 279:49
Lyubimov DV, Lyubimova TP, Cherepanov AA (2003) Dynamics of interfaces in vibration fields. Book, Moscow, p 216 (in Russian)
Mason WP (ed) (1965) Physical acoustics, vol II, Part B. Properties of polymers and nonlinear acoustics. New York, p 383
Ockendon JR, Ockendon H (1973) Resonant surface waves. J Fluid Mech 59(2):397
Rayleigh L (1896) The theory of sound, 2nd ed. London/New York, p 328
Sekerzh-Zenkovich SYa (1983) Parametric excitation of the finite amplitude waves on the interface of two fluids of different densities. Doklady AN SSSR 272(2):1083 (in Russian)
Sekerzh-Zenkovich SYa, Kalinichenko VA (1979) Excitation of internal waves in a two-layered liquid y vertical oscillations. Sov Phys Dokl 24(12):960
Sorokin VI (1957) About the effect of fountain forming on the surface of the vertically oscillating liquid. Akustich Zh 3(3):262 (in Russian)
Troyon F, Gruber R (1971) Theory of dynamic stabilization of the Rayleigh-Taylor instability. Phys Fluids 14:20692073
Wolf GH (1969) The dynamic stabilization of Rayleigh-Taylor instability and corresponding dynamic equilibrium. Z Physik B 227:291
Wolf GH (1970) Dynamic stabilization of the interchange instability of a liquid-gas interface. Phys Res Lett 24(9):444446
Wright J, Yon S, Pozrikidis C (1999) Numerical studies two-dimensional Faraday oscillations of inviscid fluids. J Fluid Mech 402:1

Numerical Simulation of the Thermal Regime of Inland Water Bodies Using the Coupled WRF and LAKE Models

D. S. Gladskikh, A. M. Kuznetsova, G. A. Baydakov, and Yu. I. Troitskaya

Abstract The work is devoted to the numerical simulation of the thermal regime of inland water bodies (lakes and reservoirs). A technique for describing the vertical temperature distribution observed under strong winds is proposed, based on the implementation of data obtained using the WRF model as input data for the LAKE model. Other methods of obtaining data on wind velocity are described, such as the results of field measurements and the use of global reanalysis data, a comparison of these methods with those proposed by the authors is given. The coupling of models presented in the work was verified on the basis of measurement data obtained at the experimental site of the Gorky reservoir in the summer of 2019.

Keywords Numerical simulation · Inland waters · Thermal regime · Models coupling · WRF · LAKE

1 Introduction

Inland water bodies, such as lakes and water reservoirs, play a significant role in the socio-economic development of regions and are the object of research for a number of scientific problems. Temperature changes in lakes and reservoirs can contribute to eutrophication processes. Blooming water can lead to mass fish mortality and deterioration of water quality. The object of the current study is the Gorky Reservoir, for which the team of the Department of Nonlinear Geophysical Research of the IAP RAS has accumulated a large array of data from field measurements in the reservoir (Kuznetsova et al. 2016a). During the field studies on the Gorky reservoir, the intensity of eutrophication varied (see Fig. 1), which is directly related to both hydrological characteristics (vertical distribution of temperature, thickness and temperature of the

D. S. Gladskikh (✉) · A. M. Kuznetsova · G. A. Baydakov · Yu. I. Troitskaya
Institute of Applied Physics RAS, Nizhny Novgorod, Russia

D. S. Gladskikh
Research Computing Center, Moscow State University, Moscow, Russia

Moscow Center for Fundamental and Applied Mathematics, Moscow, Russia

T. Chaplina (ed.), *Processes in GeoMedia—Volume IV*, Springer Geology,
https://doi.org/10.1007/978-3-030-76328-2_33

Fig. 1 The intensity of the eutrophication of the Gorky reservoir in different seasons

epilimpion), and atmospheric characteristics (intensity of short-wave radiation, wind velocity).

For correct modeling of the local climate of the regions, it is necessary to take into account the two-sided interaction of inland waters and the atmosphere. And in this regard, thermal conditions play an important role. One of the most important characteristics influencing the mixing processes in lakes is the wind velocity, which determines the characteristics of the epilimnion—the upper mixed layer of the lake. Correct data on the wind, especially at high velocities, is critical to reproduce the strong mixing regime often seen in inland waters.

2 Methods and Approaches Used for Receiving the Data on Wind Velocity

In the works of Puklakov and Grechushnikova (2001), Puklakov et al. (1999), Stepanenko et al. (2013), Abbasi et al. (2016), devoted to modeling the thermodynamics of water bodies, the results of special field measurements are often used as input data for meteorological conditions, which requires a constantly functioning equipment. At the same time, standard current observations performed at the stations of the Hydrometeorological Center located on the shore (in the case of the Gorky Reservoir, these are the Volzhskaya hydrometeorological observatory in Gorodets and the station in Yuryevets), may not demonstrate real conditions directly on the reservoir. In medium and small water bodies, such measurements may not be carried out at all.

Another approach is based on the use of NCEP/NCAR global meteorological reanalysis data. For the standard reanalysis, the time interval of meteorological data changes is 6 h, and the spatial resolution is 2.5° × 2.5°. Such a low resolution leads to the fact that even large water bodies of the European part of Russia, such as Lake Ladoga, Onega Lake or Rybinsk Reservoir, have only 1–3 cells, while the Gorky Reservoir has only one cell. In this case, the grid captures significant land areas

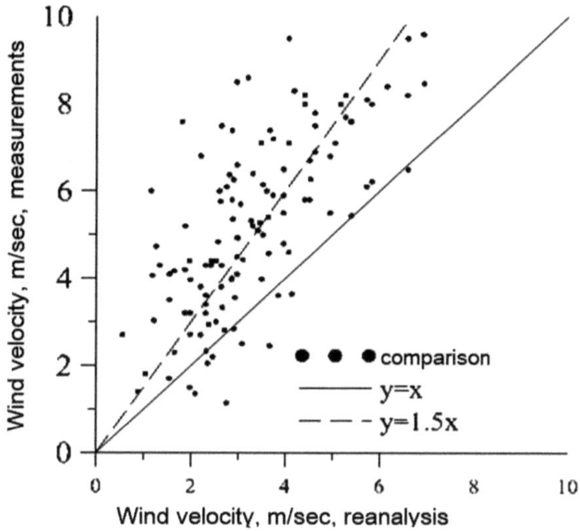

Fig. 2 Comparison of data from field measurements of wind at a height of 10 m and NCEP/NCAR reanalysis data

along with the reservoir, over which the meteorological conditions are different. Thus, due to the low spatial resolution, it is necessary to compare the reanalysis data with field measurements and introduce a special correction factor for each considered reservoir, as it was demonstrated in Sergeev et al. (2020). Figure 2 shows that for the Gorky reservoir, a correction factor (equal to 1.5) was obtained, and it is necessary to multiply the wind velocity values obtained from the reanalysis.

A more universal approach is the coupling of the atmospheric models with the one-dimensional lake models. This method is widely used in studies (Gu et al. 2015; Mallard et al. 2015; Xiao et al. 2016). In (Wang et al. 2019), the application of such a coupled model to the Nuojadu reservoir in southwestern China is considered. In this paper, the authors describe the application of the WRF model (Skamarock et al. 2019) to the region containing the water reservoir, and the implementation of the obtained atmospheric data into the LAKE model (Stepanenko et al. 2016) as boundary conditions.

3 Models Used in the Research

3.1 WRF

For a detailed description of the model, see (Skamarock et al. 2019). In this work, the simulation was carried out in 4 nested domains, reanalysis data were set as the input data, and at the output, we received wind data with a resolution of 1 km in the 4th domain (Fig. 3).

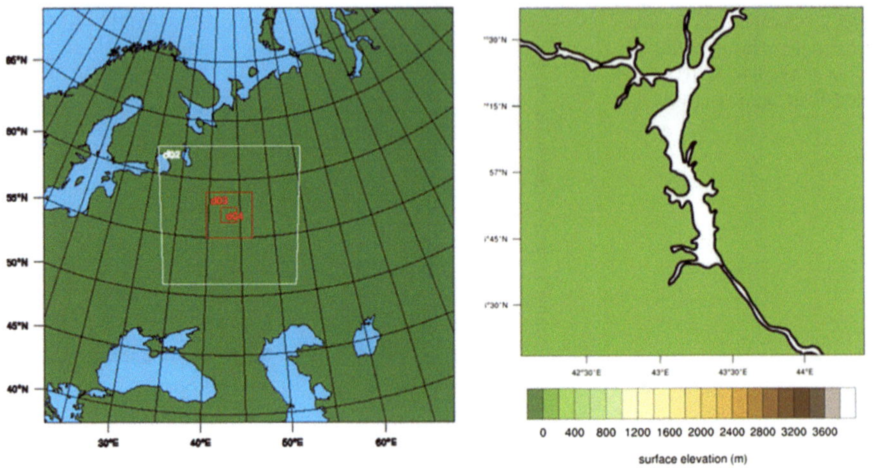

a b

Fig. 3 **a** Location of the nested domains used in WRF simulations. **b** Surface elevation (m) in the 4th domain

As the land model, MODIS LAKES (30 s) is used. For initial and boundary conditions, NCEP climate forecast system version 2 (CFSv2) 6-hourly products is used.

The WRF model settings we used were as follows. WRF single–moment 3–class scheme is chosen for microphysics option, Kain–Fritsch Scheme is used for cumulus parameterization option, Dudhia shortwave scheme and RRTM longwave scheme are used for shortwave and longwave radiation options. 5-layer thermal diffusion scheme is used for land surface option and revised MM5 scheme is used for surface layer option. Then, two cases were considered. Firstly, we used the parameterization of the PBL based on the Monin–Obukhov similarity theory, and the second case was with the LES, large-eddy simulation in the PBL. The results of the model simulation can be compared with the reanalysis data, which have the best spatial resolution nowadays, 0.2°. Figure 4 shows that the reanalysis data do not have sufficient spatial variability and do not reproduce the increase in wind velocity over the water area of the reservoir observed in the experiment.

Figure 5 demonstrates that WRF simulation reproduce much stronger winds than reanalysis data.

3.2 LAKE

For a detailed description of the model, see (Stepanenko et al. 2016). The model is based on the horizontally-averaged equations for temperature and momentum components. To describe the processes of turbulent mixing, the k-ε closure is used,

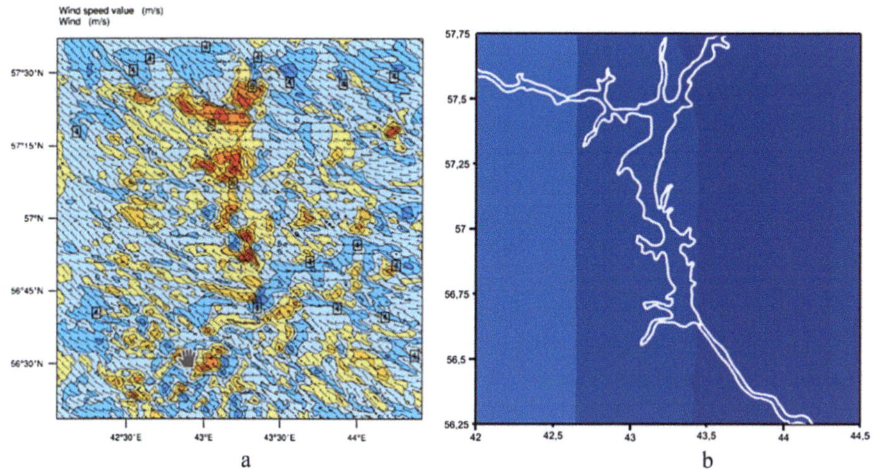

Fig. 4 Wind speed (m/s) distribution over the Gorky Reservoir using **a** WRF LES simulation, **b** CFSv2 reanalysis data

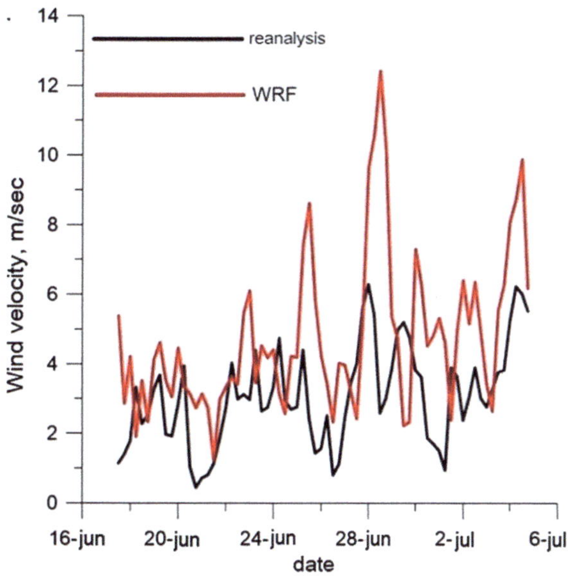

Fig. 5 Comparison of NCEP/NCAR reanalysis data on wind and WRF calculations data

based on the equations for the kinetic energy of turbulence and the rate of its dissipation. In the LAKE model, as input data for calculating the time evolution of the vertical temperature distribution, the following parameters are used: the initial temperature profile; the characteristics which do not change during the calculation (for example, parameters of the numerical scheme, constant coefficients in the equations), as well as data tables of meteorological data characterizing changing conditions at the upper (water–air) boundary during the entire counting period, for example, air temperature and humidity, short-wave radiation flux and, in particular, wind velocity components.

Meteorological data are used to set the boundary conditions for the interaction of the reservoir and the atmosphere, and in particular, the components of the friction velocity are calculated using the wind velocity values. Friction velocity is included in the boundary conditions at the upper boundary of the reservoir for the equations for calculating the dynamics of the lake: this is how the components of the current velocity of the reservoir are calculated.

For universal use in order to predict the temperature stratification of various water bodies without involving additional field measurements, a modification was performed, within which the wind velocity values obtained with the WRF are used as input data to the LAKE model. To verify the coupled models, the data of field measurements carried out at the Gorky reservoir in the summer of 2019 were used.

4 Experimental Studies of Gorky Reservoir

In the period from 2014 to the present, a number of field measurements are being carried out at the experimental site of the Gorky Reservoir. Within the framework of this study, the authors were interested in results of measuring the vertical distribution of temperature and wind velocity at a fixed point in the lake part of the reservoir (Fig. 6).

Measurements of the temperature distribution over depth were carried out using a freely sinking CTD probe, which recorded temperature values at a frequency of 6 Hz. At a descent rate of 24 cm/s, this provided a depth resolution of about 4 cm. Several measurements were made during each expedition.

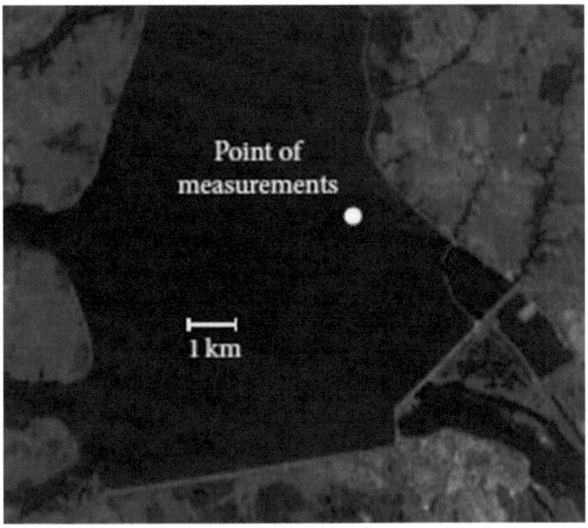

Fig. 6 Point of measurements of temperature distribution and wind velocity

A buoy station with the original design based on the oceanographic Froude buoy was used to measure the near-water wind velocity. Froude buoy is a mast submerged in water and held in a vertical position by the float close to the surface and by the load on the depth. On the buoy mast, 4 ultrasonic speed sensors (WindSonic GillInstruments Production) were located. WindSonic is two-component ultrasonic sensor with 4% measurement accuracy and velocity resolution of 0.01 m/s. Operating range of wind speed measurements 0–60 m/s includes measurements in calm conditions (Kuznetsova et al. 2016b).

5 Results and Conclusions

Within the framework of the study, calculations were carried out using the LAKE model. A typical example of such calculation is presented in this section. The starting profile, the two weeks evolution of which we traced, is shown in Fig. 7.

In one case, reanalysis data without a correction factor were used as the wind velocity. In the second case, the data from the WRF model were used. The calculation results were compared with the temperature profile measured with a CTD probe in the Gorky Reservoir. Figure 8 demonstrates that the applied technique allows us to reproduce strong mixing regime due to correct surface wind velocity data.

The calculations made it possible for the authors to draw a preliminary conclusion about the efficiency of the method and to recommend it for the application of simulation of the temperature regimes of lakes and reservoirs, particularly under strong winds conditions.

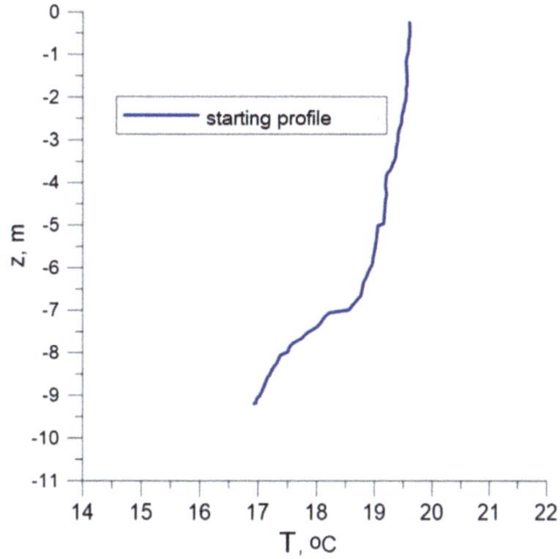

Fig. 7 Starting profile measured in the Gorky reservoir in June 2019

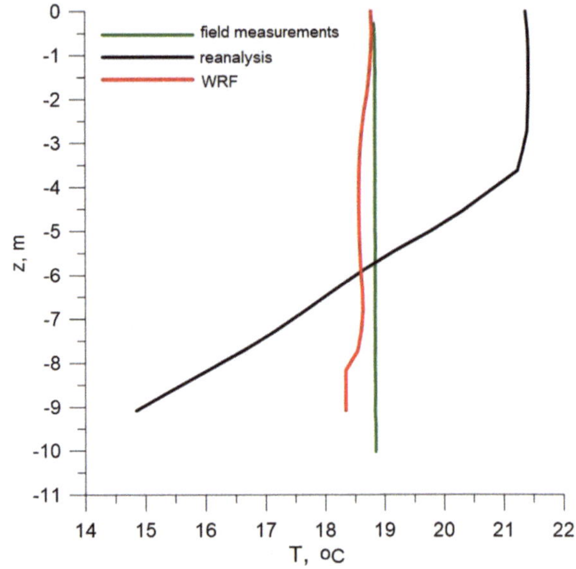

Fig. 8 Comparison of the vertical temperature distribution obtained using field measurements, with results of the LAKE model with reanalysis data and WRF calculations

Acknowledgements The work was supported by grants of the RF President's Grant for Young Scientists (MK-1867.2020.5, MD-1850.2020.5) and by the RFBR (18-05-00292, 20-05-00776).

References

Abbasi A, Annor FO, Giesen NV (2016) Investigation of temperature dynamics in small and shallow reservoirs, case study: Lake Binaba, Upper East Region of Ghana. Water 8(3):84. https://doi.org/10.3390/w8030084

Gu H, Jin J, Wu Y, Ek MB, Subin ZM (2015) Calibration and validation of lake surface temperature simulations with the coupled WRF-lake model. Clim Change 129(3–4):471–483. https://doi.org/10.1007/s10584-013-0978-y

Kuznetsova A, Baydakov G, Papko V, Kandaurov A, Vdovin M, Sergeev D et al (2016b) Adjusting of wind input source term in WAVEWATCH III model for the middle-sized water body on the basis of the field experiment. Adv Meteorol 13:2016

Kuznetsova AM, Baidakov GA, Papko VV, Kandaurov AA, Vdovin MI, Sergeev DA, Troitskaya YuI (2016) Field research and numerical simulation of wind and surface waves on medium-sized inland water bodies. Meteorol Hydrol 2:85–97. https://doi.org/10.3103/S106837391602008 [in Russian]

Mallard MS, Nolte CG, Spero TL, Bullock OR, Alapaty K, Herwehe JA, Gula J, Bowden JH (2015) Technical challenges and solutions in representing lakes when using WRF in downscaling applications. Geosci Model Dev 8:1085–1096. https://doi.org/10.5194/gmd-8-1085-2015

Puklakov VV, Grechushnikova MG (2001) Thermal regime of the Moskvoretsky reservoirs. Meteotol Hydrol 12:70–78 [in Russian]

Puklakov VV, Ershova MG, Grechushnikova MG (1999) Mathematical modeling of intra-reservoir processes in the reservoir. Prob Hydrol Hydroecol 1:302–317 [in Russian]

Sergeev DA, Gladskikh DS, Baydakov GA, Soustova IA, Troitskaya YuI (2020) On modeling the thermal regimes of inland water bodies using the data of global meteorological reanalysis (on the example of the Gorky reservoir). Meteorol Hydrol 4:104–112 [in Russian]

Skamarock WC, Klemp JB, Dudhia J, Gill DO, Liu Z, Berner J, Wang W, Powers JG, Duda MG, Barker DM, Huang X-Y (2019) A description of the advanced research WRF version 4. NCAR Tech. Note NCAR/TN-556+STR, 145 p

Stepanenko VM, Martynov A, Jöhnk KD, Subin ZM, Perroud M, Fang X, Beyrich F, Mironov D, Goyette S (2013) A one-dimensional model intercomparison study of thermal regime of a shallow, turbid midlatitude lake . Geosci Model Dev 6:1337–1352. https://doi.org/10.5194/gmd-6-1337-2013

Stepanenko V, Mammarella I, Ojala A, Miettinen H, Lykosov V, Vesala T (2016) LAKE 2.0: a model for temperature, methane, carbon dioxide and oxygen dynamics in lakes. Geosci Model Dev 9(5):1977–2006

Wang F, Ni G, Riley WJ, Tang J, Dejun Z, Sun T (2019) Evaluation of the WRF lake module (v1.0) and its improvements at a deep reservoir. Geosci Model Dev 12:2119–2138

Xiao C, Lofgren BM, Wang J, Chu PY (2016) Improving the lake scheme within a coupled WRF-lake model in the Laurentian great lakes. J Adv Model Earth Syst 8(4)

Triggering Landslides with Seismic Vibrations

A. N. Besedina, D. V. Pavlov, and Z. Z. Sharafiev

Abstract Considered is triggering landslides with seismic vibrations produced by earthquakes and explosions. Laboratory investigations have been performed of the effect of impulsive actions on slopes at a vertical and a horizontal impact set-ups designed and constructed in IDG RAS. The ultimate parameters of dynamic effects obtained in laboratory have been compared to the critical accelerations estimated on the basis of the quasi-static approach. Estimations of critical parameters of the effects of earthquakes of various magnitudes have been also performed. The results show that the minimal peak acceleration that could trigger a landslide noticeably exceeds the estimate of critical acceleration that uses the static coefficient of stability. The critical Newmark's displacements under the action of seismic vibrations of earthquakes can be reached in the case when the maximal acceleration exceeds 5–10 times the critical acceleration calculated using the quasi-static approach.

Keywords Landslide · Laboratory experiment · Model of the infinite slope · Newmark's method · Critical acceleration · Static factor of stability · Quasi-static approach

1 Introduction

Currently there are only few reliable data on the parameters of seismic vibrations that have triggered landslides. Investigations of slope phenomena emerging from seismic events usually result in determination of ultimate distances from epicenters of earthquakes of certain magnitudes that had triggered landslides (Keefer 1984). Corresponding parameters of vibrations are estimated via averaged dependencies of peak ground acceleration (PGA) and peak ground velocity (PGV) on the distance to the epicenter.

One of the widely used methods to estimate the dynamic stability of a slope is the one of Newmark (1965). It states that for a landslide to be triggered a certain critical

A. N. Besedina · D. V. Pavlov · Z. Z. Sharafiev (✉)
Sadovsky Institute for Dynamics of Geospheres, Russian Academy of Sciences, 119334 Moscow, Russia

T. Chaplina (ed.), *Processes in GeoMedia—Volume IV*, Springer Geology,
https://doi.org/10.1007/978-3-030-76328-2_34

level of internal deformations of the massif should be exceeded (those deformations more often occur along the plane where displacements localize). According to this approach deformations along the slope emerge in the case when the acceleration produced by an external load (in our case it is the seismic vibrations) exceeds the critical acceleration a_c calculated from the condition of static equilibrium.

This method was considered in detail in Wilson and Keefer (1979). Formally it can be reduced to a double integral of the dependence of acceleration on time $a(t)$ with account for the thing that sliding starts when the critical acceleration a_c is exceeded. The double integral gives the cumulative displacement D of the sliding block which is compared to the value D_N critical for landslide emergence.

Estimation of critical values of Newmark's displacement is an essential problem, but it goes beyond the scope of this work. According to Wilson and Keefer (1979) the value of $D_N = 10$ cm is the conservative estimate of the critical displacement needed for the failure to occur. So, in case the Newmark's analysis predicts a displacement of only several millimeters, the slope can be considered stable. On the other hand, if the displacement estimated via Newmark's analysis reaches several tens of centimeters, a large-scale landslide is highly possible because of a noticeable loss of strength and realization of dynamic processes.

In this work we compare in laboratory experiments the ultimate parameters of dynamic disturbances to the values of critical acceleration calculated via the quasi-static approach. Calculations of critical parameters of the effects of seismic vibrations produced by earthquakes of different magnitudes have been performed.

2 Laboratory Investigations of Slope Stability

The process of triggering slope failure with a short dynamic pulse was investigated in laboratory.

The inertia forces play the key role in the effect of seismic waves produced by explosions on a gentle slope. In many cases the stage of triggering the motion of a mass sliding relatively the base of a slope corresponds to the phase of decreasing particle velocity. At the section where the velocity grows to maximum in the time interval (t_0-$t_н$) the material of the expectant landslide moves together with the main body of the slope. At the time t_1, when the braking acceleration becomes high enough to overcome the cohesion and friction, the landslide mass gets detached and starts to move relatively the slope. In case the amplitude of displacement is sufficient for frictional weakening (Kocharyan 2016), the detached material moves further under the gravity.

Set-ups of impact type were used in laboratory experiments (Kocharyan and Spivak 2003). The dynamic effects with prevailing horizontal component was modeled at the horizontal impact set-up (HIS, Fig. 1). The maximal velocity of the container (PGV) was varied in experiments by changing the pressure in the pneumatic actuator. The required peak acceleration (PGA) was obtained by changing the duration of the impact. Modeling the effects with prevailing vertical component was

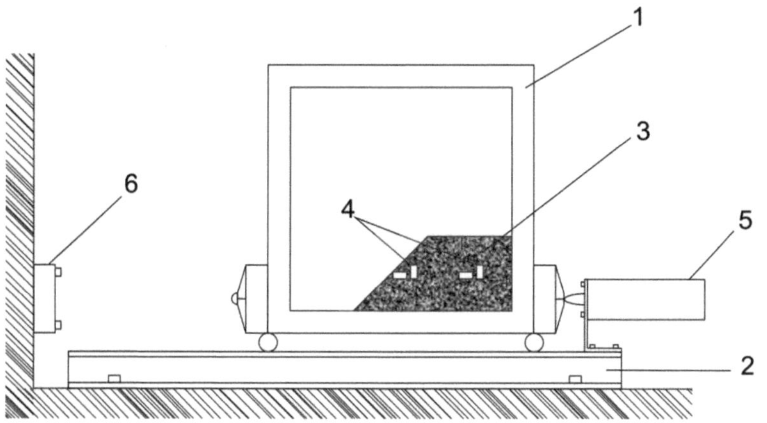

Fig. 1 A scheme of the horizontal impact set-up (HIS): 1—container with transparent side walls; 2—horizontal rails; 3—model of the slope; 4—accelerometers; 5—pneumatic actuator; 6—steel target

performed at an analogous vertical impact set-up (VIS). The container slid along vertical rails under the gravity until it hit the massive steel base. We succeeded to obtain the following peak parameters in experiments: peak ground velocity, PGV ~ 0.003 m/s to 1.3 m/s; peak ground acceleration, PGA ~ 0.01 g to 170 g.

Geomaterials with specific weights corresponding to the ones of natural materials (~1700 to 2000 kg/m^3) were used in laboratory experiments. A special technique was developed to ensure the same slope density in the same experimental series. The geomaterials were selected so that the following conditions was approximately true:

$$\begin{cases} C_m = \dfrac{l_m \gamma_m}{l_n \gamma_n} C_n \\ tg(\varphi)|_m = |tg(\varphi)|_n \end{cases} . \tag{1}$$

here C is the cohesion, φ is the angle of friction, γ is the specific weight, l is the linear size. Indexes "n" and "m" correspond to "nature" and "model".

Before each experimental series calculations of the coefficient of slope stability FS were performed—the ratio of the sum of forces holding the slope to the one destructing it. As far as the simplest model of infinite slope is considered, the coefficient of slope stability is calculated as follows (Krahn 2004):

$$FS = \frac{C}{\gamma z \cdot \cos\alpha \cdot \sin\alpha} + \frac{tg\varphi}{tg\alpha}, \tag{2}$$

here C is the coefficient of ground cohesion; γ is the ground specific weight; φ is the angle of ground internal friction; α is the slope angle; z is the vertical size of the sliding block.

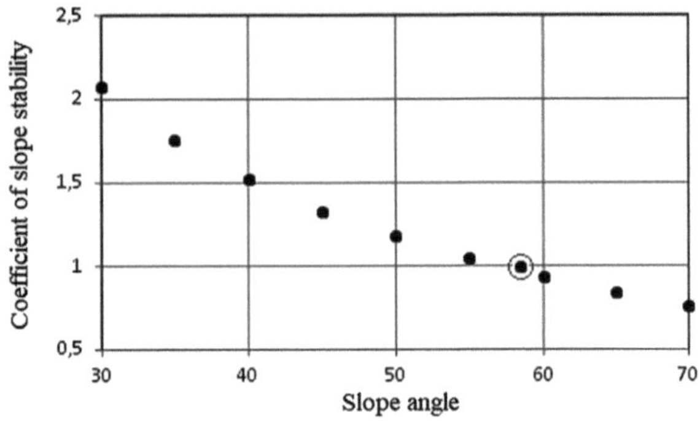

Fig. 2 Coefficient of slope stability vs slope angle. The black circle marks the critical slope angle

The coefficient of stability of the slope made of quartz sand moistened with 0.25% of glycerol of the whole mass of the sand ($C = 0.9$ kPa, $\varphi = 32.1°$) versus slope angle is shown in Fig. 2. One can see that the ultimate slope angle is about 60°.

Parameters of dynamic impact and the displacement of surface layer of the slope were controlled with accelerometers Bruel & Kjaer 4370, which were installed in the middle of the model and at the slope surface (Fig. 1). This gave us the opportunity to detect the nucleation and the start of sliding, even in cases when visually no deformations were observed. Examples of velocity waveforms obtained by integrating the recorded waveforms of acceleration are presented in Fig. 3.

The results of all the experiments were summarized at the diagrams *PGV—PGA*, which were plotted by drawing corresponding parameters of each experiment. Figure 3 gives examples of diagrams *PGV-PGA* for experiments with slopes of quartz sand with 0.25% of glycerol for the maximal and minimal slope angles at the VIS. Figure 4 shows that the minimal acceleration required to destroy the slope with the angle of 49° equals to 9g, while for the slope 59° – 3.8g.

The obtained values of a_c, as well as *FS*, *PGA* and *PGV* are presented in Table 1.

One can see that the values of minimal PGA, for which "landslide" triggering occurs exceed noticeably the values of critical acceleration a_c calculated using the static coefficient of slope stability *FS*. It should be noted that the effect of vertical impacts turns to be essentially weaker than that of the horizontal ones.

3 Results for Earthquakes of Different Magnitudes

Investigations of slope stability under the action of seismic vibrations from earthquakes of different magnitudes were performed using the program SLAMMER (Jibson et al. 2013) resting on the Newmark's method. The input data for the Newmark's analysis were the critical horizontal acceleration a_c and the waveform of acceleration. We calculated Newmark's displacements for four earthquakes with the magnitudes of M_w 6.0, 6.6, 7.1 and 7.6. Waveforms were taken from the database of

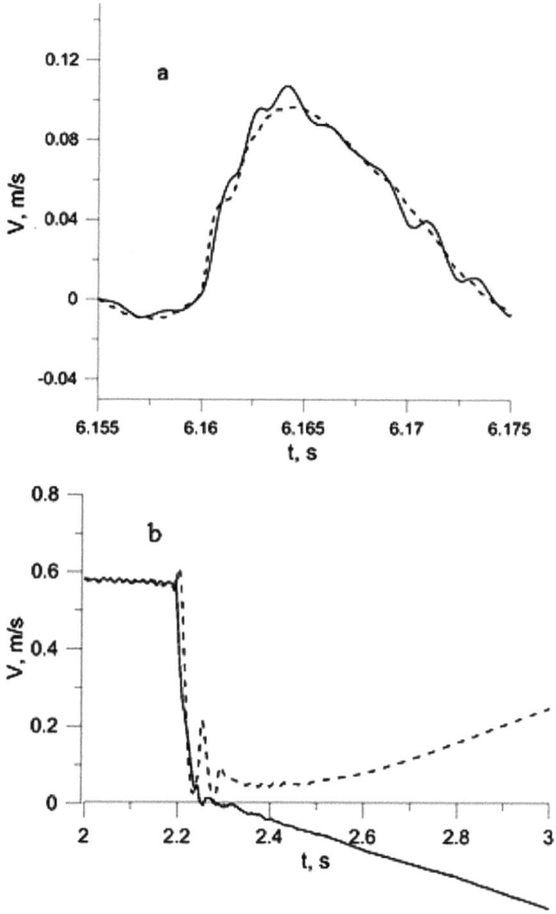

Fig. 3 Waveforms of ground velocity in a stable slope (**a**) and when a "landslide" occurs (**b**). **a** velocity component along the slope in the expectant "landslide" and in the middle of the model during a vertical impact (direction "+" corresponds to the velocity upward along the slope); **b** velocity component along the slope, "landslide" occurred at the slope of quartz sand during a horizontal impact. Solid line corresponds to the sensor in the middle of the model, dashed line—at the surface of the slope

Table 1 Critical parameters of impacts for experiments with quartz sand with 0.25% of glycerol ($C = 0.9$ kPa, $\varphi = 32.1°$)

Slope angle α, degrees	Equation of the "boundary" line	FS	a_c, g	PGV_{min}, m/s	PGA_{min}, g	Direction of impact
50	$PGV = 0.16PGA^{-0.5}$	1.26	0.17	0.35	0.6	Horizontal
60	$PGV = 0.095PGA^{-0.5}$	1.01	0.009	0.11	0.2	Horizontal
49	$PGV = 3.9PGA^{-0.5}$	1.29	0.22	0.49	9	Vertical
59	$PGV = 1.1PGA^{-0.5}$	1.03	0.03	0.26	3.8	Vertical

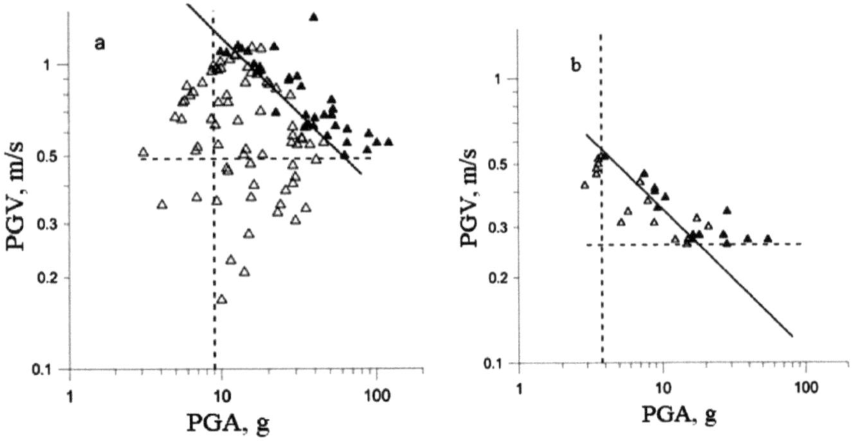

Fig. 4 Diagrams PGV—PGA for experiments with slopes of quartz sand with 0.25% of glycerol. White symbols—the slope remained stable; black symbols—"landslide" occurred. **a** slope angle $\alpha = 49°$; **b** slope angle $\alpha = 59°$. Solid straight line is the "boundary" between stability and "landslides". Dashed lines correspond to minimal values of PGV and PGA, at which "landslides" occurred

the SLAMMER program. For each earthquake waveforms with $PGA \geq 0.2$ g were taken. The results of calculations are presented in Fig. 5.

Each line in the figure corresponds to a single record registered at a station from the SLAMMER catalogue (Jibson et al. 2013). The sample includes all the stations from the catalogue for which $PGA > 0.2$ g. For each station Newmark's displacements were calculated for 10 values of critical acceleration a_c from 0.02 g to 0.2 g with the step of 0.02 g.

The results of calculations show that displacements $D_N \approx 10$ to 20 cm are reached when PGA is 5–10 times higher than a_c.

The same results in terms of PGV show that the critically dangerous displacements of $D_N \approx 10$ to 20 cm are reached at PGV ~ 10 to 40 cm/s for the earthquake $M_w6.0$ and at PGV ~ 20 to 50 cm/s for the earthquake $M_w7.6$.

4 Conclusions

1. Results of laboratory experiments show that the values of minimal PGA at which "landslides" occur noticeably exceed the values of critical acceleration calculated using the quasi-static approach.
2. For the vertical impacts the values of PGA required to trigger slope failure are approximately by an order of magnitude higher than the corresponding values for the horizontal impacts.
3. The critical Newmark's displacements DN are reached under the action of seismic vibrations from earthquakes in cases when PGA 5–10 times exceeds

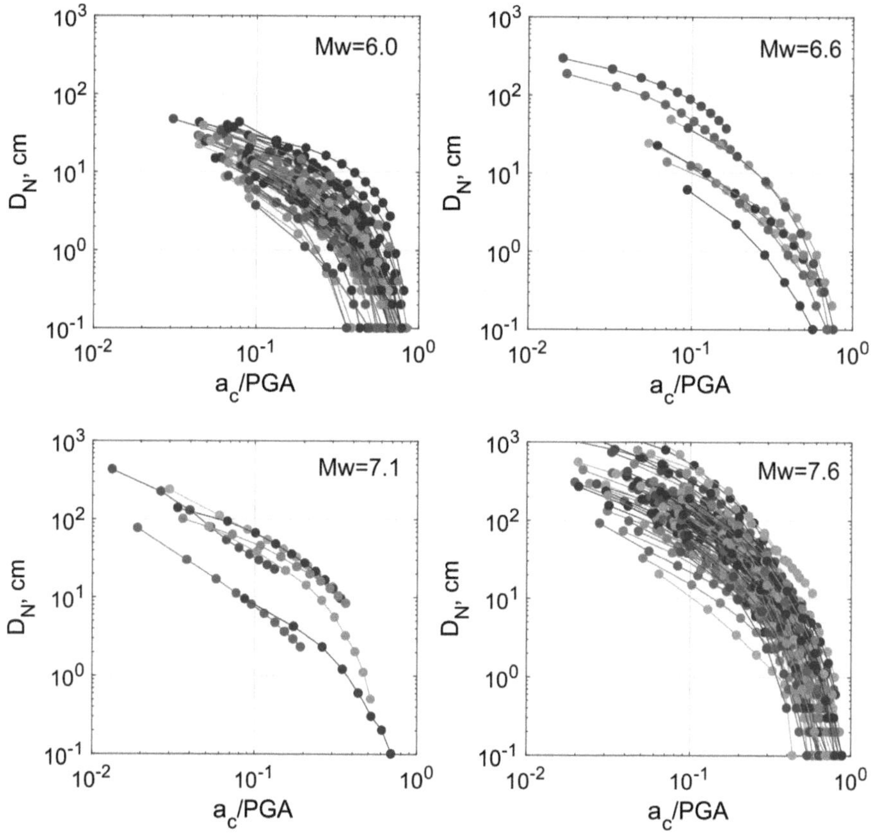

Fig. 5 Newmark's displacement vs the ratio of a_c/PGA for the earthquakes *Whittier Narrows* 1987 M_w6.0; *San Fernando* 1971 M_w6.6; *Cape Mendocino* 1992 M_w7.1; *ChiChi* 1999 M_w7.6

a_c, which is estimated using the quasi-static approach. In terms of velocity the dangerous values of displacement are reached at $PGV \sim 10$ to 40 cm/s for the earthquake M_w6.0 and at $PGV \sim 20$ to 50 cm/s for the earthquake M_w7.6.

Acknowledgements The work was partially supported by RFBR, grant # 19-05-00378 for D. V. Pavlov and Z. Z. Sharafiev. A. N. Besedina acknowledges Program No. 0146-2019-006 of the Ministry of Science and Higher Education of the Russian Federation.

References

Jibson RW, Rathje EM, Jibson MW, Lee YW (2013) SLAMMER—seismic landslide movement modeled using earthquake records (ver.1.1, November 2014): U.S. Geological Survey Techniques and Methods, book 12, chap. B1, unpaged

Keefer DK (1984) Landslides caused by earthquakes. Geol Soc Am Bullet 95(4):406–421. https://doi.org/10.1130/0016-7606(1984)95>406:LCBE<2.0.CO;2

Kocharyan GG (2016) Geomechanics of faults. GEOS, 424 p

Kocharyan GG, Spivak AA (2003) Deformation dynamic of block rock massifs. Akademkniga, 423 p

Krahn J (2004) Stability modeling with SLOPE/W. An engineering methodology: First Edition, Revision 1. GEO-SLOPE International Ltd., Calgary, Alberta , p 396

Newmark NM (1965) Effects of earthquakes on dams and embankments. Géotechnique 15(2):139–160

Wilson RC, Keefer DK (1983) Dynamic analysis of a slope failure from the 6 August 1979 Coyote Lake, California, earthquake. Bullet Seismol Soc Am 73:863–877

Modeling of the Ocean Wave System Based on Image from Satellite

S. A. Kumakshev

Abstract Modeling of the formation of an ocean wave system based on a space image taken from the International Space Station (ISS) as part of the "Uragan" (Hurricane) space experiment was performed. Due to the fact that the area of the ocean was known approximately, the simulation of wave excitation and propagation was performed for two cases: in the framework of the theory of long waves and for the ocean of infinite depth. Numerous calculations made it possible to achieve a similarity between the modeled wave system and the original one shown on the photo. This allowed us to confirm the hypothesis about the impulse nature of the initial disturbance necessary for the formation of such a phenomenon and determine its physical parameters. The model allows us to predict the development of such phenomena. Unfortunately, one static image does not allow us to judge the dynamics of the recorded phenomenon.

Keywords Waves in the ocean · Modeling of natural phenomena · Uragan (Hurricane) space experiment

1 Introduction

The Uragan space experiment involves cosmonauts conducting visual observations of various areas of the Earth's surface from the Russian segment of the ISS. At the same time, certain events of interest are recorded on video and photo equipment. First of all, this applies to natural and man-made disasters. There is an accumulation of material to find out the signs of the origin of dangerous events, and criteria for operational classification are being developed. It is necessary to identify and modelling such phenomenon. This, in turn, requires the development and application of mathematical models. Such models need to be verified in order to identify the conditions of occurrence of dangerous events and predict their formation. In the process of photographing, identifying the shooting area and creating models,

S. A. Kumakshev (✉)
Ishlinsky Institute for Problems in Mechanics RAS, Moscow, Russia
e-mail: kumak@ipmnet.ru

T. Chaplina (ed.), *Processes in GeoMedia—Volume IV*, Springer Geology,
https://doi.org/10.1007/978-3-030-76328-2_35

a team of people from various organizations was involved (Belyaev et al. 2009, 2011). Special requirements are also put forward for modeling—the speed of calculation on a personal computer. However, the analysis is carried out in conditions of limited information: only one photo is available. The archive of this space experiment contains various photographs of ocean wave systems, many of which are of interest from the point of view of operational decoding of dangerous events.

2 The Analyzed Object

The system of ocean waves shown in Fig. 1 was chosen for identification.

First, the task was to link the photo geographically. It is necessary to determine the coordinates of the ocean section shown in the photo in conditions of limited information: there is no coastline in the image. It was possible to approximately determine the area of the ocean surface (the Atlantic coast of the United States) using the Sigma program. It provides a ballistics and navigation display of the track of the ISS flight path from which the observation was taken on the Earth's surface.

The image was taken from the ISS with the following parameters: the orbit height is 350 km, the angle between the local horizon and the direction from the ISS to the object varies from 67° to 90°. This made it possible to determine the depth of the ocean in this area in the range from 200 to 2000 m.

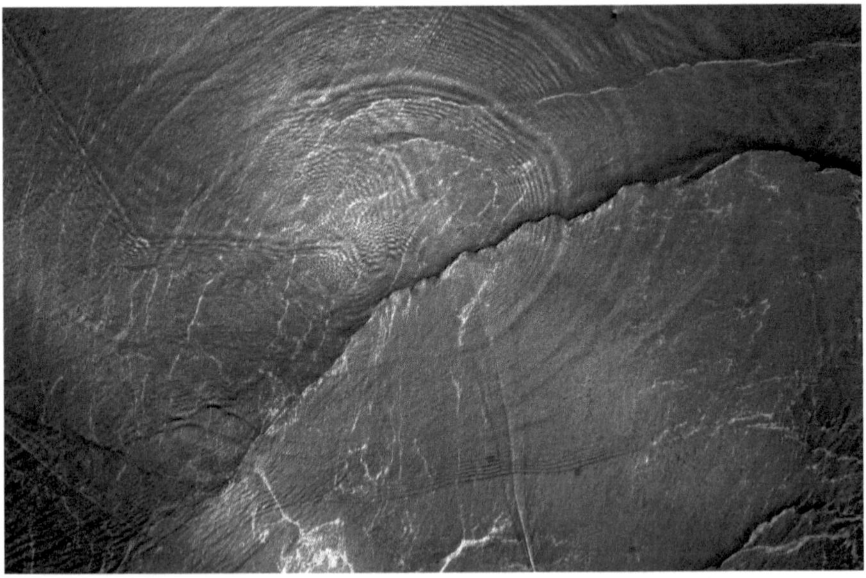

Fig. 1 The ring ocean wave system selected for identification

In the left part of the image, in the lower corner and slightly above the center, you can see two ship tracks consisting of longitudinal and transverse waves. One trace has an angle of about 150°.

The section of ocean shown in the photo covers an area of 12 by 8 km. The distance of 1 cm in the image corresponds to about 600–800 m. The largest wave system was chosen for decryption for the following reasons:

(1) This is a large and extended system and therefore can be classified as dangerous waves;
(2) The source of this wave system was quite powerful;
(3) The photo allows to evaluate the geometric parameters of the system;
(4) The crests of large ring wave packets and the structure of short wave packets are visible.

3 Pulsed Initial Disturbance

It is very likely that the wave system under study could have been formed due to a pulsed initial disturbance. As a cause, we can consider either a meteorite falling into the ocean, or a powerful explosion—underwater or surface. The latter assumption is more likely, since the area of the ocean under consideration is not very far from the major naval base Norfolk (East coast of the United States).

Based on these assumptions, a very simple physical model is developed (Mehaute and Wang 1996; Mirchina and Pelinovsky 1988). Suppose that the waves were excited as a result of the instantaneous formation of a parabolic cavity at the free surface of the liquid. We describe this assumption using the equation

$$\eta_e(r) = \begin{cases} H_e\left[2\left(\frac{r}{R}\right)^2 - 1\right], & r \le R, \\ 0, & r > R. \end{cases} \tag{1}$$

Denotes the polar radius as r, the vertical deviation of the free surface of the liquid in the cavity from its undisturbed flat horizontal position is denoted as $\eta_e(r)$; H_e is the maximum depth of the cavity. The parameters R and H_e that characterize the cavity are related to the explosion energy E by the following relations (Mirchina and Pelinovsky 1988): $R = 0.04E^{0.3}$, $H_e = 0.02E^{0.24}$. The atmospheric pressure on the free surface of the liquid was assumed to be zero. We will not take into account the viscosity of the liquid. Further consideration will be carried out in the framework of linear theory under the assumption that the height of the simulated ocean waves is many times less than their length. Since the exact depth of the ocean is unknown, two mathematical models have been proposed. One model is for shallow depth, and the second model is for infinite depth.

4 Model in the Framework of the Theory of Long Waves

Assume that the length of the excited waves is many times greater than the depth of the liquid. This is true for models describing ocean tsunamis (Bryant 2001). Choose the time interval $t \geq 0$ and at a point with a polar radius r denote by $\eta(r, t)$ the vertical deviation of the liquid surface. In the long-wave approximation, the system of hydrodynamic equations can be reduced to a single wave equation (Kochin et al. 1964). This is true because the functions describing the fluid motion do not depend on the vertical coordinate. Then, to find the profile of the free surface of the liquid $\eta(r, t)$, the following initial Cauchy problem arises

$$\begin{gathered}\frac{\partial^2 \eta(r,t)}{\partial t^2} = C_0^2 \frac{1}{r}\frac{\partial}{\partial r}\left(r\frac{\partial \eta(r,t)}{\partial r}\right),\\ \eta(r,0) = \eta_e(r),\\ \frac{\partial \eta}{\partial t}(r,0) = 0.\end{gathered} \tag{2}$$

Here g is the acceleration of gravity, $H_0 = const$ is the depth of the liquid, and $C_0 = \sqrt{gH_0}$.

5 Formula for Calculating the Height of Ocean Waves Using the Theory of Long Waves

The solution of the Cauchy problem is obtained by the Fourier–Bessel transform with respect to the variable r. The elevation of the free surface of the liquid $\eta(r, t)$ is represented as

$$\eta(r,t) = \int_0^\infty \tilde{\eta}(k,t) J_0(kr) k dk, \tag{3}$$

and $\tilde{\eta}(r, t)$ is

$$\tilde{\eta}(k,t) = \int_0^\infty \eta(r,t) J_0(kr) r dr$$

Here $J_0(kr)$ is the zero-order Bessel function of the argument (kr). Thus, we obtain the Cauchy problem for an ordinary differential equation with constant coefficients for the Fourier–Bessel image $\tilde{\eta}(k, t)$ of the function $\eta(r, t)$.

As a result of solving this problem, we find

$$\tilde{\eta}(k,t) = \tilde{\eta}_e(k)cos(C_0 tk),$$

where $\tilde{\eta}_e(k)$ is the Fourier–Bessel image of the function $\eta_e(r)$. Let us certainly find this image:

$$\tilde{\eta}_e(k) = -\frac{H_e R}{k} J_3(Rk),$$

Here $J_3(Rk)$ is the third-order Bessel function of the argument Rk. We can find the following integral representation of the solution of problem (2) by substituting intermediate images in (3):

$$\eta(r,t) = -H_e R \int_0^\infty J_3(Rk)cos(C_0 tk)J_0(rk)dk. \quad (4)$$

Unfortunately, the calculations carried out using this formula (4) did not coincide with the wave system under study.

6 Model in the Case of Infinite Depth

Now let's assume that the depth of the ocean is infinite, and the motion of the liquid is axisymmetric. Then it is not possible to escape from the vertical coordinate and it is necessary to consider the complete system of equations. We assume that the shape of the free surface of the liquid is given by Eq. (1) at the time $t = 0$. The main required value is the vertical deviation $\eta(r,t)$ of the liquid surface at time $t \geq 0$. We introduce a cylindrical coordinate system with two coordinates (r, z). We present the potential $\Phi(r,z,t)$ of the fluid velocity $\overleftarrow{v}(r,z,t)$ so that $\overleftarrow{v}(r,z,t) = \nabla\Phi(r,z,t)$..

We have an initial-boundary value problem (Kochin et al. 1964) for functions $\Phi(r,z,t)$ and $\eta(r,t)$ at $t > 0$ and $0 \leq r < \infty$, for finding $\eta(r,t)$

$$\begin{aligned}
\frac{1}{r}\frac{\partial}{\partial r}\left(r\frac{\partial\Phi(r,z,t)}{\partial r}\right) + \frac{\partial^2\Phi(r,z,t)}{\partial^2 z} &= 0,\\
\frac{\partial\eta(r,t)}{\partial t} = \left.\frac{\partial\Phi(r,z,t)}{\partial z}\right|_{z=0}&,\\
\eta(r,t) = -\frac{1}{g}\left.\frac{\partial\Phi(r,z,t)}{\partial t}\right|_{z=0}&
\end{aligned} \quad (5)$$

with initial conditions

$$\eta(r,0) = \eta_e(r), \qquad \frac{\partial\eta}{\partial t}(r,0) = 0. \quad (6)$$

Problem (5) and (6) is a mathematical model of the wave system under study for the case of infinite depth of the liquid.

7 Formulas for Calculating the Height of Ocean Waves

We obtain an exact and asymptotic formula for calculating the free surface profile of the liquid. We apply, as in the previous model, the Fourier–Bessel transform in a variable r to solve this problem. Finally, to determine the height of the waves $\eta(r, t)$, we have the exact formula

$$\eta(r,t) = -H_e R \int_0^\infty J_3(Rk) cos(\sqrt{gk}t) J_0(rk) dk, \tag{7}$$

Here $J_3(Rk)$ and $J_0(rk)$ are third- and zero-order Bessel functions with arguments (Rk) and (rk), respectively. When performing calculations using this formula, it is convenient to switch to dimensionless variables

$$\gamma = \sqrt{\frac{g}{R}}t, \quad \xi = \frac{r}{R}, \quad \kappa = kR, \tag{8}$$

Then for $\eta(\xi, \gamma)$ we have:

$$\eta(\xi,\gamma) = -H_e \int_0^\infty J_3(\kappa) cos(\gamma\sqrt{\kappa}) J_0(\xi\kappa) d\kappa. \tag{9}$$

The obtained exact formulas (7) and (9) have a simple form but, unfortunately, are suitable for operational calculations for not very large values of γ and r. However, an asymptotic formula can be obtained from (9) using the stationary phase method (Copson 1965). To do this, the following conditions must be fulfilled: $\gamma \gg 1$ and $\xi \gg 1$. In this case, the formula is convenient for calculations

$$\eta(\xi,\gamma) = -\frac{\sqrt{2}H_e}{\xi} J_3\left(\frac{\gamma^2}{4\xi^2}\right) cos\left(\frac{\gamma^2}{4\xi}\right), \tag{10}$$

Here ξ and γ is expressed in terms of dimensional values r and t by formulas (8).

8 Calculated Ocean Wave System

In the process of calculating the excitation of waves with different values of physical parameters, the task was to approximate the simulated wave system as closely as possible to the real one shown in Fig. 1 then we can estimate the energy of the explosion E and obtain a characteristic of the geometric parameters of the initial pulse perturbation, including the radius R and depth of the cavity H_e, respectively. We can also estimate the time interval between the explosion and the photo. Assessment by the photo, the distance r from the epicenter of the initial disturbance to the point farthest from it is about 6000 m. We present the wave profile, i.e., for the selected time point t, we calculate the elevation of the free surface $\eta(r, t)$ as a function of distance r (Fig. 2). The calculation is performed using the asymptotic formula (10). Values $\eta(r, t)$ and r in Fig. 2 are given in meters. As you can see, the characteristic height of the wave profile was about 0.1 m. However, if such waves form an extended and fairly regular system, they can be detected from space. The proof of this statement can be obtained from Fig. 1, which shows the waves formed from the passage of ships.

Numerous calculations have shown that the following parameters are suitable: for the explosion energy $E = 2.48 \times 10^{11}$ J; this energy corresponds to the following dimensions of the initial parabolic cavity formed on the liquid surface at the initial time: $R = 105$ m, $H_e = 10.9$ m. The time interval between the initial pulse action and photographing is $t = 800$ s.

Taking the thermal equivalent of TNT equal to 4.2 MJ/kg, we get an explosion power of 59 tons in equivalent.

Figure 2 shows that the simulated wave system consists of wave packets, similar to the ring-shaped wave system in Fig. 1.

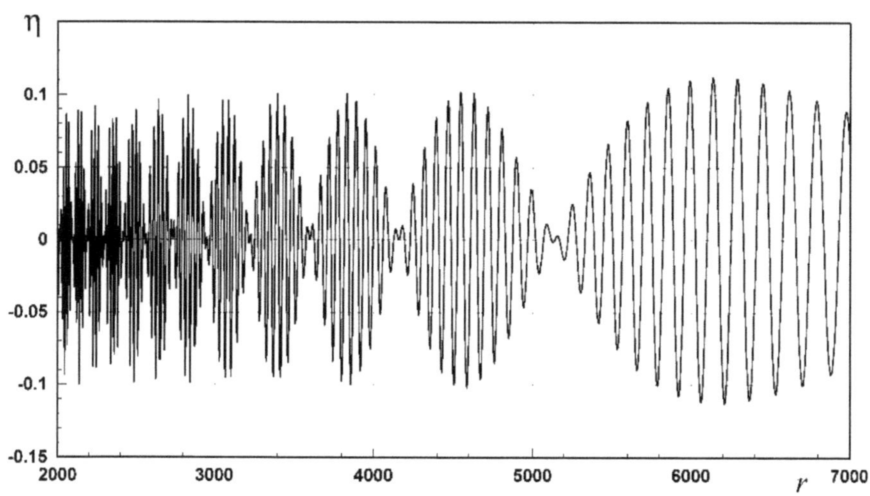

Fig. 2 Wave packets calculated using the proposed model

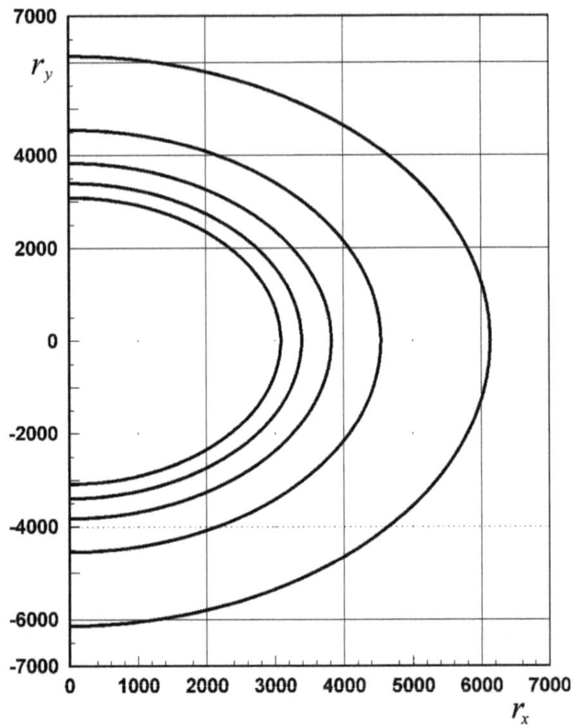

Fig. 3 Crests of wave packets calculated using the proposed model

In Fig. 1, you can see the light bars. These are the Central parts of wave packets. For comparison with these bands, we present a top view of the highest waves (i.e., the central part) of the wave packets at $t = 800$ s, $3000 < r < 7000$ m (Fig. 3). For convenience of comparison, the figure shows the right part of the wave system calculated in accordance with the proposed model.

Due to the accepted condition about the axial symmetry of the initial perturbation, the crests of the highest waves in the wave packets should look like 5 semicircles from above. Given the assumption that the camera's optical axis deviates from the vertical when taking pictures of the ocean surface, Fig. 3 shows these curves as half ellipses with a slight eccentricity.

Comparing Figs. 1 and 3, we see a good match between the halves of the ellipses and the light stripes in the original photo.

9 Conclusion

Analyzing the location of wave maxima from Fig. 2 it can be seen that the modeled wave system (its wave packets) is similar to the ring wave system in Fig. 1.

The proposed calculation algorithm quickly performs calculations on a personal computer and is easy to implement.

As a result of performing calculations based on the proposed mathematical model with different parameters of ocean wave excitation, it was possible to achieve similarity of the simulation results and the original image. As a result, it was possible to obtain estimates of the initial pulse perturbation and the time between the perturbation and photographing. The energy of the initial perturbation (explosion) was estimated in J, and the wave height is about 0.1 m. Despite the small amplitude, such waves can be registered from space. This can be seen in the original image (Fig. 1), which shows traces of ship waves.

The photograph in Fig. 1 shows a well-developed system of ocean waves. Modeling has shown that this wave system originated in a deep-water area of the ocean. It is also shown that the time between the initial perturbation and photographing is about 800 s.

Unfortunately, photographing only produces one static image. It would be interesting to have information about the development of such ocean wave systems. This would allow us to refine the mathematical model.

Acknowledgements The study was partially supported by the Government program (contract AAAA-A17-117021310387-0) and partially supported by RFBR (grant No. 18-01-00812).

References

Belyaev MY, Vinogradov PV, Desinov LV, Kumakshev SA, Sekerzh-Zen'Kovich SY (2011) Identification of a source of oceanic ring waves near Darwin's Island based on space photos. J Comput Syst Sci Int 50(1):67–80

Belyaev MYu, Desinov LV, Kumakshev SA, Sekerzh-Zen'kovich SYa, Krikalev SK (2009) Identification of a system of oceanic waves based on space imagery. J Comput Syst Sci Int 48(1):110–120

Bryant E (2001) Tsunamis. The underrated hazard. Cambridge University Press, Cambridge

Copson ET (1965) Asymptotic expansions. Cambridge University Press, Cambridge

Kochin NK, Kibel IA and Roze NV (1964) Theoretical hydromechanics. Wiley Interscience

Mehaute BLe, Wang S (1996) Water waves generated by underwater explosion. World Science, Singapore

Mirchina NR, Pelinovsky EN (1988) Estimation of underwater eruption energy based on tsunami wave data. Nat Hazards 1(3):277–283

Technology of Seismic Acoustic Detection and Research of River Paleostructures of the Sea Bottom of the Coastal Zone and Its Approbation in Blue Bay

A. L. Brekhovskikh, M. S. Klyuev, A. E. Sazhneva, A. A. Schreider, and A. S. Zverev

Abstract The technology of seismoacoustic detection and study of river paleostructures of the coastal sea zone is considered. It is proposed to use the method of vertical seismoacoustic sounding in a wide frequency range with different directional patterns and the use of high-precision GPS navigation. The principles of seismoacoustic detection and studying of bottom paleostructures have been developed. The equipment for the practical implementation of the considered methods and approaches is described. The results of their application to study the paleostructures of the Ashamba River in Blue Bay near Gelendzhik are presented. The analysis and mapping of the identified bottom paleostructures was carried out.

Keywords Coastal sea zone · Bottom geomorphology · River paleostructures · Gas flares · Seismoacoustic equipment

Bottom river paleostructures of the coastal marine zone are unique objects of marine geology that occur when deltas and riverbeds are flooded due to sea level rise or coastal land subsidence and their further evolution in marine conditions (Gulin and Kovalenko 2010; Klyuev et al. 2019a).

These structures are a kind of markers of sea level fluctuations, tectonic movements of its coastal zone, serve as accumulators of modern marine sediments and fossil organic matter, as a result of which they are of considerable interest for various fields of science—geophysics, geology, geochemistry, hydrobiology and geoecology. In practical terms, they may contain placers of valuable ore and mineral resources (gold, platinum, palladium, silver, diamonds, rubies, sapphires, etc.), which were eroded by ancient water flows or carried out from the land and accumulated in the paleo riverbed.

One of the promising approaches to their detection and study is broadband seismoacoustic profiling (Klyuev et al. 2019a; Brekhovskikh et al. 2019a, b). Its essence consists in vertical hydroacoustic sounding of the coastal water area over an area

A. L. Brekhovskikh · M. S. Klyuev (✉) · A. E. Sazhneva · A. A. Schreider
P.P. Shirshov Institute of Oceanology of RAS, Moscow, Russia

A. S. Zverev
Vernadsky Institute of Geochemistry and Analytical Chemistry of RAS, Moscow, Russia

T. Chaplina (ed.), *Processes in GeoMedia—Volume IV*, Springer Geology,
https://doi.org/10.1007/978-3-030-76328-2_36

in a wide frequency range (200 kHz–300 Hz) using combined directional patterns (narrow from 4° and wide to 180°), and providing high-precision GPS navigation.

This approach allows you to remotely study the geomorphology of the bottom surface, the structure and structure of the bottom layer, paleostructures of ancient rivers, gas emissions from the sea floor, submerged ships and objects, submerged ancient settlements, etc. To identify specific forms and types of bottom relief and objects, we use developed criteria for their recognition and original methods and algorithms based on them. As a result, it becomes possible to determine the boundaries of marine polygons with various forms of bottom relief and types of bottom objects for their further detailed study by complex geological, geochemical, hydrobiological and geoecological methods.

As shown by extensive field studies and tests (Klyuev et al. 2019a; Brekhovskikh et al. 2017a, b, 2018, 2019a, b; Schrader al. et al. 2017, 2018a, 2018b), the following equipment can be effectively used together to detect and study bottom paleostructures in the coastal zone.

high-frequency narrow-beam echo sounder (operating frequency of the order of hundreds of kilohertz, the width of the radiation pattern of the order of units of degrees);
high-frequency narrow-beam profiler (operating frequency of the order of ten kilohertz, the width of the radiation pattern of the order of units of degrees);
mid-frequency acoustic profiler with electromechanical emitter of the "boomer" (the operating frequency of the order of a few kHz, the width of the radiation pattern of the order of tens of degrees);
a low-frequency seismoacoustic profiler with a sparker-type electric spark emitter (operating frequency of the order of hundreds of Hertz, the width of the radiation pattern of the order of tens of degrees);
providing high-precision GPS navigation.

The combined use of this equipment makes it possible to obtain comprehensive information about the bottom surface and the structure of the bottom layer. Narrow-beam echo-sounder device gives information about the profile of the bottom surface. High frequency narrow-beam profiler informs about the type and structure of the upper sedimentary layer, medium. Low-frequency seismoacoustic profilograph gives information about of deeper bottom structures.

The wide frequency range of the used seismoacoustic means is due to the fact that paleostructures can be located not only on the bottom surface, but also at a depth from units to tens of meters below its surface, and the depth of penetration of the sounding signal into the bottom depends on the frequency. The lower the frequency, the more penetration but lower spatial resolution and, consequently, the higher the frequency, the less penetration but higher resolution.

Increasing the width of the radiation pattern allows you to increase the search area when the angular resolution decreases, and reducing it increases the angular resolution when productivity decreases. The formation of narrow diagrams of the radiation produced by parametric radiation.

The use of high-precision GPS navigation allows you to determine the detailed spatial structure of paleostructures and perform their detailed mapping, which increases the reliability of their identification and the accuracy of determining their parameters, and therefore, make more detailed and adequate conclusions about their origin and evolution.

Data processing uses original methods and algorithms for joint processing of information from various devices, which can significantly improve the recognition and identification of bottom paleostructures (Schreider al. et al. 2018a, b; Brekhovskikh et al. 2018).

The key issue in the detection of paleostructures is the criteria and classification features of their presence. As a result of field research and analytical study, the main criteria and classification features of their presence were determined, namely.

1. The presence of a natural "box-shaped" cross-section on the bathymetry of the sea floor and its smooth translational repetition in space.
2. The presence of a natural "box" profile filled with sediments on the seismoacoustic structure of the sea floor and its smooth translational repetition in space, including at the site of a bottom fault.
3. If the paleo riverbed is confined to a bottom fault, its transverse profile may be distorted up to V-shaped.
4. The presence of a rugged and dissected geomorphology that was not exposed to water flow indicates the absence of paleo riverbed.
5. The compliance of the position of the paleo riverbed with modern river structures of land, namely, to the riverbed, banks, valley, canyon, delta.
6. The possible presence of paleo tributaries and paleo deltas recognized by the same criteria as a paleo riverbed.
7. The presence in many cases of gas flares associated with the paleo riverbed.

The "box" profile here refers to a transverse profile in the form of a bowl with pronounced coastal slopes and a flattened bottom. It should smoothly translational repeat in the space, outlining the channel paleo riverbed. The term "box" in this context was introduced in Gulin and Kovalenko (2010) and came into wide use. If paleo riverbed confined to the bottom's fault, its cross-section may be distorted up to V-shaped.

This bowl is usually filled with weakly consolidated bottom sediments with a flat surface, since it is a local deep maximum and all sedimentary material is dumped into it.

The presence of a rugged and dissected geomorphology (rock sedimentary rocks, marl ridges, etc.), which was not exposed to the water flow, indicates the absence of riverbed in this place. As a rule, the paleo riverbed is a continuation of the modern land riverbed or its delta. Paleo tributaries can flow into the paleo riverbed, forming their confluence, and it can also be divided into branches in the form of a paleo delta.

Gas flares often accompany the paleo riverbed, since it may be confined to geological faults, and the sediments that fill it are less consolidated and more transparent for bottom gases to escape.

Thus, the technology for detecting paleostructures in the coastal zone consists of the following:

1. The use of broadband seismoacoustic sounding of the sea floor with combined directional diagrams.
2. Areal study of the polygon with accurate GPS navigation.
3. Application of criteria and classification features for the presence of paleostructures.

For the detection and study of coastal paleostructures according to the developed technology uses complex sonar profiling of the bottom surface and the upper sediment layer development P. P. Shirshov Institute of Oceanology of Russian Academy of Sciences and seismoacoustic complex "Geont-shelf" by production of the company "Spektr-geophysika", at our disposal.

The complex of P. P. Shirshov Institute of Oceanology of RAS (Klyuev et al. 2015; Schreider al. et al. 2016) includes a high-frequency narrow-beam parametric profiler (operating frequency 20 kHz, beam width 4.5°, spatial resolution about 10 cm) and a high-frequency narrow-beam echo sounder (operating frequency 200 kHz, beam width is 4°, spatial resolution is about 1–5 cm) based on the device "SeaKing DST", vector receiver of satellite GPS navigation Trimble BX982, navigation complex based on the program AquaScan 10, processing and control laptops, autonomous power supply devices, as well as data processing software.

"Geont-shelf" seismoacoustic complex (Mayev et al. 2009) produced by "Spektr-geophysika" with electrodynamic ("boomer" type, operating frequency 1–2 kHz) and electric spark ("sparker" type, operating frequency 300–800 Hz) wide-directional emitters. It includes a seismic energy storage device "SPES-270" (provides energy of 270 J at an operating voltage of up to 3000 V), emitters and a receiving oil-filled scythe (6 hydrophones, length 1 m, diameter 22 mm), a programmable amplifier with an ADC, a laptop, a GPS receiver and software for collecting, recording, reproducing and processing data, and an Autonomous power supply device.

Marine work was carried out on the board of research vessels (SIV) "Ashamba" (length 15 m, width 4 m, draft 1.5 m, displacement 27 tons.), "Professor Longinov" (MaryFisher 625, length 6.4 m, width 2.5 m,) and "Caiman" (length 3 m, width 1.6 m). The mounting of the receiving head "SeaKing DST" was carried out on a rod outside the SIV, and emitters such as "boomer", "sparker" and the receiving spit "Geont-shelf" were transported behind the ship.

The Blue (Rybatskaya) Bay near Gelendzhik town, Krasnodar Territory, where the Southern Branch of the Shirshov Institute of Oceanology of the Russian Academy of Sciences is located, was chosen as a search and studying area of coastal river paleostructures. It is a natural coastal formation that arose under the influence of various natural factors, including the formation of the mouth of the Ashamba River (Yashamba), which flows into the bay.

It is natural to assume that the geomorphology of the bottom of the Blue Bay may bear traces of the impact of the Ashamba River during the periods of ancient sea level decline, when its water area became dry. Basic studies of the geomorphology

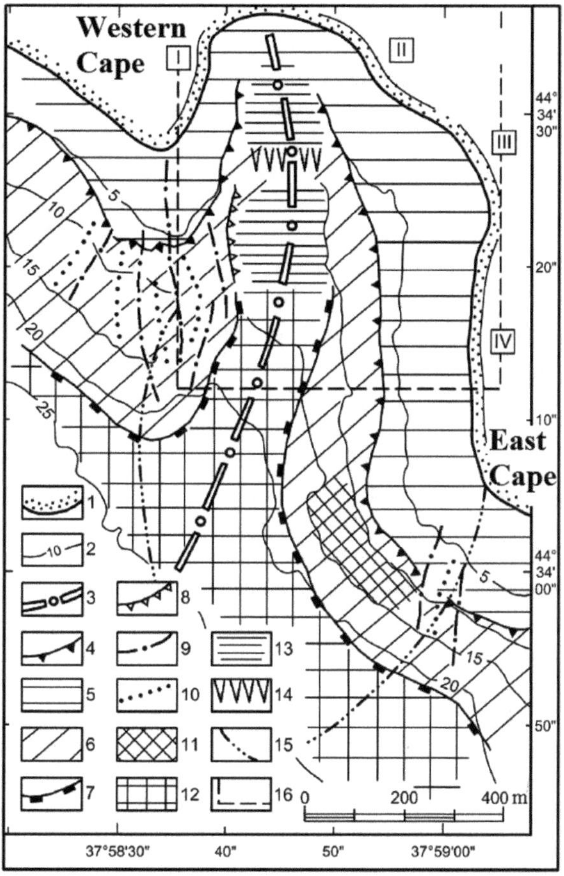

Fig. 1 Map-diagram of the geomorphology of the Blue Bay bottom (Kuklev et al. 2014). According to Yevsyukov (according to Brekhovskikh et al. 2020, modified)

of the bottom of the Blue Bay were carried out by Kuklev et al. (2014) and is shown on the diagram map in Fig. 1.

Here 1 is the coastline; 2 is isobaths; 3 is the axis of the depression; 4 is the edge of the coastal stage; 5 is the surface of the coastal stage; 6 is the slope of the depression (Graben); 7 is the foot of the slope; 8 is the foot of a steep ledge; 9 is the axes of ridges; 10 is the axes of hollows; 11 is the tectonically fragmented stage; 12 is the submerged surface of the depression; 13 is the sub horizontal surfaces at the head of the depression; 14 is the ledge separating the sub horizontal surfaces; 15 is the assumed the boundary of the outflow cone formed by the flood; 16 is the boundary of the reference polygon.

According to these studies, the configuration of the Bay has the shape of a half-rhombus. The length of the coast along its "faces" is 300 m from the West, 550 m from the Northeast, and 700 m from the East. In cross-section, the bottom of the bay is a morphologically distinct depression. Capes, conventionally called Western and Eastern, with heights of 18–20 m and 11–13 m, respectively, mark the seaward sections of the coast.

Along the perimeter, the underwater part of the Bay is outlined by a coastal shoal. Its width is from 70 to 100 m (West) to 230–260 m (East), with large values confined to the promontories. The step is bounded by brow, which is located at a depth of 4–7 m.

Below the brow of the step, the relief of the bay bottom is represented by a slope, the foot of which is located at a depth of 8–10 to 22 m. The slope is characterized by a sharp heterogeneity of morphology. On the traverse of Cape Zapadny, its width is 450 m in plan, and in the bottom relief it is composed of ridges and hollows of a sub meridional orientation. Their length is from 200 to 450 m, the amplitude is 3–5 m. In the Northern direction, this slope narrows and turns into a steep local ledge, the height of which is about 8 m.

The Eastern, relatively gentle slope of the Bay, is more pronounced in relief. At a distance of about 800 m, its width (from South to North) decreases from 230 to 100 m. On the traverse of Cape Vostochny, it is complicated by a tectonically fragmented step and sub meridional ridges (relative height 2–4 m), which are associated with hollows. At various bathymetric levels, this slope is complicated by narrow (40–70 m) steps and relatively steep (15–20°) ledges, which are 2–6 m high. Their formation seems to be related to abrasion processes. On the traverse of Cape Vostochny, the depression expands and loses its morphological expression. The same thing happens with the coastal step and the slope of the depression.

The precipitation of the Blue Bay is represented by sandy silts with shells, pebbles and marl ridges. The coastal, Central and Eastern part of the bay is mostly leveled and composed of pebbles and sand silts with shells. In the West of the bay and in the East at the exit there are marl ridges with depth differences of up to several meters.

One of the unique and interesting features of the Blue Bay, discovered and studied in 2018–2020, is the paleo riverbed of the Ashamba river and its paleo tributary, located at the bottom of the bay (Klyuev et al. 2019a; Brekhovskikh et al. 2019a, b). The research was carried out by areal profiling of the Blue Bay in the East–West and North–South directions with a step of about 30 × 30 m by using the complex of hydroacoustic profiling of the bottom surface and the upper layer of sediments developed by the Institute of Oceanology of RAS and the "Geont-shelf" seismoacoustic complex produced by "Spektr-geophysika" LLC.

Figure 2a shows the transverse profile of the paleo Ashamba riverbed in area E (Fig. 5) of the Blue Bay according to the echo sounder data (frequency $f = 200$ kHz, directional pattern $\theta = 4°$) at a depth of about 11 m. The drawing shows that the cross-section of the paleo riverbed has a "box" shape about 120 m wide with steep coastal slopes up to 6 m high and smooth, apparently, sedimentary filling at a depth of about 11 m. It can be assumed that the paleo riverbed is a channel cut by ancient streams in the rocky sedimentary base, which is partially filled with modern sediments to an almost flat surface (Figs. 3 and 4).

High-frequency parametric profilograph (frequency $f = 20$ kHz, radiation pattern $\theta = 4.5°$), as well as medium-frequency and low-frequency profilographs

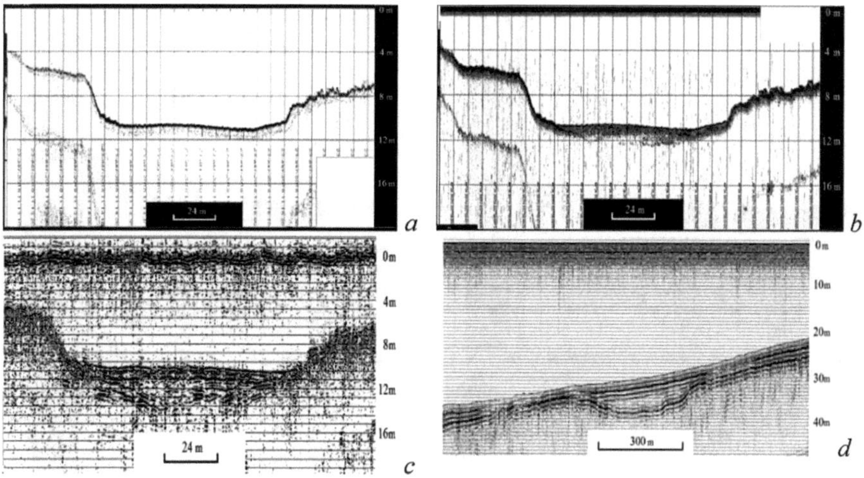

Fig. 2 The transverse profile of the Ashamba riverbed in the E region (Fig. 5) of the Blue Bay according to: **a** echo sounder (f = 200 kHz, θ = 4°), **b** high-frequency parametric profiler (f = 20 kHz, θ = 4.5°), **c** medium-frequency profiler ("boomer" type emitter, f = 1–2 kHz,, wide-directional), **d** in the S region (Fig. 5) according to the low-frequency profiler ("sparker", f = 300–800 Hz, wide-directional) (according to Brekhovskikh et al. 2020, modified)

Fig. 3 **a** The longitudinal profile of the paleo riverbed of the Ashamba river in Blue Bay; **b** the transverse profile of the paleo riverbed (left), the paleo tributary (right), and the dividing rock barrier (center). Parametric profiler data (f = 20 kHz, θ = 4.5°) (according to Brekhovskikh et al. 2020, modified)

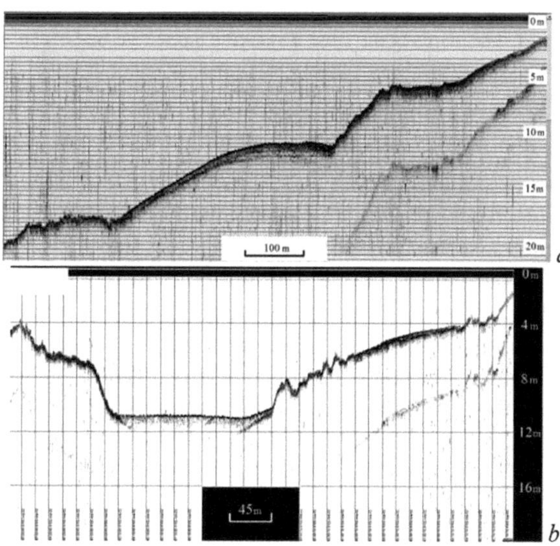

with "boomer" and "sparker" type emitters (frequencies f = 1–2 kHz and f = 300–800 Hz, wide-directional) were used to study the soils that fill the paleo riverbed and compose its coastal slopes.

Figure 2b shows the transverse profile of the Ashamba river paleo riverbed in the Blue Bay according to the data of a high-frequency parametric Profiler (frequency

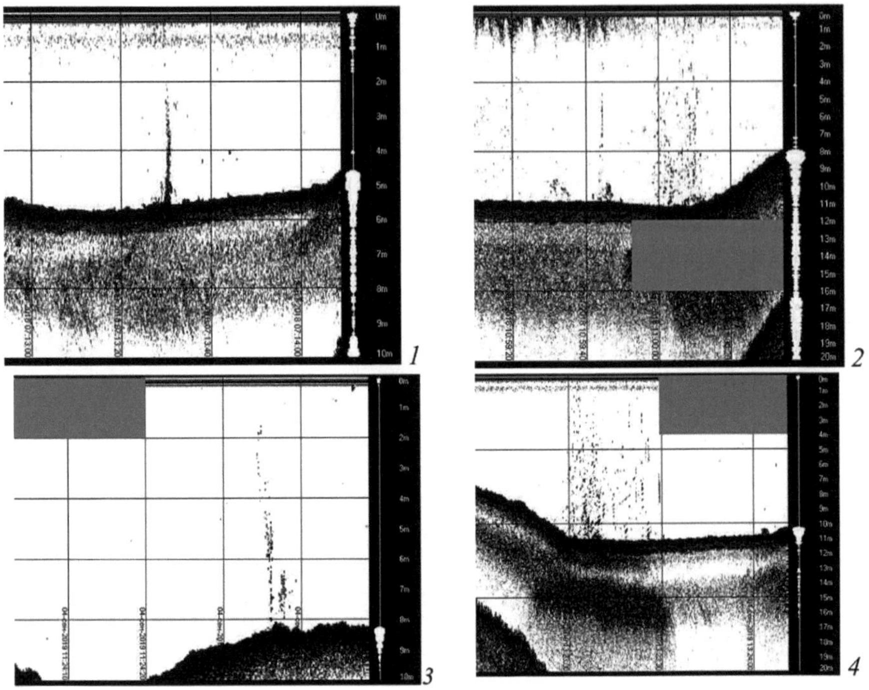

Fig. 4 Gas flares in the Blue Bay in the paleo riverbed of the Ashamba River ($f = 200$ kHz echo sounder) (according to Brekhovskikh et al. 2020, modified)

$f = 20$ kHz, radiation pattern $\theta = 4.5°$) at the same point as Fig. 2a. The figure shows that the probing pulse practically does not penetrate the coastal slopes of the paleo riverbed, which indicates their possible rock sedimentary composition, and penetrates to a depth of about 1 m deep into the soil that fills the bowl of the paleo riverbed, which indicates its possible unconsolidated sedimentary composition. It follows from the drawing that the rocky banks go under the unconsolidated sediments that fill the bowl of the paleo riverbed.

Figure 2d shows the transverse profile of the Ashamba river paleo channel in Blue Bay according to a medium-frequency profiler with a "boomer" type radiator (frequency $f = 1$–2 kHz, θ wide-directional) at the same point as Fig. 2a. The figure shows that the probing pulse still practically does not penetrate the coastal slopes of the paleo riverbed, which further confirms their possible rock sedimentary composition, and penetrates to a depth of about 4 m deep into the soil filling the paleo riverbed, which confirms its possible unconsolidated sedimentary composition. The figure clearly shows the shape of the bottom of the paleo riverbed with a maximum depth of about 4 m relative to the surface of the sediments that fill it, which, apparently, is composed of rocky sedimentary soil.

In Fig. 2c presents a cross profile of a paleo riverbed Ashamba Blue Bay according to a low-frequency profiler (emitter type "sparker", frequency $f = 300$–800 Hz, θ

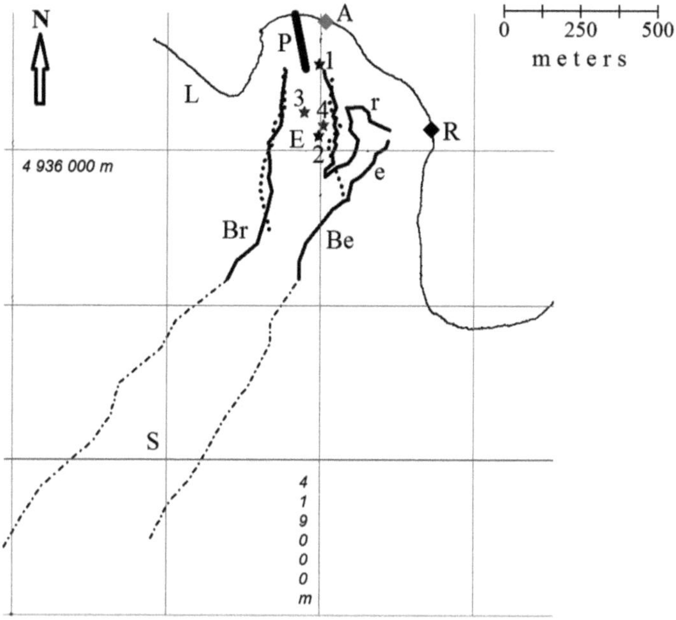

Fig. 5 Map of the paleo riverbed of the Ashamba river, the paleo tributary and gas flares in the Blue Bay, the designations are given in the text (according to Brekhovskikh et al. 2020, modified)

wide-directional) at a depth of about 30 m at point S (Fig. 5), it was found that at a distance from the shore to a depth of more than 20 m, the paleo riverbed completely goes into the thickness of bottom sediments, does not form a depression on the bottom surface and is detected only by medium-frequency and low-frequency profilers with emitters of the "boomer" type and "sparker" respectively.

Figure 3a shows the longitudinal (axial) profile of the paleo riverbed of the Ashamba river in the Blue Bay near Gelendzhik town according to the data of a high-frequency parametric profiler (frequency $f = 20$ kHz, radiation pattern $\theta = 4.5°$) approximately along the central meridian in Fig. 5. The drawing shows that at the coast at depths of about 0–12 m (right part of the picture), the paleo riverbed is buried by alluvial material (pebbles, marl fragments and sand) as a result of wave activity. At depths of about 12–17 m (the Central part of the figure), a paleo riverbed appears, filled with weakly consolidated sediments with a flat surface. At depths of more than 17 m (left part of the picture), marl rocks appear, which indicates that the paleo riverbed turns away from these rocks.

The research also revealed the paleo tributary of the Ashamba river and their confluence at a depth of about 11 m in the E region (Fig. 5) according to the parametric profiler data (frequency $f = 20$ kHz, radiation pattern $\theta = 4.5°$). Figure 3b clearly shows the dividing rock barrier (center) between the paleo riverbed (left) and the paleo tributary (right) before their confluence, which disappears on the bottom profiles after they merge into a single paleo riverbed.

In the course of research, it was also found that coastal bottom paleostructures are accompanied by gas flares (methane husky voice) (Klyuev et al. 2019b; Brekhovskikh et al. 2019c). This may be due to the fact that paleostructures are often confined to geological faults, and the sediments that fill them are less consolidated and more transparent for the release of bottom gases. Figure 4 shows gas flares in the paleo riverbed of the Ashamba River ($f = 200$ kHz echo sounder). They have both jet (1, 3) and distributed (2, 4) character.

The identified paleostructures of the Ashamba river and associated formations in the Blue Bay were summarized and mapped in flat geographical coordinates using the Global Mapper computer program in the universal transverse Mercator projection UTM zone 37 (36°E–42°E of the Northern Geosphere, meters North/South–meters West/East) in the WGS84 coordinate system (Fig. 5).

Here L is coastal line, P is dock, A modern mouth of the river Ashamba, Br is right Bank of paleo riverbed Ashamba (solid line corresponds to 20 kHz ADCP, the dotted—to "boomer" 1–2 kHz, dashed—to "sparker" 300–800 Hz), Be is the left Bank of paleo riverbed Ashamba (solid line corresponds to 20 kHz ADCP, the dotted—to "boomer" 1.0.2 kHz, dashed—to "sparker" 300–800 Hz), R is the modern mouth of the stream, r is right Bank of paleo tributary, e is the left side of paleo tributary, stars 1 through 4 are some gas torches (the number of asterisks indicates the number of the torch in Fig. 4), E is position of the sounder, profiler and "boomer" profiles (Fig. 2a, b), S is position of the sparker profile (Fig. 2c). N is North direction. Note that dashed lines ("sparker" data) show the position of a section of paleo riverbed completely buried in bottom sediments, while solid lines (profiler data) show it rising above the bottom.

Considered in this work paleostructures, from the point of view of the maps by Yevsyukov (Fig. 1) (Kuklev et al. 2014) of geomorphology of the bottom of the Blue Bay, correspond to near-horizontal surfaces in the upper depression (13), the immersed surface of the depression (12), the axis of depression (3) and the slope of depression (6), the foot of a steep ledge (8) and the foot of the slope (7). In this case, the elements 13, 12, 3, 6, 8, 7 they are not treated as components of a single geomorphological formation, in this case the paleostructures of the Ashamba River, but as separate elements of a descriptive nature.

This may be due to the fact that when drawing up the map-scheme, no seismoacoustic equipment was used to study the structure of the upper and deeper sedimentary layer of the bottom, but a single echo sounder was used.

The analysis suggests that the features we have identified are a manifestation of a single geomorphological formation, namely, a segment of the paleo riverbed of the Ashamba river with a paleo tributary flowing into it, made in marl rocks during the ancient sea level drop to 12 m about 2500 years ago during the Phanagoria regression according to Kuklev et al. (2014).

In addition, seismoacoustic studies of the Blue Bay bottom have shown that the coastal areas of the bottom (about 100 m from the shore) are partially covered with a sound-transparent (at a frequency of 20 kHz) substance (possibly weakly consolidated silts) with a thickness of about 0–30 cm, which, at a further distance from the shore, is channeled into the paleo channel of the Ashamba river and its paleo tributary, where its thickness increases to 100 cm or more. In the areas adjacent to the paleo riverbed and the paleo tributary, this thickness does not exceed the same 0–30 cm. It can be assumed that the Ashamba river paleo riverbed and its paleo tributary currently act as a kind of collector for a sound-transparent substance that, under the influence of gravity and waves on the bottom of the Bay, slowly drains from the Bay bottom area into the paleo riverbed and paleo tributary and moves along them to deeper areas of the sea floor, where it flows slowly in the surrounding silts. This assumption needs further comprehensive verification using bottom sampling methods, sensors for horizontal movement of precipitation, as well as methods for numerical mathematical modeling of precipitation dynamics in the coastal zone.

In conclusion, we note that the considered seismoacoustic methods for detecting and studying coastal bottom river paleostructures need further development and improvement. The affiliation of the identified objects to the paleostructures of the Ashamba River requires further detailed research and justification, including through detailed sampling and analysis of bottom sediments and accretions.

Acknowledgements The work was carried out within the framework of The state task No. 0149-2019-0005 of the P. P. Shirshov Institute of Oceanology of the Russian Academy of Sciences, and some methods developed under the RFBR grant No. 20-05-00089 A were used in its preparation.

References

Brekhovskikh AL, Grinberg OV, Evsenko EI, Klyuyev MS, Olkhovsky SV, Sazhneva AE, Schreider AA, Schreider al. A (2017a) On structures and objects of the bottom anthropocene in the Phanagoria GIAMZ. In: Proceedings of the XV all-Russian scientific and technical conference “MSOI-2017”: modern methods and means of Oceanological research. M., APR, vol 1, pp 251–254 (In Russia)

Brekhovskikh AL, Grinberg OV, Evsenko EI, Klyuev MS, Olkhovsky SV, Rakitin IY, Sazhneva AE, Zakharov EV, Chizhikov VV, Schreider AA, Schreider al. A (2017b) On the structure of the collapse of stones in the flooded part of the Patrei settlement according to hydroacoustic parametric profilography and its geochronology. In: Physical and mathematical modeling of processes in geomedia: third international school of young scientists, Moscow: collection of materials schools, IPMeh RAS, Moscow, pp 58–61 (In Russia)

Brekhovskikh AL, Grinberg OV, Evsenko EI, Klyuev MS, Olkhovsky SV, Rakitin IY, Sazhneva AE, Schreider AA, Schreider al. A (2018) Development of the basics of technology for studying cultural heritage objects buried in bottom unconsolidated sediments by parametric profilograph using satellite navigation data. Oceanological Res 46(2):5–14 (In Russia)

Brekhovskikh AL, Voltaire ER, Grinberg OV, Evsenko EI, Zakharov EV, Klyuev MS, Kosyan RD, Kuklev SB, Sazhneva AE, Mazurkevich AN, Olkhovsky SV, Schreider AA (2019a) Paleostructures of the Ashamba river in the geomorphology of The blue Bay bottom according to a parametric profiler with satellite navigation. In: Materials of the XVI all-Russian scientific and technical conference “MSOI-2019”: modern methods and means of oceanological research.-m.: ID of the Zhukovsky Academy, vol 1, pp 144–147 (In Russia)

Brekhovskikh AL, Voltaire ER, Grinberg OV, Evsenko EI, Zakharov EV, Zverev AS, Klyuev MS, Kosyan RD, Kuklev SB, Mazurkevich AN, Olkhovsky SV, Rakitin IY, Sazhneva AE, Schreider AA (2019b) On paleostructures of the Ashamba river in the geomorphology of the bottom of the Blue Bay according to the parametric Profiler with satellite navigation. In: Geology of the seas and oceans: materials XXIII international scientific conference (school) on Marine Geology. Volume V. M.: IO RAS, pp 72–76 (In Russia)

Brekhovskikh AL, Voltaire ER, Grinberg OV, Evsenko EI, Zakharov EV, Zverev AS, Klyuev MS, Kosyan RD, Kuklev SB, Mazurkevich AN, Olkhovsky SV, Rakitin IY, Sazhneva AE, Schreider AA (2019c) On gas husky voice of the Blue Bay near Gelendzhik according to parametric profilograph data using satellite navigation. In: Geology of the seas and oceans: materials of the XXIII international scientific conference conferences (schools) on marine geology, vol V. M.: IO RAS, pp 77–81 (In Russia)

Brekhovskikh AL, Klyuev MS, Sazhneva AE, Schreider AA, Zverev AS (2020) On the principles of seismoacoustic study of the paleostructures of the sea bottom of the coastal zone (on the example of the Blue Bay). Processes Geomedia 3(25):755–763 (In Russia)

Gulin MB, Kovalenko MV (2010) Paleo riverbeds of the Chernaya and Belbek rivers on the shelf of the South-Western Crimea—a new object of ecological research. Mar Ecol J IX(1): 23–31. Sevastopol (In Russia)

Klyuev MS, Olkhovsky SV, Fazlullin SM, Sazhneva AE, Evseenko EI, Schreider al. A (2015) On the capabilities of the parametric Profiler, echo sounder and GLONASS/GPS receiver system for complex studies of bottom anthropocene sedimentary deposits. In: Geology of the seas and oceans: proceedings of the XXI International scientific conference (school) on marine geology, vol V. M.: GEOS, pp 132–136 (In Russia)

Klyuev MS, Schreider AA, Brekhovskikh AL, Rakitin IY, Zverev AS, Voltaire ER, Olkhovsky SV, Grinberg OV, Evsenko EI, Sazhneva AE (2019) Paleo valley of the Ashamba river in the geomorphology of the Blue Bay bottom near Gelendzhik town according to parametric profilograph with satellite navigation. In: Physical and mathematical modeling of processes in geomedia: fifth international school of young scientists, Moscow: conference materials. RAS, pp 81–83 (In Russia)

Klyuev MS, Schreider AA, Brekhovskikh AL, Rakitin IY, Zverev AS, Voltaire ER, Olkhovsky SV, Grinberg OV, Evsenko EI, Sazhneva AE (2019b) Methane flares of the Blue Bay near Gelendzhik according to the parametric profilograph with satellite navigation . In: Physical and mathematical modeling of processes in geomedia: fifth international school of young scientists, Mat. Conf. – M., Moscow, IPMeh RAS, pp 83–84 (In Russia)

Kuklev S, Evsyukov Y, Rudnev V (2014) Catastrophic flooding in the Gelendzhik region. Land and sea bottom terrain transformations. Publishing house: LAP LAMBERT Academic Publishing, p 80 (In Russia)

Mayev EG, Myslivets VI, Zverev AS (2009) Structure of the upper layer of sediments and bottom relief of the Taganrog Bay of the sea of Azov. Bull Moscow Univ. Series 5: Geography. 5:78–82 (In Russia)

Schreider al. A, Klyuyev MS, Evsenko EI (2016) High-resolution geoacoustic system for geological and archaeological study of the bottom. Processes Geomedia 2:156–161 (In Russia)

Schrader al. A, Schreider AA, Galindo-Zaldivar J, Klyuev MS, Ausenco EI, Olkhovsky VS, Sazhneva AE, Zakharov EV, Chizhikov VV, Brekhovskikh AL, Rakitin YI, Grinberg OV (2017) The first data of geological and archaeological study petraske shelf of the Taman Bay of the Black sea. Processes Geomedia 2:557–562 (In Russia)

Schreider al. A, Schreider AA, Klyuev MS, Sazhneva AE, Brekhovskikh AL, Olkhovsky SV, Zakharov EV, Chizhikov VV, Evsenko EI, Rakitin IY, Grinberg OV (2018a) Features of the technology of using parametric hydroacoustic means for searching, identifying and monitoring objects in the bottom layer. Processes Geomedia 2:920–927 (In Russia)

Schreider al. A, Schreider A, Klyuev MS, Sazhneva AE, Brekhovskikh AL, Olkhovsky SV, Zakharov EV, Chizhikov VV, Evsenko EI, Rakitin IY, Grinberg OV (2018b) Technological features of using a parametric profiler to study the bottom layer. Processes Geomedia 4(18):1249–1252 (In Russia)

Changes in Sea Surface Roughness in Light Wind

I. P. Shumeyko, A. Yu. Abramovich, and V. M. Burdyugov

Abstract Correct interpretation and application of remote sensing data from spacecraft requires detailed information about the sea surface. The structure and variability of the sea surface are analyzed under conditions that are most favorable for remote monitoring of processes occurring near the ocean–atmosphere boundary. Such conditions occur when the wind speed U does not exceed 5–7 m/s. Under these conditions, areas of ripples and slicks may simultaneously exist on the sea surface. Slick-ripple contrasts are clearly visible on radar and optical images of the sea surface. It is shown that linear regression equations describing the dependence of the variance of the sea surface slopes, constructed for a wide range of wind speeds, overestimate the variance of the slopes in the region of low speeds. At a wind speed of less than 5–7 m/s, the slope dispersion increases faster with increasing wind than at higher wind speeds. The simultaneous presence of slicks and ripples on the sea surface is a factor limiting the accuracy of altimetric determination of wind speed.

Keywords Remote sensing · Sea surface · Slope · Slick · Ripple · Light wind · Wind waves

1 Introduction

During remote sensing of the ocean from spacecraft, information about the processes that occur near the ocean–atmosphere boundary is obtained by registering electromagnetic waves reflected from the sea surface. One of the main characteristics of the sea surface that determines the signal recorded on the spacecraft is its slopes. Surface waves of all scales contribute to the slope dispersion, but the contribution of short waves is predominant (Zapevalov et al. 2021). The energy of short waves depends on many factors: wind, currents, the presence of surfactants, etc. (Monin

I. P. Shumeyko (✉) · A. Yu. Abramovich
Sevastopol State University, Sevastopol, Russian Federation

V. M. Burdyugov
Marine Hydrophysical Institute RAS, Sevastopol, Russian Federation

T. Chaplina (ed.), *Processes in GeoMedia—Volume IV*, Springer Geology,
https://doi.org/10.1007/978-3-030-76328-2_37

and Krasitskiy 1985). When the wind speed is less than 5–7 m/s, its role is not dominant, so conditions when the wind does not exceed the specified value are optimal for remote observation of phenomena on the sea surface.

With a weak wind, slicks are observed on the sea surface. Slices are areas of the sea surface where short waves are partially or completely suppressed. The appearance of slicks is caused by: the impact of internal waves on the sea surface, a decrease in wind speed below the threshold value (calm slicks), Langmuir circulation, biogenic films and others (Apel et al. 1975; Konstantinov and Novotryasov 2013; Bulatov et al. 2003; Serebryany 2012). The main cause of anthropogenic slicks is the oil spill (Brekke and Solberg 2005).

When sounding the sea surface at small incidence angles, the normalized radar cross section of radio waves is described by the expression (Bass and Fuks 1979).

$$\sigma(\theta) = \pi \sec^4\theta\,|R_0|^2 P(\xi_x, \xi_y)\big|_{\xi_x = tg\theta,\, \xi_y = 0} \quad (1)$$

where θ is the incidence angle, R_0 is the Fresnel coefficient for normal incidence, P is a two-dimensional probability density function for orthogonal components of sea surface slopes, ξ_x and ξ_y there are slopes in orthogonal directions. Equation (1) is obtained for the case when the slope ξ_x is oriented in the sensing direction. The accuracy of calculating the radio emission and sunlight reflected from the sea surface depends on how correctly the characteristics of its slopes are specified. The accuracy of the calculation of radio waves and sunlight reflected from the sea surface depends on how correctly the characteristics of sea surface slopes are specified.

Sea waves are weakly nonlinear. Measurements show that the deviations of the slope distribution as well as the elevation distribution from the Gaussian are small (Zapevalov and Ratner 2003; Babanin and Polnikov 1995; Khristoforov et al. 1992). However, they are important in forming the field of electromagnetic waves scattered by the sea surface (Khristoforov et al. 1994; Bréon and Henriot 2006).

The aim of this work is to generalize and analyze the results of studies of the structure of the sea surface in low winds, carried out by different research groups with different types of equipment (in situ measurements and remote sensing).

2 Sea Surface Slope Measurements

Various types of equipment are used to measure sea surface slopes (Zapevalov et al. 2021). Such equipment is a wave buoy (Longuett-Higgins et al. 1963), an array of wave gauge sensors (Kalinin and Leikin 1988; Zapevalov et al. 2009), a laser slopemeter (Zapevalov 2002), a radar (Karaev et al. 2008), a device that probes a surface in the optical range (Bréon and Henriot 2006; Cox and Munk 1954). The minimum wavelength that contributes to the variance of the measured slopes depends on the technical characteristics of the equipment. For wave buoys, it is determined by the size of the buoy body (meters), for wave gauge sensors it is determined by the distance between the gauge sensors (decimeters), for laser slopemeter it is determined by the size of the laser spot on the sea surface (millimeters), for radar it is determined by the length of the radio wave (centimeters and decimeters). The length of the electromagnetic wave in the optical range is much smaller than the length of any sea wave, which makes it possible to measure all waves present on the surface.

Next, we will consider the sum of slope variances oriented along ξ_u and across ξ_c the wind direction $\overline{\xi^2} = \overline{\xi_u^2} + \overline{\xi_c^2}$. The variance of slopes determined by various types of equipment can be represented as

$$\overline{\xi^2} = \int_0^{k_0} \Upsilon_u(k)\,dk + \int_0^{k_0} \Upsilon_c(k)\,dk, \tag{2}$$

where k is the wave number of the surface wave; $\Upsilon_u(k)$ and $\Upsilon_c(k)$ are the spectrum of the upwind and crosswind components of the slopes. The upper limit of integration k_0 is determined by the minimum wavelength λ_0, which depends on the spatial resolution of the given type of equipment. The lower limit of integration equal to zero indicates that there is no limitation on the maximum length of the waves.

3 Dependences of the Variance of the Sea Surface Slopes on the Wind Speed

The regression dependences of slope variance on wind speed obtained from the sounding data are shown in Fig. 1. Curves 1 and 2 are based on slope data obtained from aerial photographs of the sun glitter (Cox and Munk 1954). The results of this work published in 1954 have until now been widely used to simulate radio waves reflected from the sea surface. Curve 1 corresponds to pure water, curve 2

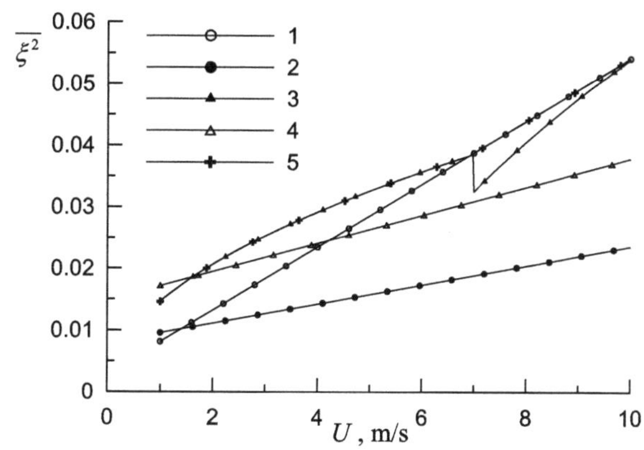

Fig. 1 Dependence of sea surface slope variance $\overline{\xi^2}$ on wind speed U

corresponds to an artificial slick created by spilling a mixture of surfactants. Recently, the regression relationships $\overline{\xi^2} = \overline{\xi^2}(U)$ were confirmed by remote sensing data obtained from the spacecraft (Bréon and Henriot 2006). The characterization of slope was carried out using an optical scanner POLDER (POLarization and Directionality of the Earth Reflectances). The NSCAT scatterometer (NASA Scatterometer) was used to determine the wind speed. The regressions $\overline{\xi^2} = \overline{\xi^2}(U)$ obtained by two different methods coincide up to the standard deviation. Curves 4 and 5 are plotted according to the data of radiometric measurements (Li et al. 2007) and laser sounding from spacecraft (Hu et al. 2008). Different estimates of the sea surface roughness level corresponding to curves 1, 4, and 5 are explained by the fact that it was created by waves of different scales (Zapevalov et al. 2021).

As a rule, dependencies $\overline{\xi^2} = \overline{\xi^2}(U)$ are described by linear equations. In [Wu], the data describing changes in the variance of sea surface slopes presented in [Cox] were analyzed once again. It was found that at wind speeds $U \leq 7$ m/s and $U > 7$ m/s, variance $\overline{\xi^2}$ changes in different ways. The proposed relationship in Fig. 2 corresponds to curve 3. According to this dependence, the rate of change with a strong wind is higher than with a weak one.

The result obtained by [Wu] contradicts existing ideas about the behavior of the spectrum of surface waves in the short-wave range. The energy of short waves that make the predominant contribution to slope dispersion increases rapidly at low wind speeds. As the wind increases, the rate of growth of short wave energy slows down (Apel 1994; Elfouhaily et al. 1997; Cheng et al. 2006).

A similar nature of the change in wave energy (rapid growth with a weak wind, slow growth with a strong wind) was confirmed by direct measurements of the sea surface slopes (Khristoforov et al. 1992). The measurements were carried out using a laser slopemeter. The measurements were carried out under meteorological conditions varying from calm to strong wind. The slope modulus of the sea surface was studied as a parameter characterizing the level of roughness $\xi_m = \sqrt{\xi_u^2 + \xi_c^2}$.

The dependence of the mean value of the slope modulus $\overline{\xi_m}$ on the wind speed obtained in field experiments is described by the equations (Khristoforov et al. 1992)

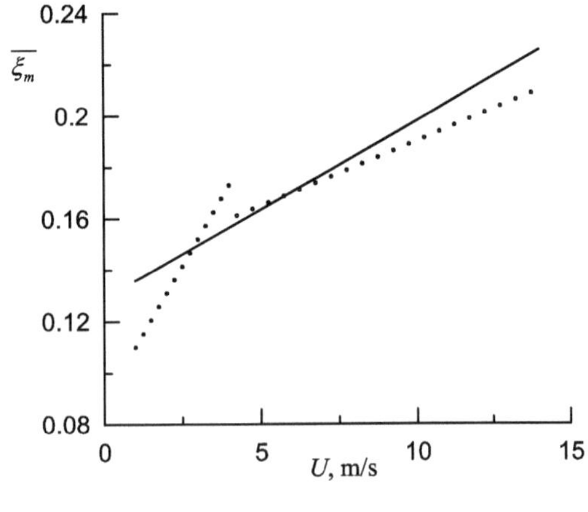

Fig. 2 Regression dependencies $\overline{\xi_m} = \overline{\xi_m}(U)$. The solid line is the calculation for the entire range of wind speeds, dotted lines are calculations for ranges of 1–4 m/s and 4–14 m/s

$$\overline{\xi_m} = \begin{cases} 0.089 + 0.021\, U\, npu\, U \leq 4\,\mathrm{m/s} \\ 0.14 = 0.005\, U\, npu\, U > 4\,\mathrm{m/s} \end{cases} \quad (3)$$

Dependence (3) is shown in Fig. 2.

Linear regression calculated from the same measurement data used for regression (3) over the entire range of wind speeds is described by the equation

$$\overline{\xi_m} = 0.129 + 0.0069\, U. \quad (4)$$

From the regressions shown in Fig. 2 it follows that the calculation of the sea surface roughness level over a wide range of wind speed leads to its overestimation in the domain of low speed and underestimation in the domain of high speed.

Another method for monitoring the roughness of the sea surface is based on laser sensing from a low height (Lebedev et al. 2016). When laser sounding into the nadir, only the glitter formed by facet, the local slope of which is less than a certain critical value, enter the photodetector aperture $\xi_m < \xi_K$. The value ξ_K is determined by the aperture diameter D of the photodetector and the distance from the photodetector to the reflecting surface H, $\xi_K = D/(4H)$. During continuous laser sounding of the sea surface, the recorded signal is a sequence of electrical pulses corresponding to the mirror reflections of the laser beam, which is further analyzed. Typically, the frequency of glitter F, the average duration of glitter τ, and the average power of the reflected signal I are analyzed.

Data from laser sounding of the sea surface indirectly confirm the nonlinearity of changes in the parameters that characterize its roughness. The variability of pulses recorded under different meteorological conditions is shown in Fig. 3 (Zapevalov 2000). The measurements were carried out with a lidar mounted at a height of 5 m. In the range of wind speeds from calm to 10 m/s, the frequency of recorded glitter increases linearly.

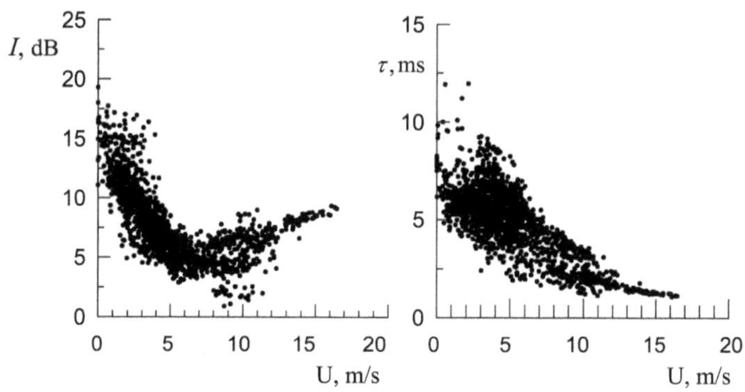

Fig. 3 Dependence of the reflected signal characteristics I and τ on wind speed U

The relationship between the parameters of the recorded glitter and the wind speed is stochastic. The spread of parameters characterizing roughness is small at high wind speeds, when the wind is the main factor determining the wave spectrum.

4 The Roughness of the Sea Surface in Slicks and Ripples

In a weak wind, areas of slicks and ripples can simultaneously exist on the sea surface. Instruments with high spatial resolution are used to study the topographic structure in these areas.

Field measurements carried out using string sensors showed (Khristoforov 1982) that in slicks, generated by internal waves, the energy of the wave spectral components decreases in a wide frequency range from 2 to 30 Hz. Such a wide range is probably due to two mechanisms: the modulation of the surface flow by internal waves and changes in the concentration of surfactants. Especially strong changes in the spectrum occur in the frequency range of 6–8 Hz, which corresponds to the transition from the shortest gravitational waves to gravitational-capillary waves.

It was also found that short surface waves with periods of 0.3–0.6 s become more "sinusoidal" in slicks, and more "trochoidal" in the ripple area. This change in the shape of short waves indicates that their distribution of surface elevations in slicks is closer to the Gaussian distribution than in the ripple region.

Measurements carried out by a laser slopemeter showed that the transition from a smooth surface (slick) to a rough surface (ripples) causes a sharp change in the slope dispersion (Khristoforov et al. 1992). There is also a change in the older statistical moments. The regression dependences of the slope variance determined in a weak wind are graphically shown in Fig. 4. It can be seen that from the stage of complete calm, in which slope variance are minimal and remain approximately at the same level in a certain range of low wind speeds of $0.8 < U < 1.7$ m/s, to the stage of pure wind waves, the state of the sea surface can be transformed in two ways. Both paths lead to the same result, since when the wind speed exceeded the level of 5 m/s, slicks from internal waves disappeared. When the wind speed decreases, the process of changing the topographic structure of the sea surface goes in the opposite direction

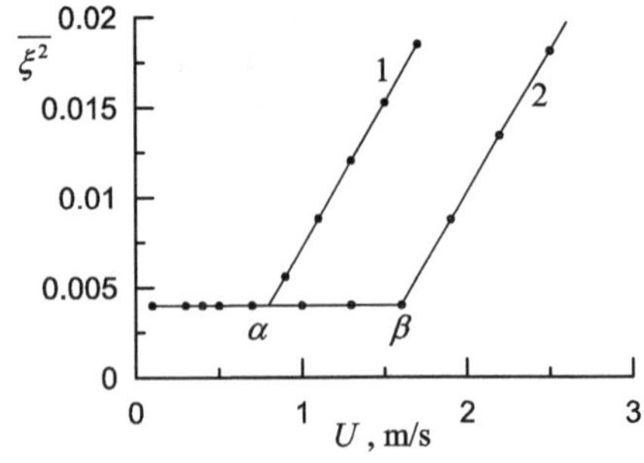

Fig. 4 Schematic description of changes in slope variance $\overline{\xi^2}$ in the ripple zone (curve 1) and in the slick zone (curve 2)

according to one of the specified diagrams. The calm slick state is achieved regardless of what the intermediate states were.

In (Khristoforov et al. 1994), changes in the structure of the sea surface were analyzed on the basis of low-altitude laser sensing data. Changes in glitter were studied when a laser beam hits smoothed surface areas in two main natural slick situations: in slicks associated with internal waves and in Langmuir slicks. The slicks created by the internal waves and Langmuir circulation are visually very distinct. A comparison of these data obtained in these situations reveals the following. In slicks created by internal waves, the frequency F decreases from 10–12 Hz to 5–6 Hz, the average power I of the reflected signal increases by 4–8 dB, and the average duration of the flares decreases τ by ~20%. In Langmuir slicks, the variations in the parameters I and τ are of the same order, but the changes in the frequency F are much smaller.

When comparing the two situations, you should take into account the significant difference in background conditions. Slicks from internal waves oriented parallel to their crests are observed at wind speeds of less than 4–5 m/s, and when the wind increases to speeds of more than 5–6 m/s, they disappear. Instead of slicks from internal waves, there are wind-stretched smoothing bands created by the Langmuir circulation.

5 Conclusion

The problem of remote sensing of the ocean has two components. The first component is how atmospheric processes and processes in the ocean are manifested on the sea surface. The second component is how these manifestations are displayed during remote sensing. In this paper, the first component is considered.

The analysis of experimental studies of the structure of the sea surface in low wind is carried out. The conditions arising in light wind are most favorable for remote monitoring of processes occurring near the ocean–atmosphere boundary. At the same time, the wind speed must exceed a certain level (~1 m/s), so that ripple appear on the sea surface. The shape of the slicks, their speed of movement, their lifetime, and other characteristics make it possible to identify the processes that create slicks.

Currently, there is a clear lack of information about the structure of the sea surface in light winds. It is necessary for the development of algorithms for recognizing the physical mechanisms that create slicks. Also, information on changes in slope variance and short wave energy is needed to refine the geophysical model function.

Acknowledgements The authors are grateful to A. S. Zapevalov, for a helpful discussion of the problem considered in this paper. This work was carried out as part of a state assignment on the topic No. 0827-2018-0003.

References

Apel JR (1994) An improved model of the ocean surface wave vector spectrum and its effects on radar backscatter. J Geophys Res 99(C8):16269–16291

Apel JR, Byrne HM, Proni JR, Charnell RL (1975) Observations of oceanic internal and surface waves from the earth resources technology satellite. J Geophys Res 80(6):865–881

Babanin AV, Polnikov VG (1995) On the non-Gaussian nature of wind waves. Phys Oceanogr 6(3):241–245

Bass FG, Fuks IM (1979) Wave scattering by statistically rough surface. Pergamon, p 540

Brekke C, Solberg AHS (2005) Oil spill detection by satellite remote sensing. Rem Sens Environ 95:1–13

Bréon FM, Henriot N (2006) Spaceborne observations of ocean glint reflectance and modeling of wave slope distributions. J Geoph Res Oceans 111(C06005). https://doi.org/10.1029/2005JC003343

Bulatov MG, Kravtsov YuA, Lavrova OYu, Litovchenko KT, Mityagina MI, Raev MD, Sabinin KD, Trokhimovskiĭ YG, Churyumov AN, Shugan IV (2003) Physical mechanisms of aerospace radar imaging of the ocean. Phys Usp 46(1):63–79

Cheng Y, Liu Y, Xu Q (2006) A new wind-wave spectrum model for deep water. Indian J Mar Sci 35(3):181–194

Cox C, Munk W (1954) Measurements of the roughness of the sea surface from photographs of the sun glitter. J Optical Soc America 44(11):838–850

Elfouhaily T, Chapron B, Katsaros K, Vandemark D (1997) A unified directional spectrum for long and short wind-driven waves. J Geophys Res 102:15781–15786

Hu Y, Stamnes K, Vaughan M, Pelon J, Weimer C, Wu D, Cisewski M, Sun W, Yang P, Lin B, Omar A, Flittner D, Hostetler C, Trepte C, Winker D, Gibson G, Santa-Maria M (2008) Sea surface wind speed estimation from space-based lidar measurements. Atmos Chem Phys 8:3593–3601

Kalinin SA, Leikin I (1988) Measurement of the slopes of wind waves in the Caspian Sea. Izvestiya akademii nauk SSSR. Fizika atmosfery i okeana 24(11):1210–1217

Karaev VYu, Kanevsky MB, Meshkov EM, Titov VI, Balandina GN (2008) Measurement of the variance of water surface slopes by a radar: verification of algorithms. Radiophys Quantum Electron 51(5):360–371

Khristoforov GN (1982) Changes in the structure of sea wind waves in the area of surface slick. In: Impact of large-scale internal waves on the sea surface. Institute of applied physics RAS: Gorky, pp 189–208 [in Russian]

Khristoforov GN, Zapevalov AS, Babii MV (1992) Measurements of sea-surface roughness parameters during the transition from calm to wind generated wave. Izvestiya Akademii Nauk Fizika Atmosfery i Okeana 28(4):424–431

Khristoforov GN, Zapevalov AS, Smolov VE, Feldman YuR (1994) Lidar sensing of topographic inhomogeneities on the rough sea surface. Phys Oceanogr 5(6):461–468

Konstantinov OG, Novotryasov VV (2013) Surface manifestations of internal waves observed using a land-based video system. Izv Atmos Ocean Phys 49(3):334–338

Lebedev NE, Aleskerova AA, Plotnikov EM (2016) The development of optical methods for sea surface slope measurement. Sovremennye Problemy Distantsionnogo Zondirovaniya Zemli Iz Kosmosa. 13(3):136–149 [in Russian]

Li L, Fukushima H, Suzuki K, Suzuki N (2007) Optimization of Cox and Munk sun-glint model using ADEOS/II GLI data and SeaWinds data. In: Coastal Ocean remote sensing. Proceedinngs of SPIE, vol 6680, p 668006

Longuett-Higgins MS, Cartwrighte DE, Smith ND (1963) Observation of the directional spectrum of sea waves using the motions of the floating buoy. In: Proceedings of conference Ocean wave spectra. Prentice Hall, Englewood Cliffs. N. Y., pp 111–132

Monin AS, Krasitskiy VP (1985) Phenomena on the surface of the ocean. Leningrad: Gidrometeoizdat, p 375 [in Russian]

Serebryany AN (2012) Slick- and suloy generating processes in the sea. Internal waves. Sovremennye problemy distantsionnogo zondirovaniya Zemli iz kosmosa 9(2) 275–286 [in Russian]
Zapevalov AS (2000) Variability of lidar signal features when remotely sensing the sea surface. Oceanology 40(5):742–746
Zapevalov AS (2002) Statistical characteristics of the moduli of slopes of the sea surface. Phys Oceanogr 12(1):24–31
Zapevalov AS, Ratner YuB (2003) Analytic model of the probability density of slopes of the sea surface. Phys Oceanogr 13(1):1–13
Zapevalov AS, Bol'shakov AN, Smolov VE (2009) Studying the sea surface slopes using an array of wave gauge sensors. Oceanology 49(1):31–38
Zapevalov A, Pokazeev K, Chaplina T (2021) The distribution of the variance of the sea surface slopes on the spatial scales creating their waves. In: Simulation of the sea surface for remote sensing. Springer, Cham, pp 157–172

Far Internal Gravity Waves Fields Generated by a Sources Distributed on a Moving Plane

V. V. Bulatov and Yu. V. Vladimirov

Abstract The problem of the internal gravity waves far fields generated by a moving non-local source of disturbances is considered. The integral representation of the solution has the form of a sum of wave modes. The perturbation method was used to study the characteristics of the corresponding dispersion relations that determine the behavior of far wave fields. The asymptotics of solutions for individual wave modes are obtained, which make it possible to calculate wave fields far from arbitrary non-local sources of disturbances. A general scheme of mathematical modeling of the internal gravity waves far fields generated by a moving non-local sources is proposed.

Keywords Stratified medium · Internal gravity waves · Source distribution · Far fields

Perturbations of various physical nature are important mechanisms of generation of internal gravity waves (IGW) in stratified media (ocean, atmosphere of the Earth); they include natural sources (moving disturbances of the atmospheric pressure, flows over uneven bottom topography, leeward perturbations of flows near mountains, tidal generation) and anthropogenic influence (marine technological structures, collapse of turbulent mixing regions, underwater explosions) (Lighthill 1978; Miropol'skii and Shishkina 2001; Mei et al. 2017; Morozov 2018). Generation and propagation of IGW in real oceanic conditions are essentially nonlinear. It is possible to linearize the system describing the wave dynamics under some reasonable assumptions. A combination of model concepts and semi-empirical, and possibly, even experimental data may become interesting (Adcroft and Campin 2011; Elizarova 2009; Bulatov and Vladimirov 2012; 2019). Solutions of linear equations of IGW generated by local sources of disturbances were studied in detail in (Lighthill 1978; Miropol'skii and Shishkina 2001; Gray et al. 1983; Kallen 1987; Borovikov et al. 1995). Therefore, knowing such solutions, it is possible to write a solution for non-local sources distributed vertically and horizontally. The following approach is also possible. Let

V. V. Bulatov (✉) · Yu. V. Vladimirov
Ishlinsky Institute for Problems in Mechanics RAS, Vernadskogo ave. 101-1, 119526 Moscow, Russia

T. Chaplina (ed.), *Processes in GeoMedia—Volume IV*, Springer Geology,
https://doi.org/10.1007/978-3-030-76328-2_38

us assume that there is a nonlocal source of disturbances that leaves behind a turbulent wake (a region of eddies or a mixing zone) that generates IGW by means of a specific mechanism. We locate an imaginary plane parallel to the trajectory of motion of the source of perturbations; the initial data on the plane can be determined both experimentally and by numerical simulations. Then it is possible to solve the problem of the propagation of IGW in a linear formulation using the data on the plane as the boundary conditions. Much real information can be stored under these a priori conditions (Elizarova 2009; Robey 1997; Voisin 2007; Wang et al. 2017a, b). A linear theory based on these conditions can give quite satisfactory results far from the regions of nonlinearity. This approach seems to be quite reasonable, since good results should be expected from the linear theory far from the turbulence regions, mixing zones, eddy formations, and other nonlinear phenomena (Bulatov and Vladimirov 2012, 2019, 2020). In this paper, we consider an example illustrating this approach, which makes it possible to calculate properties of IGW far from nonlocal sources of perturbations, the distribution of which is given on a certain plane in a layer of stratified medium.

Vertical component W of velocity of linear IGW, which are generated by a point source of perturbations moving in an inviscid, incompressible, stratified medium of depth H with velocity V on plane $y = 0$ in the Boussinesq approximation satisfies equation (Lighthill 1978; Miropol'skii and Shishkina 2001; Bulatov and Vladimirov 2012, 2019)

$$LW(x + Vt, y, z, z_0) = 0, L = \frac{\partial^2}{\partial t^2}\left(\Delta + \frac{\partial^2}{\partial z^2}\right) + N^2(z)\Delta \tag{1}$$

$$\Delta = \frac{\partial^2}{\partial x^2} + \frac{\partial^2}{\partial y^2}, N^2(z) = -\frac{g}{\rho_0(z)}\frac{d\rho_0(z)}{dz},$$

$$W = \delta(x + Vt)\delta(z - z_0)(y = 0), \tag{2}$$

$$W = 0\,(z = 0, -H), \tag{3}$$

where $N^2(z)$ is squared Brunt-Väisälä frequency (buoyancy frequency), g is acceleration due to gravity, and $\rho_0(z)$ is unperturbed density of the stratified medium. Solution $U(x + Vt, y, z)$ with arbitrary boundary condition at $y = 0$ of the following form $U(x + Vt, 0, z) = f(x + Vt, z)$ is determined by the corresponding convolution

$$U(x + Vt, y, z) = \int_{-H}^{0} dz_0 \int_{-\infty}^{\infty} W(x + Vt - \lambda, y, z - z_0) f(\xi, z_0) d\lambda \tag{4}$$

Uniqueness of the solution is provided by the radiation condition, which can be set as follows. We consider function $\eta(t, x, y, z)$ that satisfies Eq. (1), boundary conditions (3), and boundary condition at $y = 0$ written as

$$\eta = \delta(x + Vt)\delta(z - z_0)\exp(\varepsilon t) \quad (y = 0) \text{ at } \varepsilon > 0 \tag{5}$$

If we seek function η as: $\eta = \exp(\varepsilon t)\xi(x + Vt, y, z)$ and introduce additional condition for decreasing $\xi \to 0$ $(y \to \infty)$, then ξ is determined uniquely. Then, solution W of problem (1)–(3) is determined as the limit: $W(x + Vt, y, z) = \lim\limits_{\varepsilon\to 0} \xi$ Solution of problem (1), (3), (5) can be reasonably sought by separating variables. Let w be a solution of Eq. (1) satisfying boundary conditions (3), and boundary condition at $y = 0$

$$w = \exp(i(\mu x + \Omega t))\delta(z - z_0) \quad (y = 0),\ \Omega = \omega - i\varepsilon. \tag{6}$$

Then, solution of problem (1), (3), (5) at $t > 0$ is written in terms of w as:

$$\eta(t, x, y, z) = \frac{1}{4\pi^2 i}\int\limits_{-\infty}^{\infty} d\mu \int\limits_{-\infty - i\varepsilon}^{\infty - i\varepsilon} \frac{w d\omega}{\omega - \mu V}.$$

We shall seek for the solution to problem (1), (3), (6) using the method of separation of variables and consider the spectral problem

$$\Omega^2 \frac{\partial^2 \phi}{\partial z^2} + (\mu^2 + \nu^2)(N^2(z) - \Omega^2)\phi = 0,$$
$$\phi = 0 \ (z = 0, -H), \tag{6}$$

where, μ is a free parameter, ν is a spectral parameter. This problem has a set of eigen values $\pm\nu_n(\mu)$ and eigen functions $\phi_n(z, \mu)$; these functions are complete and orthonormal with the weight $(N^2(z) - \Omega^2)$ (Lighthill 1978; Miropol'skii and Shishkina 2001; Bulatov and Vladimirov 2012, 2019). Therefore, the following expansion is possible:

$$\delta(z - z_0) = \sum_n (N^2(z_0) - \Omega^2)\, F_n(z, z_0, \mu),$$
$$F_n(z, z_0, \mu) = \phi_n(z, \mu)\phi_n(z_0, \mu).$$

If we assume that ν_n is a root with positive imaginary part, solution of problem (1), (3), (5) will be written as:

$$w = \sum_n (N^2(z) - \Omega^2)\exp(i(\mu x + \Omega t + \nu_n y))\, F_n(z, z_0, \mu).$$

Then, we can present solution of problem (1), (3), (4) as:

$$\eta(t,x,y,z)=\sum_n\int_{-\infty}^{\infty}\frac{(N^2(z)-(\mu V-i\varepsilon)^2)}{2\pi}\exp(\varepsilon t+i\nu_n(\mu)y$$
$$+i\mu(x+Vt))F_n(z,z_0,\mu)d\mu,$$

where $\nu_n(\mu)$ are eigen numbers of spectral problem:

$$(\mu V-i\varepsilon)^2\frac{\partial^2\phi_n}{\partial z^2}+\left(\mu^2+\nu_n^2(\mu)\right)\left(N^2(z)-(\mu V-i\varepsilon)^2\right)\phi_n=0,$$
$$\phi_n=0\,(z=0,-H). \tag{6}$$

Let us consider properties of functions $\nu_n(\mu)$ assuming that $\varepsilon=0$. This function is either real or imaginary depending on the sign of $\nu_n^2(\mu)$. If $\mu V>N_{\max}$, where $N_{\max}$ is the maximum value of the Brunt-Väisälä frequency, then ν_n^2 is negative: $\nu_n^2=-\gamma_n^2(\mu)$ and we have to take values $\nu_n=i\gamma_n(\mu)$. A transition to small $\varepsilon>0$ does not change the sign of the imaginary part of ν_n. If $\mu V<N_{\max}$, then ν_n^2 is positive and can be determined from equation: $\omega_n(\kappa)=\mu V$, where $\kappa=\sqrt{\mu^2+\nu^2}$, and $\omega_n(\kappa)$ is determined from the solution of the main spectral problem of internal gravity waves (Lighthill 1978; Miropol'skii and Shishkina 2001; Morozov 2018) (we neglect rotation of the Earth)

$$\frac{\partial^2\phi_n(z,k)}{\partial z^2}+k^2\left(\frac{N^2(z)}{\omega_n^2(k)}-1\right)\phi_n(z,k)=0,$$
$$\phi_n=0\,(z=0,-H)$$

The main properties of functions $\mu=\mu_n(\nu)$ are studied in detail in (Bulatov and Vladimirov 2012; Bulatov and Vladimirov 2019; Gray et al. 1983; Borovikov et al. 1995). These functions monotonously increase at $\nu\to\infty$ and tend to a finite limit equal to $N_{\max}/V\equiv M$. Function $\mu_n(\nu)$ monotonously increases at $\nu\to\infty$ and tends to a finite limit equal to M. Therefore, the inverse function monotonously increases and tends to infinity at $\mu\to M$. At the given values of μ the sign of $\nu_n(\mu)$ is found from the condition that coefficient at $\exp(\pm i\nu_n(\mu)y)$ decreases when $y\to+\infty$. If we consider domain $y<0$, we select the opposite sign. Let us calculate the imaginary addition to $\nu_n(\mu)$. We present $\nu_n(\mu)$ as $\nu_n=\nu_{n0}+i\varepsilon\nu_{n1}$, then using the theory of perturbations we can get from Eq. (7) that

$$\nu_{n1}=-\frac{1}{\nu_{n0}}\left(\frac{1}{\mu V(\nu_{n0}^2+\mu^2)}+\mu V\left(\nu_{n0}^2+\mu^2\right)\int_{-H}^{0}\phi_n^2(z,\mu)dz\right).$$

Therefore, at $y>0$ we select the minus sign for the coefficient at $\exp(\pm i\nu_n(\mu)y)$, and positive sign if the values of y are negative. Let us now consider the case when $\mu>M$. In this case $\nu_n^2(\mu)$ is negative and correspondingly $\nu_n(\mu)$ is purely imaginary;

its absolute value is not smaller than M, which follows from the existence of solution to the Sturm–Liouville problem: $\frac{d^2\phi}{dx^2}+q(x)\phi=0$, only if $q(x)$ is not smaller than 0 at all x. At $\mu > M$ we get that $N^2(z)-\mu^2V^2 < 0$ at all z, therefore, $\mu^2+\nu_n^2(\mu) > 0$ at $\mu > M$. Therefore, $\nu_n^2(\mu) < -M^2$ at all $\mu > M$. Hence, the contribution of the imaginary part of function $\nu_n(\mu)$ is an exponentially small value. Inverse function $\nu = \nu_n(\mu)$ increases and tends to infinity at $\mu \to M$. At small μ, the following expansion is possible: $\nu_n(\mu) = \alpha_n\mu + j_n\mu^3 + ...$, where $\alpha = q_n, \quad j_n = \beta_n q_n^4, \quad q_n, \beta_n$ are coefficients of expansion $\mu_n(\nu) = q_n^{-1}\nu - \beta_n\nu^3 + ...$ at zero. At $\varepsilon = 0$ and $\mu < M$. Equation (6) determines two families of dispersion curves $\nu = \pm\nu_n(\mu)$; among these two branches we select the one whose imaginary part is positive at $\varepsilon < 0$. This branch can be easily chosen using the method of perturbations; therefore, we take $\nu = -\nu_n(\mu)$. Now we can make a transition to the limit at $\varepsilon \to 0$, and as a result get a solution of problem (1)–(3) in the form

$$W = \frac{1}{2\pi}\sum_n \left(I_n^0 + I_n^+ + I_n^-\right),$$

$$I_n^0 = \int\limits_{-M}^{M} A_n(z, z_0, \mu)\exp(i\mu(x+Vt) - i\nu_n(\mu)y)d\mu,$$

$$I_n^- = \int\limits_{-\infty}^{-M} A_n(z, z_0, \mu)\exp(i\mu(x+Vt) - \gamma_n(\mu)y)d\mu,$$

$$I_n^+ = \int\limits_{M}^{\infty} A_n(z, z_0, \mu)\exp(i\mu(x+Vt) - \gamma_n(\mu)y)d\mu,$$

$$A_n(z, z_0, \mu) = \left(N^2(z_0) - \mu^2V^2\right)F_n(z, z_0, \mu).$$

We assume in integrals I_n^0 that $\nu_n(\mu)$ is a monotonously increasing function of μ determined from (6). Integrals $I_n^{\pm}$ are exponentially small at $y \gg 1$. Therefore, at $y \gg 1$ (at large distances from plane $y = 0$) we get

$$W = \sum_n W_n$$

$$W_n = \frac{1}{2\pi}\int\limits_{-M}^{M} A_n(z, z_0, \mu)\exp(i\mu(x+Vt) - i\nu_n(\mu)y)d\mu.$$

This expression differs from integral presentations of individual wave modes considered in (Bulatov and Vladimirov 2012; Bulatov and Vladimirov 2019; Gray et al. 1983; Borovikov et al. 1995) by the coefficient at the exponent. Therefore, asymptotics of the components of W_n at $y \gg 1$ can be found similarly. At $x + Vt < q_n y$ (where $q_n = \sqrt{V^2c_n^{-2} - 1}$, and c_n is the maximum group velocity of wave mode

n), W_n exponentially decreases. At $x + Vt > q_n y$, asymptotics of W_n can be found using the stationary phase method:

$$W_n \approx 2\frac{N^2(z_0) - (\mu_n^*)^2 V^2}{\sqrt{2\pi y \left|\nu_n''(\mu_n^*)\right|}} \cos(\Phi_n)\, F_n\left(z, z_0, \mu_n^*\right),$$

$$\Phi_n = \mu_n^*(x + Vt) - \nu_n(\mu_n^*)y + \pi/4 \tag{8}$$

where μ_n^* is a solution of equation $\frac{x+Vt}{y} = \frac{\partial \nu_n(\mu)}{\partial \mu}$. At $x + Vt \to q_n y$, stationary points $\mu = \pm\mu_n^*$ merge, and then near the wave front of individual mode $x + Vt = q_n y$, asymptotics of W_n at $y \gg 1$ are expressed from the Airy function (Bulatov and Vladimirov 2012; Bulatov and Vladimirov 2019; Borovikov et al. 1995):

$$W_n \approx \frac{N^2(z_0)\, F_n(z, z_0, 0)}{q_n \sqrt[3]{3\beta_n q_n y}} Ai\left(\frac{x + Vt - q_n y}{q_n \sqrt[3]{3\beta_n q_n y}}\right),$$

$$Ai(\tau) = \frac{1}{2\pi} \int\limits_{-\infty}^{\infty} \cos\left(\tau u - u^3/3\right) du. \tag{9}$$

Let us qualitatively describe propagation of far field of IGW from a nonlocal source of perturbations: at $x + Vt < q_1 y$, the wave field is negligibly small, at $x + Vt = q_1 y$, the front of the first wave mode arrives to the fixed point of observations; in the vicinity of the front the field is described by formulas (4), (9) through the Airy function convolution; at $q_1\, y < x + Vt < q_2\, y$, the far wave field consists of one wave packet, which is determined by convolution from formulas (4), (8); at $x + Vt = q_2 y$, the second wave packet arrives to the point of observations; in the vicinity of this packet, the wave field is also described by Airy functions (4), (9); at $q_1\, y < x + Vt < q_3\, y$, the wave field consists of two terms determined by the corresponding convolutions (4), (8); at $x + vt = q_3\, y$, the third wave packet arrives to the point of observations; the total wave field consists of three terms, and so on. Numerical simulations for realistic distributions of buoyancy frequency observed in the ocean show that the main contribution to the far wave field of IGW is done only by a few first wave modes. Therefore, in the majority of the practically important oceanographic problems the applicable accuracy of simulations of far wave fields can be provided by not more than 3–5 wave modes (Bulatov and Vladimirov 2012, 2019, 2020; Borovikov et al. 1995).

The asymptotics of the solutions obtained in this paper make it possible to calculate wave fields far from arbitrary nonlocal sources of perturbations. A general scheme for modeling the far fields of IGW from moving nonlocal sources of disturbances can be presented as follows. Main characteristics of the wave field parameters (velocity components, density, pressure) are determined from the numerical solution of the complete system of hydrodynamic equations. Their distribution can be specified on a given plane. If we assume that a linear model of wave dynamics of stratified media is adequate, far fields of IGW can be calculated from formulas (4), (8), (9) far from the sources of perturbations. The results of calculations show that under real hydrological conditions of the ocean, only the first few wave modes make the main contribution to the far fields. Therefore, the results obtained in this work make it possible not only to efficiently calculate far wave fields, but also to carry out a qualitative assessment of the solutions obtained.

Acknowledgements The work is carried out with financial support from the Russian Foundation for Basic Research, project 20-01-00111A.

References

Adcroft A, Campin J-M (2011) MITgcm user manual. MIT, Cambridge

Borovikov VA, Bulatov VV, Vladimirov YV (1995) Internal gravity waves excited by a body moving in a stratified fluid. Fluid Dyn Res 5:325–336

Bulatov VV, Vladimirov YuV (2012) Wave dynamics of stratified mediums. Nauka, Moscow

Bulatov VV, Vladimirov YV (2019) A general approach to ocean wave dynamics research: modelling, asymptotics, measurements. OntoPrint Publishers, Moscow

Bulatov V, Vladimirov Y (2020) Generation of internal gravity waves far from moving non-local source. Symmetry 12(11):1899

Elizarova TG (2009) Quasi-gas dynamic equations. Springer, Berlin

Gray EP, Hart RW, Farrel RA (1983) The structure of the internal Mach front generated by a point source moving in a stratified fluid. Phys Fluids 26:2919–2931

Kallen E (1987) Surface effects of vertically propagation waves in a stratified fluid. J Fluid Mech 148:111–125

Lighthill J (1978) Waves in fluids. Cambridge University Press, Cambridge

Mei CC, Stiassnie M, Yue DK-P (2017) Theory and applications of ocean surface waves. In: Advanced series of ocean engineering, vol 42. World Scientific Publishing

Miropol'skii YZ, Shishkina OV (2001) Dynamics of internal gravity waves in the ocean. Kluwer Academic Publishers, Boston

Morozov TG (2018) Oceanic internal tides. In: Observations, analysis and modeling. Springer, Berlin

Robey H (1997) The generation of internal waves by a towed sphere and its wake in a thermocline. Phys Fluids 9:3353

Voisin B (2007) Lee waves from a sphere in a stratified flow. J Fluid Mech 574:273–292

Wang H, Chen K, You Y (2017a) An investigation on internal waves generated by towed models under a strong halocline. Phys Fluids 29:065104

Wang J, Wang S, Chen X, Wang W, Xu Y (2017b) Three-dimensional evolution of internal waves rejected from a submarine seamount. Phys Fluids 29:106601

Application of the Salome Software Package for Numerical Modeling Geophysical Tasks

V. P. Pakhnenko

Abstract As a software package for mathematical modeling of processes in geophysics, the open, integrable platform Salome was chosen. The choice of this package was due to the openness of the package and the ability to solve the following problems: strength, filtration calculations, heat conduction problems in a solid, problems associated with the deformation of the computational grid, as well as related problems. The features of using the Salome package, its structure, purpose and characteristics of the main functional blocks are considered. The results of the analysis of its capabilities for solving the problems of assessing the distribution of temperatures, mechanical stresses, thermoelasticity, porosity in the simulated object are presented.

Keywords Numerical simulation · Salome · Software modules · Visualization · Poroelasticity

1 Introduction

The rapid development of computer technology has led to the use of numerical methods for solving a wide range of problems in geophysics, mechanics, and the theory of elasticity. The leading role among numerical methods is currently occupied by the finite element method, which reduces the solutions of systems of partial differential equations to the representation of the object under study in the form of some discrete grid and approximation of the system of piecewise continuous functions to be solved.

There are many programs designed to solve numerical simulation problems. The most well-known and widespread are universal combine packages, suitable for solving a variety of mathematical problems. In terms of functionality, they are generally divided into two categories: packages intended mainly for numerical calculations (for example MatLab) and computer algebra systems (Computer Algebra System), which include Mathematica, Maple, MathCad—they are also called symbolic or

V. P. Pakhnenko (✉)
Ishlinsky Institute for Problems in Mechanics RAS, Vernadskogo ave., 101-1, 119526 Moscow, Russia

T. Chaplina (ed.), *Processes in GeoMedia—Volume IV*, Springer Geology,
https://doi.org/10.1007/978-3-030-76328-2_39

analytical computing systems (Symbolic Manipulation Program). These are the most versatile mathematical packages capable of solving both numerical and analytical problems. For solving problems in the field of geophysics, software packages designed for solving partial differential equations by the finite element method are of greatest interest. Among the most famous software products created for these purposes are ANSYS, Comsol Multiphysics, OpenFoam, Salome, etc. Aubry (2013).

As a software package for mathematical modeling of processes in geophysics, the open, integrable platform Salome was chosen. The choice of this package was due to the openness of the package and the possibility of solving the following problems: strength, filtration calculations, heat conduction problems in a solid, problems associated with the deformation of the computational mesh, as well as related problems. The report examines the features of using the Salome package, its structure, purpose and characteristics of the main functional blocks. The results of the analysis of its capabilities for solving problems of assessing the distribution of temperatures, mechanical stresses, thermoelasticity, porosity in a modeled object are presented.

The aim of this work is to analyze the possibility of using the Salome software package for solving the poroelastic problem. We will assume that the poroelasticity problem is formulated for a bounded domain Ω with boundary Γ. Then the mathematical model contains equations for displacement and pressure, which are respectively obtained from the equation of equilibrium and continuity. Let us denote the displacement vector of the porous medium through u, and the pressure of the filtering liquid through p. In the quasi-stationary approximation, the stress–strain state of a porous medium is described using the equilibrium equation (Biot 1941):

$$div\sigma_t(u, p) = 0 \tag{1}$$

where σ_t—general stress tensor, which is specified through the generalized Hooke's law for an isotropic body (Biot 1941; Geertsma 1957):

$$\sigma_t = \sigma - \alpha pI = 2\mu\varepsilon(u) + \lambda\varepsilon_v I - \alpha pI \tag{2}$$

where σ—effective stress tensor, I—unit tensor, α—Biot's coefficient characterizing the relationship between displacement and pressure, ε—strain tensor:

$$\varepsilon(u) = \frac{1}{2}\left(grad(u) + grad\left(u^T\right)\right) \tag{3}$$

and μ, λ—Lame coefficients or elastic moduli. Note that the coefficient μ is also called the shear modulus. They can be expressed in terms of Young's modulus E and Poisson's ratio ν (Biot 1941):

$$\mu = \frac{E}{2(1+v)}, \lambda = \frac{vE}{(1+v)(1-2v)}.$$

The volumetric strain ε_v is defined as the sum of the elements of the main diagonal of the strain tensor ε:

$$\varepsilon_v = tr(\varepsilon) = div(u) \tag{4}$$

Let's define the volumetric stress σ_v as:

$$\sigma_v = \frac{1}{3} tr(\sigma_t)$$

Then from the generalized Hooke's law:

$$\sigma_v = K\varepsilon_v - \alpha p \tag{5}$$

where K—module of volumetric compression:

$$K = \lambda + \frac{2}{3}\mu = \frac{E}{3(1 - 2v)}$$

Note that of the five considered elastic constants μ, λ, E, v and K, only two are independent, that is, knowing any two elastic constants, one can find the other three.

In Biot (1941) and Geertsma (1957) it is shown that the mathematical model of poroelasticity is determined by the formulas:

$$div(\sigma_t) - \alpha grad(p) = 0 \tag{6}$$

$$\alpha \frac{\partial div(u)}{\partial t} + \beta \frac{\partial p}{\partial t} - div\left(\frac{k}{\eta} grad(p)\right) = f(x, t) \tag{7}$$

Equations (6) and (7) are supplemented with initial and boundary conditions.

$$p(x, 0) = p_0(x), x \in \Omega \tag{8}$$

$$div\sigma(u_0) - \alpha grad(p) = 0, x \in \Omega \tag{9}$$

When $p_0 = 0$, the initial condition for displacement is $u_0 = 0$.

Thus, to solve the problem of poroelasticity, it is necessary to evaluate the distribution of mechanical stresses and strains arising in the object under study, taking into account the process of fluid filtration in a porous medium (Cook et al. 2007; Karev et al. 2017). Taking into account Eq. (2), this problem is in many respects similar to the problem of thermoelasticity, in solving which it is also necessary to determine the distributions of mechanical stresses and strains, but taking into account the temperature distribution. At the same time, the problems in the field of thermoelasticity are more elaborated than the problems of poroelasticity. In this case, the problem of thermoelasticity can be decomposed into even simpler problems: mechanical (assessment of stresses and deformations) and thermal (assessment of temperature distribution).

In accordance with the above, the optimal option for constructing the work is to first consider mechanical and thermal problems separately, then consider their joint implementation, namely the Lamé problem and further assess the consideration of the process of fluid filtration in a porous medium to solve the poroelasticity problem.

A feature of the Salome package is that it opensource, that is, all the sources of this package are in the public domain, that is, it provides the ability to view the implementation of the package's functionality. Logically, the work of the package is structured as follows, first, a description of the investigated physical object, its parameters, geometric dimensions and other characteristics is performed, then this description is formalized in the form of a discrete grid, on the basis of which the model under study is calculated and further visualization of the results obtained.

Let's take a closer look at the structure of the Salome package and the purpose of its main modules:

- Geometry
- Mesh
- AsterStudy
- ParaVIS.

The Geometry module creates a geometric image of the modeled object. Its shape is formed, various parameters are set, which are sequentially displayed in the geometry tree. Also in this module 4 types of groups are created: a group of points, a group of faces (lines), a group of planes and a volumetric group. The creation of these groups is an important part of this software package because all the simulated forces and conditions of a specific task are set specifically for these groups. After creating a geometric object, you need to build a mesh, for this there is a Mesh module. It contains a variety of mesh options that can be tailored to suit your specific modeling needs. The given grid parameters are written into its algorithm and the hypothesis into the grid tree. In this module, you can also form grid groups, for further convenience of applying the conditions of the task to them. As a result of the Mesh module, the mesh is calculated, which must be transferred to the main processing module AsterStudy. The entire calculation of the problem under consideration in the Salome package is performed by the AsterStudy module. This module is based on the Code_Aster software product. It appeared in 1991. The developer of this complex was the French company "EDFR & D", which is engaged in research in the field of nuclear energy. Since 2001 the Code_Aster software product is distributed under the terms of the GNUFDL license (www.gnu.org/copyleft/fdl.html). Code_Aster software complex works under the majority of modern 64-bit Linux operating systems. Note that the company "EDFR & D" does not officially support the work of Code_Aster under the MSWindows operating system. The main part of the Code_Aster calculation modules is written in the FORTRAN algorithmic language. The organization of the computing process as a whole is entrusted to the Code_Aster part, which in the technical documentation for Code_Aster was called the supervisor. The supervisor is implemented in the algorithmic Python language. The library of the Code_Aster software package includes more than 400 different types of finite elements.

Among the main types of finite elements are:

- Flat finite elements;
- Axisymmetric finite elements;
- Volumetric finite elements;
- Bar finite elements for calculating trusses;
- Bar finite elements for frame design;
- Finite elements for plate calculation;
- Finite elements for calculating membranes;
- Finite elements for the calculation of shells;
- Finite elements for calculating flexible threads;
- Non-deformable finite elements.

The AsterStudy module contains the following functional modules: Mesh, Model Definition, Material, Function and List, BC and LOADS (boundary conditions and load), Pre Analysis, Analysis, post Processing, Fracture and Fatique, Output. Mesh—Loads a previously defined mesh. Model Definition defines three model classes: Acoustic, Mechanic, Thermic. For each of the classes of models, there is a wide range of predefined analysis methods. Next, the Material module sets the material that is used in the model. First, the properties of the material from which the analyzed object is made are determined, and then these properties are assigned to the geometric object. The next important module is BC and LOADS, it sets the boundary conditions that are used in the model. They apply to those groups that were previously defined in the Geometry and Mesh module. The next step is to select the analysis method used for the model under study, it is set in the Analysis module (for example, Static Mechanical Analysis—mechanical analysis for the isotropic case, or Linear Thermal Analysis—linear temperature analysis, etc.). Then it is necessary to specify the list of results for visualization and the way of their presentation. This function is performed by the Output module. In fact, this module specifies what data and in what format they should be received. It also specifies the type of output. For example, you only need to get data on stresses and temperatures, or both. It should also be noted that the AsterSudy module of the Salome package contains only methods of mechanical, acoustic and thermal treatment and cannot be used to simulate poroelasticity problems directly due to the lack of tools for analyzing liquid filtration.

To visualize the results of the model calculation, the resulting file must be opened using the ParaVIS module. This module provides a graphical visualization of the simulation done, and the functions of the ParaVIS module allow you to change various parameters for a more visual representation. The module also allows displaying different results on one model. For example, for a specific task, it is possible to display first the temperature distribution, then the voltage, or view the results of stress action along each of the coordinate axes separately. For the practical mastering of the Salome software package and gaining skills in working with it, several geophysical problems were modeled. Let us consider the results obtained in more detail. The first challenge simulated was to build a tensile model for a steel anchored plate. The Geometry module was used to define the geometry of an anchored steel plate in a 3-D tree. For this, the Box Consrtuction object was used, which creates a three-dimensional plate with specified dimensions. Then, using the Group object, two groups of planes are specified, on which the boundary conditions are applied. Accordingly, this is a plane along the OX axis, which is rigidly fixed, and a plane along the OY axis, to which a force is applied. The mesh module then creates a mesh of the steel anchored plate. Note that it can have from 0 to 3 coordinate axes. There are several preset types of grids, which differ in sampling parameters. The Mesh module provides the ability to define both different mesh sizes and different types of mesh cells, which can be not only quadrangular.

After the mesh is drawn, a tree of named objects is generated in the Mesh module, and a named mesh object containing the generation algorithm is added to the already created Box Construction object. At the next stage, the discretization operation is performed in accordance with the specified parameters, that is, the specified mesh is applied to the specified geometric object. Note that the mesh is isotropic, the default number of segments created is the same for all created axes. Since in practical problems this is usually not the case and the layer thickness and surface size are significantly different, then along one of the axes the number of mesh layers should be different. To solve this problem, the Group tool is used. Group can be set by points, by lines, by layers (planes) and by volume. After the mesh of the steel fixed plate is built, go to the AsterStudy module and load the calculated mesh. Then in the AsterStudy module, using the Model Definition tool, we define the model class—mechanical. The next step is to set the material used in the model. In this case, first, the properties of the material from which the analyzed object is made are set, and then these properties are assigned to the geometric object. In an isotropic medium, two parameters are set Young's modulus = 2.1E11 (E—Young's modulus) and Polsson's ratio = 0.3 (NU—Poisson's ratio). To account for gravity, the density parameter must be specified.

In the next step, boundary conditions and mechanical loads are set, which is done using the tools of the BC and LOADS module. A structural element that is rigidly fixed and the coordinates of its position are specified. Next, a structural element is selected to which a tensile force is applied and its value, direction and point of application are set. All of the listed subassemblies have been defined in the Geometry module using the Group tool. The next step is to choose the analysis method that is used for the model under study. In this case, this is Static Mechanical Analysis. Then we define the type and presentation of the simulation results. At the final stage, the file with the simulation results is opened using the ParaVIS module. As a result of the work done, the stress distribution on the fixed steel plate is obtained.

In this paper we simulated two cases with two different sizes of steel plates. Both plates were fixed along the OX axis and different stresses were applied along the OY axis.

First case: DX = 200, DY = 400, DZ = 20, U = 1500 H (Fig. 1).

Second case: DX = 200, DY = 200, DZ = 20, U = 2000 H (Fig. 2).

Fig. 1 Static Linear Mechanic Analysis (Plate 1, U = 1500 H)

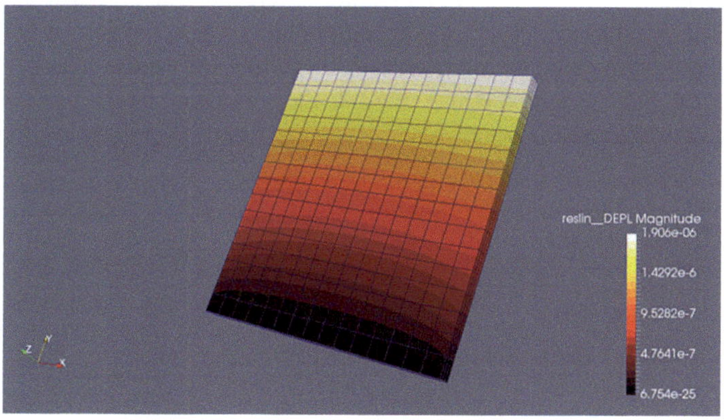

Fig. 2 Static Linear Mechanic Analysis (Plate 2, U = 2000 H)

The corresponding simulation results are shown in Figs. 1 and 2. Simulation of two cases with different parameters is due to the fact that in the course of the first case it turned out that the problem of changing the geometric parameters in the Salome package is not a trivial task and cannot be solved by simply changing the values of these parameters ... When at least one parameter is changed, the conditions for forming the mesh are violated, so the model has to be rebuilt. Therefore, to acquire skills in working with the package, we had to solve the same problem for various geometric parameters. On the whole, the simulation showed that the mechanical task was effectively implemented. The results shown in Figs. 1 and 2 show that, despite some inconveniences, the Salome package provides the ability to change the parameters of the modeled object.

The next completed task was the thermal task. Unlike a mechanical problem, a thermal one simulates the temperature distribution and the results of thermal action. The AsterStudy module provides the ability to solve a thermal problem in two cases:

Table 1 Boundary conditions

Temperature function of coordinates and time	TEMP_IMPO
Linear relationship between temperatures at grid points	LIAISON_DDL LIAISON_GROUP LIAISON_MAIL

stationary and transient. For them in the BC and LOADS module, the boundary conditions are specified, which are given in Tables 1 and 2.

Geometry and mesh creation is done similarly to the mechanical case described above. In the work, two different thermal problems were simulated. In the first case, a small square was heated on one of the surfaces. Accordingly, in the Geometry module, groups of planes were defined, which on the same plane separated areas with different initial temperatures. Next, thermal analysis was applied to the model, with all the specified parameters of temperatures and material properties. The results of numerical simulation of the considered thermal problem are shown in Fig. 3.

In the second simulated thermal problem, the dimensions and boundary conditions were changed. In the Group module, two side faces of a rectangular plate of different temperatures were fixed. Further work was carried out according to the algorithm

Table 2 Loadings

Natural convection	ECHANGE	$\lambda(T)\frac{dT}{dn} = h(t) \cdot (Text - T)$
Heat transfer between walls	ECHANGE_PAROI	$\lambda_1 \frac{dT_1}{dT_2} = h(T_1 - T_2)$
Heat flux: constant or a function of time and coordinates	FLUX_REP	$\lambda(T)\frac{dT}{dn} = f(t, x)$
Non-linear heat flux: a function of temperature	FLUX_NL RAYONNEMENT	$\lambda(T)\frac{dT}{dn} = f(T)$
Heat source	SOURCE	S(x, t)

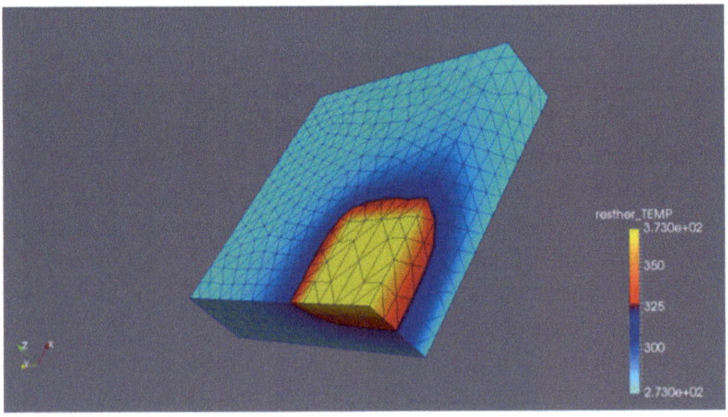

Fig. 3 Linear Thermal Analysis (small square $T_1 = 373$ K, plane $T_2 = 273$ K)

described above. The simulation results for the second version of the thermal problem are shown in Fig. 4.

At the next step in the work, a combination of mechanical and thermal problems was simulated, that is, the thermoelastic problem or the Lame problem was considered. A rectangular steel plate with a cut-out hole in the middle was chosen as the test object. Due to the central symmetry, a quarter of such a plate was simulated in the work (Fig. 5). For this, a group of points was created in the Geometry module that characterizes the hole and the group of planes. Further, the formed mesh was used both mechanical and thermal analysis methods. Two faces of the plate were rigidly fixed, and a force was applied to the third. The result of stress simulation is shown in Fig. 5.

Also in this task it is possible to simulate the stresses at the extreme points of the cut hole (Fig. 6).

The work performed has shown that the Salome package contains the necessary functionality for modeling thermoelasticity problems. Since the problems of thermoelasticity and poroelasticity are in many respects similar, as shown above. Then there are two ways to implement the poroelastic problem: the first is the calculation of

Fig. 4 Linear Thermal Analysis (left fixed edge $T_1 = 450$ K, K, right fixed edge $T_2 = 300$ K)

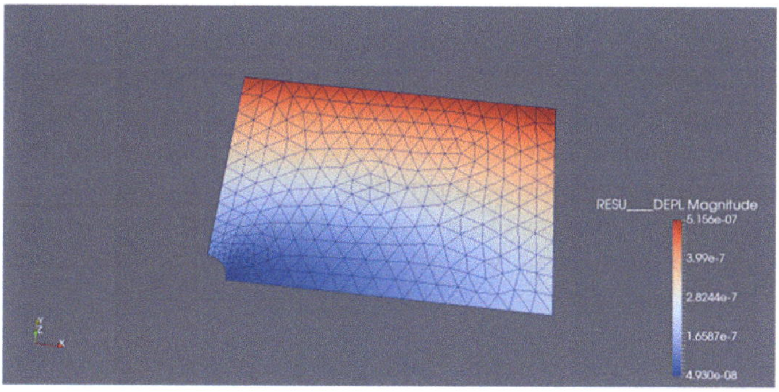

Fig. 5 Thermo-Mechanic Analysis (Plate 3, U = 1000 H)

Fig. 6 Thermo-Mechanic Analysis (stress on hole)

poroelasticity by changing the coefficients specified in the thermoelastic problem; the second is the development of our own analysis methods for the AsterStudy module, which will take into account the process of liquid filtration in a porous medium, which is the direction of further work.

2 Conclusion

As a result of the work carried out, it was found that the Salome package provides the ability to numerically simulate a wide range of mechanical, thermal and acoustic problems. It contains quite effective means of setting the parameters of the simulated physical object. In this case, the geometry and its parameters are described in the form of discrete grids. The package contains a fairly advanced functionality that provides the ability to load and unload such meshes from other packages.

The AsterStudy module based on the Code_Aster software product is actually the core of the Salome package and is responsible for obtaining the main results of numerical modeling. The visualization of the obtained results is carried out by the ParaVIS module, which visually displays the obtained results.

The main advantage of the Salome package is the free access to the source code of all the components of the package. But this is also a disadvantage of the package, because anyone can write their own module and make it publicly available, which in turn creates confusion with the documentation and package integrity.

The performed calculations of mechanical, thermal and thermoelastic problems give confidence in the possibility of using this package for numerical modeling in the field of poroelasticity.

Acknowledgements The work is support in part by the Russian Science Foundation (project 21-11-00029).

References

Aubry J-P (2013) Beginning with Code_Aster. A practical introduction to finite element method using Code_Aster, Gmsh and Salome. Framasoft, p 357

Biot MA (1941) General theory of three dimensional consolidation. J Appl Phys 12(2):155–164

Cook J, Frederiksen RA, Hasbo K et al (2007) On the importance of mechanical properties of rocks: laboratory verification of geomechanical data. Oil Gas Rev 44–69 (in Russian)

Geertsma J (1957) The effect of fluid pressure decline on volumetric changes of porous rocks. J Pet Technol 210:331–340

Karev VI, Kovalenko YF, Ustinov KB (2017) Modeling deformation and failure of anisotropic rocks nearby a horizontal well. J Min Sci 53(3):425–433

MIX
Papier aus verantwortungsvollen Quellen
Paper from responsible sources
FSC® C105338

If you have any concerns about our products,
you can contact us on
ProductSafety@springernature.com

In case Publisher is established outside the EU,
the EU authorized representative is:
Springer Nature Customer Service Center GmbH
Europaplatz 3, 69115 Heidelberg, Germany

Printed by Libri Plureos GmbH
in Hamburg, Germany